Panel Data Econometrics with R

Panel Data Econometrics with R

Yves Croissant
Professor of Economics
CEMOI
Faculté de Droit et d'Economie
Université de La Réunion
France

Giovanni Millo
Senior Economist
Group Insurance Research, Assicurazioni Generali S.p.A.
Trieste, Italy

WILEY

This edition first published 2019

Registered Offices
John Wiley & Sons, Inc., 111 River Street, Hoboken, NJ 07030, USA
John Wiley & Sons Ltd, The Atrium, Southern Gate, Chichester, West Sussex, PO19 8SQ, UK

Editorial Office
9600 Garsington Road, Oxford, OX4 2DQ, UK

For details of our global editorial offices, customer services, and more information about Wiley products visit us at www.wiley.com.

Wiley also publishes its books in a variety of electronic formats and by print-on-demand. Some content that appears in standard print versions of this book may not be available in other formats.

Library of Congress Cataloging-in-Publication Data

Names: Croissant, Yves, 1969- author. | Millo, Giovanni, 1970- author.
Title: Panel data econometrics with R / Yves Croissant, Giovanni Millo.
Description: First edition. | Hoboken, NJ : John Wiley & Sons, 2019. |
Includes index. |
Identifiers: LCCN 2018006240 (print) | LCCN 2018014738 (ebook) | ISBN 9781118949177 (pdf) | ISBN 9781118949184 (epub) | ISBN 9781118949160 (cloth)
Subjects: LCSH: Econometrics. | Panel analysis. | R (Computer program language)
Classification: LCC HB139 (ebook) | LCC HB139 .C765 2018 (print) | DDC 330.0285/5133–dc23
LC record available at https://lccn.loc.gov/2018006240

Cover Design: Wiley
Cover Image: ©Zffoto/Getty Images

Set in 10/12pt WarnockPro by SPi Global, Chennai, India

10 9 8 7 6 5 4 3 2 1

To Agnès, Fanny and Marion, to my parents
- Yves

To the memory of my uncles, Giovanni and Mario
- Giovanni

Contents

Preface *xiii*
Acknowledgments *xvii*
About the Companion Website *xix*

1 Introduction *1*
1.1 Panel Data Econometrics: A Gentle Introduction *1*
1.1.1 Eliminating Unobserved Components *2*
1.1.1.1 Differencing Methods *2*
1.1.1.2 LSDV Methods *2*
1.1.1.3 Fixed Effects Methods *2*
1.2 R for Econometric Computing *6*
1.2.1 The Modus Operandi of R *7*
1.2.2 Data Management *8*
1.2.2.1 Outsourcing to Other Software *8*
1.2.2.2 Data Management Through Formulae *8*
1.3 plm for the Casual R User *8*
1.3.1 R for the Matrix Language User *9*
1.3.2 R for the User of Econometric Packages *10*
1.4 plm for the Proficient R User *11*
1.4.1 Reproducible Econometric Work *12*
1.4.2 Object-orientation for the User *13*
1.5 plm for the R Developer *13*
1.5.1 Object-orientation for Development *14*
1.6 Notations *17*
1.6.1 General Notation *18*
1.6.2 Maximum Likelihood Notations *18*
1.6.3 Index *18*
1.6.4 The Two-way Error Component Model *18*
1.6.5 Transformation for the One-way Error Component Model *19*
1.6.6 Transformation for the Two-ways Error Component Model *20*
1.6.7 Groups and Nested Models *20*
1.6.8 Instrumental Variables *20*
1.6.9 Systems of Equations *20*
1.6.10 Time Series *21*
1.6.11 Limited Dependent and Count Variables *21*
1.6.12 Spatial Panels *21*

2	**The Error Component Model**	*23*
2.1	Notations and Hypotheses	*23*
2.1.1	Notations	*23*
2.1.2	Some Useful Transformations	*24*
2.1.3	Hypotheses Concerning the Errors	*25*
2.2	Ordinary Least Squares Estimators	*27*
2.2.1	Ordinary Least Squares on the Raw Data: The *Pooling* Model	*27*
2.2.2	The *between* Estimator	*28*
2.2.3	The *within* Estimator	*29*
2.3	The Generalized Least Squares Estimator	*33*
2.3.1	Presentation of the GLS Estimator	*34*
2.3.2	Estimation of the Variances of the Components of the Error	*35*
2.4	Comparison of the Estimators	*39*
2.4.1	Relations between the Estimators	*39*
2.4.2	Comparison of the Variances	*40*
2.4.3	Fixed *vs* Random Effects	*40*
2.4.4	Some Simple Linear Model Examples	*42*
2.5	The Two-ways Error Components Model	*47*
2.5.1	Error Components in the Two-ways Model	*47*
2.5.2	Fixed and Random Effects Models	*48*
2.6	Estimation of a Wage Equation	*49*
3	**Advanced Error Components Models**	*53*
3.1	Unbalanced Panels	*53*
3.1.1	Individual Effects Model	*53*
3.1.2	Two-ways Error Component Model	*54*
3.1.2.1	Fixed Effects Model	*55*
3.1.2.2	Random Effects Model	*56*
3.1.3	Estimation of the Components of the Error Variance	*57*
3.2	Seemingly Unrelated Regression	*64*
3.2.1	Introduction	*64*
3.2.2	Constrained Least Squares	*65*
3.2.3	Inter-equations Correlation	*66*
3.2.4	SUR With Panel Data	*67*
3.3	The Maximum Likelihood Estimator	*71*
3.3.1	Derivation of the Likelihood Function	*71*
3.3.2	Computation of the Estimator	*73*
3.4	The Nested Error Components Model	*74*
3.4.1	Presentation of the Model	*74*
3.4.2	Estimation of the Variance of the Error Components	*75*
4	**Tests on Error Component Models**	*83*
4.1	Tests on Individual and/or Time Effects	*83*
4.1.1	F Tests	*84*
4.1.2	Breusch-Pagan Tests	*84*
4.2	Tests for Correlated Effects	*88*
4.2.1	The Mundlak Approach	*89*
4.2.2	Hausman Test	*90*
4.2.3	Chamberlain's Approach	*90*

4.2.3.1 Unconstrained Estimator *91*
4.2.3.2 Constrained Estimator *93*
4.2.3.3 Fixed Effects Models *93*
4.3 Tests for Serial Correlation *95*
4.3.1 Unobserved Effects Test *95*
4.3.2 Score Test of Serial Correlation and/or Individual Effects *96*
4.3.3 Likelihood Ratio Tests for AR(1) and Individual Effects *99*
4.3.4 Applying Traditional Serial Correlation Tests to Panel Data *101*
4.3.5 Wald Tests for Serial Correlation using *within* and First-differenced Estimators *102*
4.3.5.1 Wooldridge's *within*-based Test *102*
4.3.5.2 Wooldridge's First-difference-based Test *103*
4.4 Tests for Cross-sectional Dependence *104*
4.4.1 Pairwise Correlation Coefficients *104*
4.4.2 CD-type Tests for Cross-sectional Dependence *105*
4.4.3 Testing Cross-sectional Dependence in a pseries *107*

5 Robust Inference and Estimation for Non-spherical Errors *109*
5.1 Robust Inference *109*
5.1.1 Robust Covariance Estimators *109*
5.1.1.1 Cluster-robust Estimation in a Panel Setting *110*
5.1.1.2 Double Clustering *115*
5.1.1.3 Panel Newey-west and SCC *116*
5.1.2 Generic Sandwich Estimators and Panel Models *120*
5.1.2.1 Panel Corrected Standard Errors *122*
5.1.3 Robust Testing of Linear Hypotheses *123*
5.1.3.1 An Application: Robust Hausman Testing *125*
5.2 Unrestricted Generalized Least Squares *127*
5.2.1 General Feasible Generalized Least Squares *128*
5.2.1.1 Pooled GGLS *129*
5.2.1.2 Fixed Effects GLS *130*
5.2.1.3 First Difference GLS *132*
5.2.2 Applied Examples *133*

6 Endogeneity *139*
6.1 Introduction *139*
6.2 The Instrumental Variables Estimator *140*
6.2.1 Generalities about the Instrumental Variables Estimator *140*
6.2.2 The *within* Instrumental Variables Estimator *141*
6.3 Error Components Instrumental Variables Estimator *143*
6.3.1 The General Model *143*
6.3.2 Special Cases of the General Model *145*
6.3.2.1 The *within* Model *145*
6.3.2.2 Error Components Two Stage Least Squares *146*
6.3.2.3 The Hausman and Taylor Model *146*
6.3.2.4 The Amemiya-Macurdy Estimator *147*
6.3.2.5 The Breusch, Mizon and Schmidt's Estimator *147*
6.3.2.6 Balestra and Varadharajan-Krishnakumar Estimator *147*
6.4 Estimation of a System of Equations *154*
6.4.1 The Three Stage Least Squares Estimator *155*

6.4.2 The Error Components Three Stage Least Squares Estimator *156*
6.5 More Empirical Examples *158*

7 Estimation of a Dynamic Model *161*
7.1 Dynamic Model and Endogeneity *163*
7.1.1 The Bias of the OLS Estimator *163*
7.1.2 The within Estimator *164*
7.1.3 Consistent Estimation Methods for Dynamic Models *165*
7.2 GMM Estimation of the Differenced Model *168*
7.2.1 Instrumental Variables and Generalized Method of Moments *168*
7.2.2 One-step Estimator *169*
7.2.3 Two-steps Estimator *171*
7.2.4 The Proliferation of Instruments in the Generalized Method of Moments Difference Estimator *172*
7.3 Generalized Method of Moments Estimator in Differences and Levels *174*
7.3.1 Weak Instruments *174*
7.3.2 Moment Conditions on the Levels Model *175*
7.3.3 The System GMM Estimator *177*
7.4 Inference *178*
7.4.1 Robust Estimation of the Coefficients' Covariance *178*
7.4.2 Overidentification Tests *179*
7.4.3 Error Serial Correlation Test *181*
7.5 More Empirical Examples *182*

8 Panel Time Series *185*
8.1 Introduction *185*
8.2 Heterogeneous Coefficients *186*
8.2.1 Fixed Coefficients *186*
8.2.2 Random Coefficients *187*
8.2.2.1 The Swamy Estimator *187*
8.2.2.2 The Mean Groups Estimator *190*
8.2.3 Testing for Poolability *192*
8.3 Cross-sectional Dependence and Common Factors *194*
8.3.1 The Common Factor Model *195*
8.3.2 Common Correlated Effects Augmentation *196*
8.3.2.1 CCE Mean Groups vs. CCE Pooled *198*
8.3.2.2 Computing the CCEP Variance *199*
8.4 Nonstationarity and Cointegration *200*
8.4.1 Unit Root Testing: Generalities *201*
8.4.2 First Generation Unit Root Testing *204*
8.4.2.1 Preliminary Results *204*
8.4.2.2 Levin-Lin-Chu Test *205*
8.4.2.3 Im, Pesaran and Shin Test *205*
8.4.2.4 The Maddala and Wu Test *206*
8.4.3 Second Generation Unit Root Testing *207*

9 Count Data and Limited Dependent Variables *211*
9.1 Binomial and Ordinal Models *213*
9.1.1 Introduction *213*

9.1.1.1 The Binomial Model *213*
9.1.1.2 Ordered Models *214*
9.1.2 The Random Effects Model *214*
9.1.2.1 The Binomial Model *214*
9.1.2.2 Ordered Models *217*
9.1.3 The Conditional Logit Model *219*
9.2 Censored or Truncated Dependent Variable *223*
9.2.1 Introduction *223*
9.2.2 The Ordinary Least Squares Estimator *223*
9.2.3 The Symmetrical Trimmed Estimator *225*
9.2.3.1 Truncated Sample *225*
9.2.3.2 Censored Sample *226*
9.2.4 The Maximum Likelihood Estimator *226*
9.2.4.1 Truncated Sample *226*
9.2.4.2 Censored Sample *227*
9.2.5 Fixed Effects Model *227*
9.2.5.1 Truncated Sample *227*
9.2.5.2 Censored Sample *229*
9.2.6 The Random Effects Model *233*
9.2.6.1 Truncated Sample *233*
9.2.6.2 Censored Sample *234*
9.3 Count Data *236*
9.3.1 Introduction *236*
9.3.1.1 The Poisson Model *236*
9.3.1.2 The NegBin Model *237*
9.3.2 Fixed Effects Model *237*
9.3.2.1 The Poisson Model *237*
9.3.2.2 Negbin Model *239*
9.3.3 Random Effects Models *239*
9.3.3.1 The Poisson Model *239*
9.3.3.2 The NegBin Model *240*
9.4 More Empirical Examples *243*

10 Spatial Panels *245*
10.1 Spatial Correlation *245*
10.1.1 Visual Assessment *245*
10.1.2 Testing for Spatial Dependence *246*
10.1.2.1 CD p Tests for Local Cross-sectional Dependence *247*
10.1.2.2 The Randomized W Test *247*
10.2 Spatial Lags *250*
10.2.1 Spatially Lagged Regressors *251*
10.2.2 Spatially Lagged Dependent Variables *253*
10.2.2.1 Spatial OLS *254*
10.2.2.2 ML Estimation of the SAR Model *254*
10.2.3 Spatially Correlated Errors *255*
10.3 Individual Heterogeneity in Spatial Panels *258*
10.3.1 Random versus Fixed Effects *258*
10.3.2 Spatial Panel Models with Error Components *260*
10.3.2.1 Spatial Panels with Independent Random Effects *260*

10.3.2.2 Spatially Correlated Random Effects *261*
10.3.3 Estimation *261*
10.3.3.1 Spatial Models with a General Error Covariance *262*
10.3.3.2 General Maximum Likelihood Framework *263*
10.3.3.3 Generalized Moments Estimation *267*
10.3.4 Testing *269*
10.3.4.1 LM Tests for Random Effects and Spatial Errors *269*
10.3.4.2 Testing for Spatial Lag vs Error *272*
10.4 Serial and Spatial Correlation *277*
10.4.1 Maximum Likelihood Estimation *277*
10.4.1.1 Serial and Spatial Correlation in the Random Effects Model *277*
10.4.1.2 Serial and Spatial Correlation with KKP-Type Effects *278*
10.4.2 Testing *281*
10.4.2.1 Tests for Random Effects, Spatial, and Serial Error Correlation *281*
10.4.2.2 Spatial Lag vs Error in the Serially Correlated Model *284*

Bibliography *285*

Index *297*

Preface

While R is the software of choice and the undisputed leader in many fields of statistics, this is not so in econometrics; yet, its popularity is rising both among researchers and in university classes and among practitioners. From user feedback and from citation information, we gather that the adoption rate of panel-specific packages is even higher in other research fields outside economics where econometric methods are used: finance, political science, regional science, ecology, epidemiology, forestry, agriculture, and fishing.

This is the first book entirely dedicated to the subject of doing panel data econometrics in R, written by the very people who wrote most of the software considered, so it should be naturally adopted by R users wanting to do panel data analysis within their preferred software environment. According to the best practices of the R community, every example is meant to be replicable (in the style of package vignettes); all code is available from the standard online sources, as are all datasets. Most of the latter are contained in a dedicated companion package, *pder*. The book is supposed to be both a reasonably comprehensive reference on R functionality in the field of panel data econometrics, illustrated by way of examples, and a primer on econometric methods for panel data in general.

While we have tried to cover the vast majority of basic methods and much of the more advanced ones (corresponding roughly to graduate and doctoral level university courses), the book is still less exhaustive than main reference textbooks (one for all, Baltagi, 2013) the *a priori* being that the reader should be able to apply all the methods presented in the book through available R code from *plm* and related, more specialized packages.

One should note from the beginning that, from a computational viewpoint, the average R user tends to be more advanced than users of commercial statistical packages. R users will generally be interested in interactive statistical programming whereby they can be in full control of the procedures they use and eventually be looking forward to write their own code or adapt the existing one to their own purposes. All that said, despite its reputation, R lends itself nicely to standard statistical practice: issuing a command, reading output. Hence the potential readership spans an unusually broad spectrum and will be best identified by subject rather than by level of technical difficulty.

Examples are usually written without employing advanced features but still using a fair amount of syntax beyond what would be the plain vanilla "estimate, print summary" procedure sketched above; the reader replicating them will therefore be exposed to a number of simple but useful constructs—ranging from general purpose visualization to compact presentation of results—stemming from the fact that she is using a full-featured programming language rather than a canned package.

The general level is introductory and aimed at both students and practitioners. Chapters 1–2, and to some extent 4–5, cover the basics of panel data econometrics as taught in undergraduate econometrics classes, if at all. With some overlapping, the main body of the book (Ch. 3–6) covers the typical subjects of an advanced panel data econometrics course at graduate level. Nevertheless, the coverage of the later chapters (especially 7–10) spans fields typical of current applied research; therefore it should appeal particularly to graduate students and researchers. For all this, the book might play two main roles: companion to advanced textbooks for graduate students taking a panel data course, with Chapters 1–7 covering the course syllabus and 8–10 providing more cutting-edge material for extensions; and reference text for practitioners or applied researchers in the field, covering most of the methods they are ever likely to use, with applied examples from recent literature. Nevertheless, its first half can be used in an undergraduate course as well, especially considering the wealth of examples and the possibility to replicate all material. Symmetrically, the last chapters can appeal to researchers wanting to employ cutting-edge methods—for which there is usually around only quite unfriendly code written in matrix language by methodologists—with the relative user-friendliness of R. As an example, Ch. 10 is based on the R tutorials one of the authors gives at the Spatial Econometrics Advanced Institute in Rome, the world-leading graduate school in applied spatial econometrics.

Econometrics is a late comer to the world of R, although of course much of basic econometrics employs standard statistical tools, which were present in base R. Typical functionality, addressing the emphasis on model assumptions and testing, which is characteristic of the discipline, started to appear with the lmtest package and the accompanying paper of Zeileis & Hothorn (2002); a review paper on the use of R in econometrics, focused on teaching, was published at about the same time (Racine & Hyndman, 2002). This was followed by further dedicated packages extending the scope of specialized methods to structural equation modeling, time series, stability testing, and robust covariance estimation, to name a few; while despite the availability of some online tutorials, no dedicated book would appear in print until Kleiber & Zeileis (2008).

In the wake of any organized and comprehensive R package for panel data econometrics, Yves Croissant started developing plm in 2006, presenting one early version of the software at the 2006 useR! Conference in Vienna. Giovanni Millo joined the project as coauthor shortly thereafter. Two years later, an accompanying paper to *plm* (Croissant & Millo, 2008) featured prominently in the econometrics special issue of the *Journal of Statistical Software* testifying the improved availability of econometric methods in R and the increased relevance of the R project for the profession.

More recently, Kevin Tappe has become the third author. Liviu Andronic, Arne Henningsen, Christian Kleiber, Ott Toomet, and Achim Zeileis importantly contributed to the package at various times. Countless users provided feedback, smart questions, bug reports, and, often, solutions.

Estimating the user base is no simple task, but the available evidence points at large and growing numbers. The 2008 paper describing an earlier version of the package has since been downloaded almost 100,000 times and peaked on Goggle Scholar's list as the 25th most cited paper in the *Journal of Statistical Software*, the leading outlet in the field, before hitting the five-year reporting limit. At the time of writing, it counts over 400 citations on Google Scholar, despite the widespread bad habit of not citing software papers. The monthly number of package downloads from a leading mirror site has been recently estimated at 6,000.

Chapters 2, 3, 6, 7, and 8 have been written by Yves Croissant; 1, 5, 9 (except the first generation unit root testing section), and 10 by Giovanni Millo, chapter 4 being co-written.

The book has been produced through Emacs+ESS (Rossini et al., 2004) and typeset in LaTeX using *Sweave* (Leisch, 2002) and later *knitr* (Xie, 2015). Plots have been made using *ggplot2* (Wickham, 2009) and *tikz* (Tantau, 2013).

The companion package to this book is *pder* (Croissant & Millo, 2017); the methods described are mainly in the *plm* package (Croissant & Millo, 2008) but also in *pglm* (Croissant, 2017) and *splm* (Millo & Piras, 2012). General purpose tests and diagnostics tools of packages *car* (Fox & Weisberg, 2011), *lmtest* (Zeileis & Hothorn, 2002), *sandwich* (Zeileis, 2006b), and *AER* (Kleiber & Zeileis, 2008) have been used in the code, as have some more specialized tools available in *MASS* (Venables & Ripley, 2002), *censReg* (Henningsen, 2017), *nlme* (Pinheiro et al., 2017), *survival* (Therneau & Grambsch, 2000), *truncreg* (Croissant & Zeileis, 2016), *pcse* (Bailey & Katz, 2011), and *msm* (Jackson, 2011). *dplyr* (Wickham & Francois, 2016) has been used to work with data.frames and *Formula* with general formulas. *stargazer* (Hlavac, 2013) and *texreg* (Leifeld, 2013) were used to produce fancy tables, the *fiftystater* package (Murphy, 2016) to plot a United States map. The packages presented and the example code are entirely cross-platform as being part of the R project.

Acknowledgments

We thank Kevin Tappe, now a coauthor of "plm," for his invaluable help in improving, checking and extending the functionality of the package. It is difficult to overstate the importance of his contribution.

Achim Zeileis, Christian Kleiber, Ott Toomet, Liviu Andronic, and Nina Schoenfelder have contributed code, fixes, ideas, and interesting discussions at different stages of development. Too many users to list here have provided feedback, good words of encouragement, and bug reports. Often those reporting a bug have also provided, or helped in working out, a solution.

We thank the authors of all the papers that are replicated or simply cited here, for their inspiring research and for making their datasets available. Barbara Rossi (editor) and James MacKinnon (maintainer of the data archive) of the *Journal of Applied Econometrics* (*JAE*) are thanked together with the original authors for kindly sharing the *JAE* data archive datasets.

Personal thanks

Yves Croissant
The first drafts of several chapters of the book have been written while giving a panel data course in the applied economics master of the University of La Reunion. I thank the students of this course for their useful feedback, which helped improving the text. I've been working with Fabrizio Carlevaro on several projects for about 20 years. During this collaboration, he shared with me his deep knowledge of econometrics, and the endless discussions we had were an invaluable source of inspiration for me.

Giovanni Millo
I thank my parents, Luciano and Lalla, for lifelong support and inspiration; Roberta, for her love and patience; my uncle Marjan, for giving me my first electronic calculator—a TI30—when I was a child, sparking a lasting interest for automatic computing; my mentors Attilio Wedlin, Gaetano Carmeci, and Giorgio Calzolari, for teaching me econometrics; and Davide Fiaschi, Angela Parenti, Riccardo "Jack" Lucchetti, Eduardo Rossi, Giuseppe Arbia, Gianfranco Piras, Elisa Tosetti, Giacomo Pasini, and other friends from the "small world" of Italian econometrics—again, too many to list exhaustively here—for so many interesting discussions about econometrics, computing with R, or both.

About the Companion Website

This book is accompanied by a companion website:

www.wiley.com/go/croissant/data-econometrics-with-R

The website includes code for reproducing all examples in the book, which can be found below:

Examples Ch.1
Examples Ch.2
Examples Ch.3
Examples Ch.4
Examples Ch.5
Examples Ch.6
Examples Ch.7
Examples Ch.8
Examples Ch.9
Examples Ch.10

The datasets are to be found in the *pder* package in the below link:
https://cran.r-project.org/web/packages/pder/index.html

Scan this QR code to visit the companion website.

1

Introduction

This book is about doing panel data econometrics with the R software. As such, it is aimed at both panel data analysts who want to use R and R users who endeavor in panel data analysis. In this introductory chapter, we will motivate panel data methods through a simple example, performing calculations in base R, to introduce panel data issues to the R user; then we will give an overview of econometric computing in R for the analyst coming from different software packages or environments.

1.1 Panel Data Econometrics: A Gentle Introduction

In this section we will introduce the broad subject of panel data econometrics through its features and advantages over pure cross-sectional or time-series methods. According to Baltagi (2013), panel data allow to control for individual heterogeneity, exploit greater variability for more efficient estimation, study adjustment dynamics, identify effects one could not detect from cross-section data, improve measurement accuracy (micro-data instead of aggregated), use one dimension to infer about the other (as in panel time series).

From a statistical modeling viewpoint, first and foremost, panel data techniques address one broad issue: unobserved heterogeneity, aiming at *controlling for unobserved variables possibly biasing estimation.*

Consider the regression model

$$y = \alpha_o + \beta_o x + \gamma_o z + \epsilon_o$$

where x is an observable regressor and z is unobservable. The feasible model on observables

$$y = \alpha + \beta x + \epsilon$$

suffers from an omitted variables problem; the OLS estimate of $\hat{\beta}$ is consistent if z is uncorrelated with either y or x: otherwise it will be biased and inconsistent.

One of the best-known examples of unobserved individual heterogenetiy is the agricultural production function by Mundlak (1961) (see also Arellano, 2003, p. 9) where output y depends on x (labor), z (soil quality) and a stochastic disturbance term (rainfall) so that the data-generating process can be represented by the above model; if soil quality z is known to the farmer, although unobservable to the econometrician, it will be correlated with the effort x and hence $\hat{\beta}_{\text{OLS}}$ will be an inconsistent estimator for β.

This is usually modeled with the general form:

$$y_{nt} = \alpha + \beta^\top x_{nt} + (\eta_n + \nu_{nt}) \tag{1.1}$$

Panel Data Econometrics with R, First Edition. Yves Croissant and Giovanni Millo.

Companion website: www.wiley.com/go/croissant/data-econometrics-with-R

where η_n is a time-invariant, generally unobservable characteristic. In the following we will motivate the use of panel data in the light of the need to control for unobserved heterogeneity. We will eliminate the individual effects through some simple techniques. As will be clear from the following chapters, subject to further assumptions on the nature of the heterogeneity there are more sophisticated ways to control for it; but for now we will stay on the safe side, depending only on the assumption of time invariance.

1.1.1 Eliminating Unobserved Components

Panel data turn out especially useful if the unobserved heterogeneity z is (can be assumed) time-invariant. Leveraging the information on time variation for each unit in the cross section, it is possible to rewrite the model (1.1) in terms of observables only, in a form that is equivalent as far as estimating β is concerned. The simplest one is by subtracting one cross section from the other.

1.1.1.1 Differencing Methods

Time-invariant individual components can be removed by first-differencing the data: lagging the model and subtracting, the time-invariant components (the intercept and the individual error component) are eliminated, and the model

$$\Delta y_{nt} = \beta^{\top} \Delta x_{nt} + \Delta v_{nt} \tag{1.2}$$

(where $\Delta y_{nt} = y_{nt} - y_{n,t-1}$, $\Delta x_{nt} = x_{nt} - x_{nt-1}$ and, from (1.1), $\Delta \epsilon_{nt} = \epsilon_{nt} - \epsilon_{n,t-1}$ for $t = 2, \ldots, T$) can be consistently estimated by pooled OLS. This is called the *first-difference*, or FD estimator.

1.1.1.2 LSDV Methods

Another possibility to account for time-invariant individual components is to explicitly introduce them into the model specification, in the form of individual intercepts. The second dimension of panel data (here: time) allows in fact to estimate the η_ns as further parameters, together with the parameters of interest β. This estimator is referred to as *least squares dummy variables*, or LSDV. It must be noted that the degrees of freedom for the estimation do now reduce to $NT - N - K$ because of the extra parameters. Moreover, while the $\hat{\beta}$ vector is estimated using the variability of the full sample and therefore the estimator is NT-consistent, the estimates of the individual intercepts $\hat{\eta}_n$ are T-consistent, as relying only on the time dimension. Nevertheless, it is seldom of interest to estimate the individual intercepts.

1.1.1.3 Fixed Effects Methods

The LSDV estimator is adding a potentially large number of covariates to the basic specification of interest and can be numerically very inefficient. A more compact and statistically equivalent way of obtaining the same estimator entails transforming the data by subtracting the average over time (individual) to every variable. This, which has become the standard way of estimating fixed effects models with individual (time) effects, is usually termed *time-demeaning* and is defined as:

$$y_{nt} - \bar{y}_{n.} = (x_{nt} - \bar{x}_{n.})\beta + (v_{nt} - \bar{v}_{n.}) \tag{1.3}$$

where $\bar{y}_{n.}$ and $\bar{x}_{n.}$ denote individual means of y and X.

This is equivalent to estimating the model

$$y_{nt} = \alpha_n + x_{nt}\beta + v_{nt},$$

i.e., leaving the individual intercepts free to vary, and considering them as parameters to be estimated. The estimates $\hat{\alpha}_n$ can subsequently be recovered from the **OLS** estimation of time-demeaned data.

Example 1.1 individual heterogeneity – `Fatalities` data set
The Fatalities dataset from Stock and Watson (2007) is a good example of the importance of individual heterogeneity and time effects in a panel setting.

The research question is whether taxing alcoholics can reduce the road's death toll. The basic specification relates the road fatality rate to the tax rate on beer in a classical regression setting:

$$frate_n = \alpha + \beta beertax_i + \epsilon_n.$$

Data are 1982 to 1988 for each of the continental US states.

The basic elements of any estimation command in R are a `formula` specifying the model design and a dataset, usually in the form of a `data.frame`. Pre-packaged example datasets are the most hassle-free way of importing data, as needing only to be called by name for retrieval. In the following, the model is specified in its simplest form, a bivariate relation between the death rate and the beer tax.

```
data("Fatalities", package="AER")
Fatalities$frate <- with(Fatalities, fatal / pop * 10000)
fm <- frate ~ beertax
```

The most basic step is a cross-sectional analysis for one single year (here, 1982). One proceeds first creating a model object through a call to `lm`, then displaying a `summary.lm` of it. Printing to screen occurs when interactively calling an object by name. Notice that subsetting can be done inside the call to `lm` by feeding an expression that solves into a logical vector to the `subset` argument: data points corresponding to `TRUE`s will be selected, `FALSE`s discarded.

```
mod82 <- lm(fm, Fatalities, subset = year == 1982)
summary(mod82)

Call:
lm(formula = fm, data = Fatalities, subset = year == 1982)

Residuals:
   Min     1Q Median     3Q    Max
-0.936 -0.448 -0.107  0.230  2.172

Coefficients:
            Estimate Std. Error t value Pr(>|t|)
(Intercept)    2.010      0.139   14.46   <2e-16 ***
beertax        0.148      0.188    0.79     0.43
---
Signif. codes:
0 '***' 0.001 '**' 0.01 '*' 0.05 '.' 0.1 ' ' 1

Residual standard error: 0.67 on 46 degrees of freedom
Multiple R-squared:  0.0133,   Adjusted R-squared:  -0.00813
F-statistic: 0.621 on 1 and 46 DF,  p-value: 0.435
```

The beer tax turns out statistically insignificant. Turning to the last year in the sample (and employing `coeftest` for compactness):

```
mod88 <- update(mod82, subset = year == 1988)
library("lmtest")

coeftest(mod88)

t test of coefficients:

             Estimate Std. Error t value Pr(>|t|)
(Intercept)     1.859      0.106   17.54   <2e-16 ***
beertax         0.439      0.164    2.67    0.011 *
---
Signif. codes:
0 '***' 0.001 '**' 0.01 '*' 0.05 '.' 0.1 ' ' 1
```

the coefficient is significant *and positive*! Similar results appear for any single year in the sample.

Pooling all cross sections together, without considering any form of individual effect, can be done using the regular `lm` function or, equivalently, `plm`; in this second case, for reasons which will be clearer in the following, this is not the default behavior, so the optional `model` argument has to be specified, setting it to `'pooling'`.

Drawing on this much enlarged dataset does not change the qualitative result:

```
library("plm")
poolmod <- plm(fm, Fatalities, model="pooling")
coeftest(poolmod)

t test of coefficients:

             Estimate Std. Error t value Pr(>|t|)
(Intercept)    1.8533     0.0436   42.54  < 2e-16 ***
beertax        0.3646     0.0622    5.86  1.1e-08 ***
---
Signif. codes:
0 '***' 0.001 '**' 0.01 '*' 0.05 '.' 0.1 ' ' 1
```

Taxing beer would seem to increase the number of deaths from road accidents so that, extending this line of reasoning far beyond what the given evidence supports, i.e., far outside the given sample, one could even argue that free beer might lead to safer driving. Similar results, contradicting the most basic intuition, appear for any single year in the sample.

Panel data analysis will provide a solution to the puzzle. In fact, we suspect the presence of unobserved heterogeneity: in specification terms, we suspect the restriction $\alpha_n = \alpha \forall n$ in the more general model

$$frate_{nt} = \alpha_n + \beta beertax_{nt} + \epsilon_{nt}$$

to be invalid. If omitted from the specification, the individual intercepts – but for a general mean – will end up in the error term; if they are not independent of the regressor (here, if unobserved state-level characteristics are related to how the local beer tax is set) the OLS estimate will be biased and inconsistent.

As outlined above, the simplest way to get rid of the individual intercepts is to estimate the model in differences. In this case, we consider differences between the first and last years in the sample. A limited amount of work on the dataset would be sufficient to define a new variable $\Delta_5 y_{nt} = y_{nt} - y_{n,t-5}$ but, as it turns out, for reasons that will become clear in the following chapters, the `diff` method well-known from time series does work in the correct way when applied to panel data through the **plm** package, i.e., `diff(y, s)` is correctly calculated as $y_{nt} - y_{nt-s}$:

```
dmod <- plm(diff(frate, 5) ~ diff(beertax, 5), Fatalities, model="pooling")
coef(dmod)
     (Intercept) diff(beertax, 5)
        -0.02524         -0.95554
```

Estimation on five-year differences finally yields a sensible result: after controlling for state heterogeneity, higher taxation on beer is associated with a lower number of fatalities.

As discussed, another way to control for time-invariant unobservables is to estimate them out explicitly. Separate intercepts could be easily added in plain R using the formula syntax:

```
lsdv.fm <- update(fm, . ~ . + state - 1)
lsdvmod <- lm(lsdv.fm, Fatalities)
coef(lsdvmod)[1]
beertax
-0.6559
```

The estimate is numerically different but supports the same qualitative conclusions.

Fixed effects (*within*) estimation yields an equivalent result in a more compact and efficient way. Specifying `model='within'` in the call to `plm` is not necessary because this estimation method is the default one.

```
library("plm")
femod <- plm(fm, Fatalities)
coeftest(femod)

t test of coefficients:

        Estimate Std. Error t value Pr(>|t|)
beertax   -0.656      0.188   -3.49  0.00056 ***
---
Signif. codes:
0 '***' 0.001 '**' 0.01 '*' 0.05 '.' 0.1 ' ' 1
```

The fixed effects model, requiring only minimal assumptions on the nature of heterogeneity, is one of the simplest and most robust specifications in panel data econometrics and often the benchmark against which more sophisticated, and possibly efficient, ones are compared and judged in applied practice. Therefore it is also the default choice in the basic estimating function `plm`.

Example 1.2 no heterogeneity – `Tileries` data set
There are cases when unobserved heterogeneity is not an issue. The `Tileries` dataset contains data on output and labor and capital inputs for 25 tileries in two regions of Egypt, observed over 12 to 22 years. We estimate a production function. The individual units are rather homogeneous, and the technology is standard; hence, most of the variation in output is explained by the observed inputs. Here, a pooling specification and a fixed effects one give very similar results, especially if restricting the sample to one of the two regions considered:

```
data("Tileries", package = "pder")
coef(summary(plm(log(output) ~ log(labor) + machine, data = Tileries,
             subset = area == "fayoum")))
            Estimate Std. Error    t-value   Pr(>|t|)
log(labor) 0.9174031    0.04661 19.681312 2.933e-45
machine    0.0001074    0.01244  0.008638 9.931e-01
```

```
coef(summary(plm(log(output) ~ log(labor) + machine, data = Tileries,
             model = "pooling", subset = area == "fayoum")))
            Estimate Std. Error t-value  Pr(>|t|)
(Intercept) 0.173423    0.07054  2.4584 1.493e-02
log(labor)  0.964845    0.03818 25.2705 3.992e-60
machine     0.002243    0.01000  0.2242 8.228e-01
```

Notice that we have employed yet another way of compactly looking at the coefficients' table only, instead of printing the whole model summary: the `coef.plm` extractor method, applied to a `summary.plm` object.

By the object orientation of R, applying `coef` to a *model* or to the *summary of a model* – in object terms, to a `plm` or to a `summary.plm` – will yield different results. The curious reader might want to try it himself.

In the following chapters we will see how to test formally for the absence of significant individual effects. For now let us concentrate on how to get things done in R, and the relation to how you would in some other environments.

1.2 R for Econometric Computing

R is widely considered a powerful tool with a relatively steep learning curve. This is true only up to a point as far as econometric computing with R is considered. In fact, rather than complicated, R is *scalable*: it can adapt to the level of difficulty/proficiency adequate for the current user. One might say that R is a "complicated" statistical tool in the same way as a drill is a more complicated tool than a hammer, or a screwdriver. Just like a drill, nevertheless, R *can* actually turn screws: although it can also do so much more.[1]

In a sense, R encompasses most other econometric software, with the exception of that based exclusively on a graphical user interface. While the effective way to use R for econometric computing is to take advantage from its peculiarities, e.g., leveraging the power of object orientation,

1 A drill can be used in place of a hammer for driving nails too, although with limited efficiency. So can R; but this is another story.

it is in fact possible to mimic in R both the *modus operandi* of procedural statistical packages and of course the functionality of other matrix languages.

In the following we will briefly hint at effective ways to perform econometric computing in R, referring the reader to Kleiber and Zeileis (2008) for a more complete treatment; then, in order to provide a friendly introduction to users of different software, we will show how R can be employed the way one would use a "canned" statistical package, or a "hard-boiled" matrix language.

1.2.1 The Modus Operandi of R

R can be used interactively, issuing one command at a time and reading the results from the session log; or it can be operated in batch mode, writing and then executing an R script. The two modes usually mix up, in that even if one writes commands in an editor, it is customary to execute them one by one, or possibly in small groups.

An edited `.R` file has a number of advantages, first of all that the whole session will be completely reproducible as long as the original data are available. There are nevertheless ways to recover all statements used from a session log, which can be turned into an executable .R script with a reasonable amount of editing, or even more easily from the command history, so that if one starts loosely performing some exploratory calculation and then changes his or her mind, perhaps because of some interesting result, nothing is lost. In short, after an interactive session, one can save:

- the session log in a text file (`.txt`)
- the command history in a text file (`.Rhistory`)
- the whole workspace, or a selection of objects, in a binary file (`.Rdata` or, respectively, `.rda`)

From a structured session's approach, there are two competing approaches to the preservation of a reproducible statistical analysis, like one that led to writing a scientific paper: either "the data are real,", or "the commands are real." In the first case, one saves all the objects that have been created during the work session: perhaps the original data, as read from the original source into a `data.frame` but most importantly the model, and possibly test, objects produced by the statistical procedures so that each one can be later (re)loaded, inspected, and printed out, yielding the needed scientific results. In the second case, the original data are kept untransformed, next to plain text files containing all the R statements necessary for full reproduction of the given analysis. This can be done by simply conserving the data file and one or more .R files containing the procedures; or in more structured formats like the popular Sweave framework and utility (Leisch, 2002), whereby the whole scientific paper is dynamically reproducible.

The "commands are real" approach has the advantage of being entirely based on human-readable files (supposing the original data are also, as is always advisable, kept in human-readable format), and its clarity is hard to surpass. Any analysis is reproducible on every platform where R can be compiled, and any file is open to easy inspection in a text editor, should anything go wrong, while binary files, even from Open Source software like R, are always potentially prone to compatibility problems, however unlikely. But considerations on computational and storage demands also play a role.

Computations are performed just once in the first case – but for the (usually inexpensive) extraction of results from already estimated model objects – and at each reproduction in the second; so that the "real data" approach can be preferable, or even the only practical alternative, for computationally heavy analyses. By contrast, the "real commands" approach is much more parsimonious from the viewpoint of storage space, as besides the original data one only needs to archive some small text files.

1.2.2 Data Management

1.2.2.1 Outsourcing to Other Software

In the same spirit, although R is one of the best available tools for managing data, users with only a casual knowledge of it can easily preprocess the data in the software of their choice and then load them into R. The **foreign** package (R Core Team, 2017) provides easy one-step import from a number of popular formats. Gretl (Cottrell and Lucchetti, 2007) took it one step further, providing the ability to call R from inside Gretl and to send to it the current dataset. In general, passing through a conversion into tab- (or space-, or comma-) delimited text and a call to the `read.table` function will solve most import problems and provide an interface between R and anything else, including spreadsheets.

1.2.2.2 Data Management Through Formulae

Even at this level one should notice, however, that R formulae are very powerful tools, accepting a number of transformations that can be done "on the fly" eliminating most of the need for data pre-processing. An obvious example are logs, lags, and differences or, as seen above, the inclusion of dummy variables. Power transformations and interaction terms can also be specified inside formulae in a very compact way. A limited investment of time can let even the casual user discover that most of his usual pre-processing can be disposed of, leaving a clean process from the original raw dataset to the final estimates.

Perhaps the use of formulae in R is the first investment an occasional user might want to do, for all the time and errors it saves by streamlining the flow between the original data and the final result.

1.3 plm for the Casual R User

This book is best for readers with familiarity with the basics of R. Nevertheless, using R interactively – the way econometric software is usually employed – to perform most of the analyses presented here requires very few language-related concepts and only three basic abilities:

- how to import data,
- which commands to issue to obtain estimates,
- optionally, how to save the output to a text file or render it toward LaTeX (but one could as well copy results from the active session).

This corresponds to the typical work flow of a statistician using specialized packages, where one issues one single high-level command, possibly of a very rich nature and with lots of switches, performing some complicated statistical procedure in batch mode, and gets the standard output printed out on screen.

Distinctions are of course sharper than this, and the boundaries between specialized packages, where macro commands perform batch procedures, and matrix languages, where in principle estimators have to be written down by the user, are blurred. In fact, and with time, packages have grown proprietary programming features and sometimes matrix languages of their own, so that much development on the computational frontier of econometric methods can be done by the users in interpreted language, just as happens in the R environment, rather than provided in compiled form by the software house. A notable example of this convergence is Gretl (Cottrell and Lucchetti, 2007), a GUI-based open-source econometric package with full-featured scripting capabilities, entirely programmable and extensible. Some well-known commercial offerings have also taken similar paths.

From the other end of the spectrum, matrix languages have built up huge libraries of ready-made, high-level functions performing complex procedures in one go.

In the following, for the sake of exposition, we will stick to cliché and assume that users of procedural languages expect to run a regression issuing one single command, although perhaps with a lot of arguments, and obtain a lengthy and very comprehensive output containing all the estimation results and diagnostics they might ever need, while matrix language users seek to perform regressions from scratch as $\hat{\beta} = (X^\top X)^{-1} X^\top y$, and obtain any post-estimation diagnostics in the same fashion.

1.3.1 R for the Matrix Language User

The latter viewpoint in our stylized world is that of die-hard econometricians-programmers, who do anything by coding estimators in matrix language. Understandably, the transition toward R is easier done in this case, as it too is a matrix language in its own right. Armed with some cheat sheet providing the translation of basic operators, users of matrix languages can be up and running in no time, learning the important differences in syntax and the language idiosyncrasies of R along the way. As for the moment, here is how linear regression "from scratch" is done in R:

Example 1.3 linear regressions – `Fatalities` data set

In order to perform linear regression "by hand" (*i.e.*, without resorting to a higher level function than simple matrix operators), we have to prepare the y vector and the X matrix, intercept included and then use them in the R translation of the least squares formula:

```
y <- Fatalities$frate
X <- cbind(1, Fatalities$beertax)
beta.hat <- solve(crossprod(X), crossprod(X,y))
```

Notice the use of the numerically efficient operators `solve` and `crossprod` instead of the plain syntax `solve(t(X) %*% X) %*% t(X) %*% y`, which – up to the numerically worst conditioned cases – would produce identical results. (Notice also that we do not need to explicitly make a vector of ones: binding by column (`cbind`-ing) the scalar 1 to a vector of length N, the former is recycled as needed.)

Next, we check that our hand-made calculation produces the same coefficients as the higher-level function `lm`:[2]

```
beta.hat
         [,1]
[1,] 1.8533
[2,] 0.3646
mod <- lm(frate ~ beertax, Fatalities)
coef(mod)
(Intercept)     beertax
     1.8533      0.3646
```

2 Notice that although the coefficients produced by the two methods are numerically the same, from a software viewpoint they are two different object types: the former a 2×1 `matrix`, the latter a (named) `numeric` vector.

It is less straightforward to perform an LSDV or a fixed effects analysis. In the former case, one must create a matrix of state dummy variables: this is cumbersome to do in plain matrix language but is much easier if leveraging the features of R's formulae: in the latter case, it is enough to add the individual index under form of a `factor`: *i.e.*, the R type for qualitative variables.[3]

```
LSDVmod <- lm(frate ~ beertax + state - 1, Fatalities)
coef(LSDVmod)["beertax"]
beertax
-0.6559
```

Estimation is also relatively easy in the fixed effects case, provided that a peculiar feature of R without an obvious counterpart in other matrix languages steps in: ragged arrays. In the following snippet, the `mean` function is *applied* along the individual index to obtain the time means for each individual, which are then replicated along the length of the time dimension. The vectors of time averages are then subtracted from the original vectors to obtain the time-demeaned data, on which plain OLS can be applied (`attach` and `detach` are used to bring the contents of the `data.frame` to user level, to avoid having to point at each variable through the `Fatalities$`... prefix).

```
attach(Fatalities)
frate.tilde <- frate - rep(tapply(frate, state, mean),
                           each = length(unique(year)))
beertax.tilde <- beertax - rep(tapply(beertax, state, mean),
                           each = length(unique(year)))
lm(frate.tilde ~ beertax.tilde - 1)

Call:
lm(formula = frate.tilde ~ beertax.tilde - 1)

Coefficients:
beertax.tilde
       -0.656
detach(Fatalities)
```

This simple example already gives an idea of the small computational complications arising from LSDV or fixed effects estimation. For example, it would not work for unbalanced panels *as is*. The simple modification required to generalize the above snippet to the unbalanced case is left as an exercise for the willing reader.

1.3.2 R for the User of Econometric Packages

The opposite vision is to resort to macro commands. At a bare minimum, users who are familiar with procedural languages can obtain the same result with R:

- issue estimation command,
- get printed output

3 Text labels like state names would be automatically converted, while numerical codes would not. In the latter case, one would use `as.factor(state)` within the `formula`.

despite the logical separation between the steps of creating a model object, summarizing it, and printing the summary, which can a) be executed separately but can also b) be nested inside the same statement, exploiting the functional logic of R, by which "inner" arguments are evaluated first, (implicitly) printing the summary of a model object which is estimated on the fly inside the same statement.[4] Easier done than said:

```
summary(plm(fm, Fatalities))
Oneway (individual) effect Within Model

Call:
plm(formula = fm, data = Fatalities)

Balanced Panel: n = 48, T = 7, N = 336

Residuals:
     Min.   1st Qu.    Median   3rd Qu.      Max.
-0.58696 -0.08284 -0.00127  0.07955  0.89780

Coefficients:
        Estimate Std. Error t-value Pr(>|t|)
beertax   -0.656      0.188   -3.49  0.00056 ***
---
Signif. codes:
0 '***' 0.001 '**' 0.01 '*' 0.05 '.' 0.1 ' ' 1

Total Sum of Squares:    10.8
Residual Sum of Squares: 10.3
R-Squared:        0.0407
Adj. R-Squared: -0.12
F-statistic: 12.1904 on 1 and 287 DF, p-value: 0.000556
```

The construct `summary(myestimator(myformula, mydata, ...))` will generally work, displaying estimation results to screen, for most estimators. Diagnostics will often have a `formula` method so that a statement along the lines of `mytest(myformula, mydata, ...)` will produce the desired output, or, at most, they will require the trivial task of making a "model" object before applying the desired test to it: which can as well happen in one single statement, like `mytest(myestimator(myformula, mydata, ...))`. In this sense, R is a good substitute of procedural languages, at least those that require text input from the command line; despite the fact of also being so much more.

If one is not scared of typing, we might even say that inputting the above statement is not far from the level of difficulty of using a point-and-click GUI. Sure it is not any more difficult to read output from the above R command than that of the standard regression in a GUI package.

1.4 plm for the Proficient R User

A better knowledge of R will disclose a wealth of possibilities streamlining the production process of empirical research. Actually, while R might look difficult or unfriendly to the beginner,

4 Intentionally convoluted sentence. This is what actually happens under the bonnet, but the user need not necessarily worry about it.

for the proficient user the overall workload when producing a piece of scientific research may turn out to be much lower than with competing solutions. The convenient features that allow for a more advanced management of research activity with respect to the usual paradigm "analyze the data – save the results – write the paper around them" can also be seen in the light of producing reproducible econometric research.

1.4.1 Reproducible Econometric Work

Performing econometric work in R, possibly in conjunction with LaTeX through literate statistical tools like Sweave (Leisch, 2002) and knitr (Xie, 2015), satisfies desirable standards of reproducibility.

Following Peng (2011), "[an] important barrier [to reproducible research] is the lack of an integrated infrastructure for distributing [it] to others." Yet such infrastructures have recently emerged in statistics and have been proposed for econometric practice. As advocated by Koenker and Zeileis (2009), one way of ensuring the complete reproducibility of one's research is to provide a self-contained Sweave file – "a tightly coupled bundle of code and documentation" – including all the text as well as the code generating the results of the paper so that, given the original data, the complete document can be reproduced exactly by anybody, on practically any computing platform.

Three aspects of R are worth highlighting in this context: object orientation; code availability, documentation, and management; and reproducible econometric research through literate programming functionalities. The latter two, in particular, help situate econometric work (properly) done with R toward the better end of the reproducibility spectrum in Peng (2011), the "gold standard" of full replication, as providing "a detailed log of every action taken by the computer," which can be replicated by anyone with any type of machine and an Internet connection. In this sense, R code is *linked* and *executable* (Peng, 2011, Fig.1) without the need for either proprietary software or particular hardware/operating system, with the only possible limit of computing power.

As for *availability*, R is open-source software (OSS); hence, all code can be used, inspected, copied, and possibly modified at will. Source code, in the words of Koenker and Zeileis (2009), is "the ultimate form of documentation for computational science," and being accessible it can more easily be subjected to critical scrutiny (on the subject, see also Yalta and Lucchetti, 2008; Yalta and Yalta, 2010).

Besides accessibility, being OSS has important consequences on numerical accuracy (see Yalta and Yalta, 2007) and, what matters most here, on the particular aspect of reproducibility. The R project encourages (in a sense, enforces) documentation of code through its packaging system: in order for a package to build, every (user-level) function inside it must be properly documented, with valid syntax and working examples, as checked by automated scripts. Reliability levels are explicit too: the main distribution site, the Comprehensive R Archive Network (cran.r-project.org) accepts stable versions of packages, subject to a further validation step; earlier versions of code, labeled according to development status (from "Planning" to "Mature"), are to be found on collaborative development platforms of which R-Forge (r-forge.r-project .org/) (Theußl and Zeileis, 2009) is a prominent example. The latter, although typically containing very recent methods, are subject to all the above mentioned quality controls but also allow for immediate patching of code; all changes are tracked inside the system's version history and are open to inspection from any user.

Lastly, and perhaps most importantly here, R explicitly encourages reproducibility of research through utilities like Sweave (Leisch, 2002), which implements literate programming techniques weaving together code and documentation in a dynamic document, as discussed in Meredith and Racine (2009) and Koenker and Zeileis (2009, 2.5). Convenient interfaces for weaving together R and LaTeX are available, from Emacs + ESS (Rossini et al., 2004) to the more recent RStudio (Racine, 2012). This book has in fact been prepared as a dynamic LaTeX document, using the Emacs editor in ESS mode.

1.4.2 Object-orientation for the User

R has object-orientation features. Beside their user-friendliness, such features have a role of their own in reproducibility: simplifying the code makes it more readable and using modular, high-level components with sensible defaults for the different objects is generally safer, especially for the accident-prone data manipulations and transformations typical of panel data.

Methods for extracting (individual, average, or pooled) coefficients, standard errors and measures of fit from model objects of different kinds work with the same syntax, although with different internals, transparently for the user. Formulae with compact representations of lags and differences can be supplied to panel estimators, where the above operators will automatically adjust to the particular context of panel data. Moreover, compact formulations of dynamic models can be indexed, as in `lag(x, 1:i)` for $x_{t-1}, \ldots, x_{t-i}$, and used inside flow control structures, simplifying the making of large tables. Preliminary data manipulation can often be avoided altogether, calculating lags, differences, logs, or more specific panel operations, such as averaging or demeaning over the time or individual dimension, inside the model formula. As observed before, this generally allows to maintain only two files: the original data source and the procedures, with obvious benefits to reliability and replicability of results.

The flexibility object-orientation features provide is highlighted when considering that the R workspace can contain objects of many different kinds at the same time: in this instance, panel or simple models, model formulae, matrices or lists of weights for representing spatial dependence, and, differently from some widespread econometric packages, datasets of various dimensions at the same time. Such flexibility is particularly useful in research work that blends methods from different lines of research together, in order to avoid having to use different software environments for the tasks at hand, and the common pitfalls of not saving the code relative to preliminary data manipulations, or that which combines the results together (see Peng, 2011, p. 1226).

1.5 plm for the R Developer

The last frontier for **plm** users is to become developers. The operation of **plm** is based on a specific data infrastructure able to deal with the peculiar aspects of panel data: basically, their double indexing feature, the possibility of *unbalancedness*, and the frequent need for transformations along one (or both) dimension(s). This mid-level functionality for (panel) data transformation is in general accessible at user level and can be very handy for those developing new methods, e.g., involving estimation over transformed data. It is in fact already in use by a number of other packages: in particular, but not only, some packages aimed at more specific needs presented in this book (**pglm**, **splm**), which are based on this infrastructure and are mostly compliant with **plm**'s conventions and syntax.

Just as the econometric estimation of a fixed effects model proceeds through applying standard OLS to demeaned data, so does the implementation in **plm**, like many others. Yet, unlike many other software packages, here these steps can be readily performed in an explicit fashion.

Example 1.4 explicit *within* transformation – `Fatalities` data set
In order to demonstrate *within* regression, we apply the transformation functions directly in the model formula, excluding *a priori* the intercept (which has been transformed out):

```
w.mod <- plm(Within(frate) ~ Within(beertax) - 1, data=Fatalities,
             model = "pooling")
coef(w.mod)
Within(beertax)
        -0.6559
```

(If trying this at home, remember that, unlike the coefficient, the standard error from this model's output would have to be adjusted by the degrees of freedom to match that of the canned *within* routine. See the discussion in the next chapter, 2.2.3.)

As often happens with R, "ideas are turned into software" (Chambers, 1998) in a natural way, the computational approach following the conceptual flow of the statistical reasoning. Moreover, while all of the software tools provided, being open-source, can ultimately be inspected by the skilled programmer, in the case of **plm** much of the infrastructure is available at user level, conveniently packaged with help and examples, both for instruction purposes and as a building block for further development.

1.5.1 Object-orientation for Development

One last observation is in order, whose scope is not limited to **plm** or panel data econometrics. For a developer, working inside the R project has the huge benefit that she is able to access a majority of all available statistical techniques from *inside* her preferred computing environment, by simply loading the relevant package. In our particular field, this means that one can leverage functionality from, say, general statistics, such as, e.g., using principal components analysis to approximate common factors (see Chapter 8); or from quantitative geography, such as calculating distances between the centroids of regions to make spatial weights matrices (see Chapter 10). This has to do with the *functional* orientation of R, by which complex (statistical) tasks are *abstracted* into functions and therefore made available irrespective of the internals (what happens under the hood).

Another side of abstraction is *object-orientation*: generic methods are often provided, which particularize into different actual computations depending on the object they are fed. Simple examples are `summary` and `plot`, which will produce different outcomes if applied to, say, a `numeric` or an `lm`.

A related, relevant feature of R, and in general of the S language (Chambers, 1998), for the developer is that *functions are a data type*. This means that a function (the abstraction of a statistical procedure) can be passed on to another statistical procedure simply calling it by name. A simple example is the case of the Wald test for generic linear restrictions of the form $R\gamma = r$ on the parameter vector γ:

$$\text{Wald}(R\hat{\gamma} - r)^\top [R^\top \hat{\text{V}} R]^{-1} (R\hat{\gamma} - r) \tag{1.4}$$

Taking the OLS estimate of the linear model as an example, the standard – or "classical" – covariance matrix $\hat{\text{V}}_{\text{OLS}} = \frac{\sum_{n=1}^{N} \hat{e}^2}{N-(K+1)}(Z^\top Z)^{-1}$ will only be appropriate if the errors are independent and identically distributed. If heteroscedasticity is present, the parameter estimates $\hat{\gamma}_{OLS}$ are still consistent, but $\hat{V}_{OLS}$ is not. The test can then be robustified employing a heteroscedasticity-consistent covariance estimator in place of $\hat{V}_{OLS}$ (Zeileis, 2006a).

The R counterpart of the Wald test is the `linearHypothesis` function, aliased by the abbreviation `lht` (Fox and Weisberg, 2011). Mimicking the relevant statistical procedure, the latter will use `coef` and – by default – `vcov` methods to extract $\hat{\gamma}$ and $\hat{V}$ from the estimated model, plugging them into (1.4). By default, an `lm` object will contain $\hat{V}_{OLS}$ but the user can, optionally, provide *a different way to calculate the covariance* under form of the function argument `vcov`.

Example 1.5 Wald test with user-supplied covariance – `Tileries` data set
As previously seen, the production function model in the `Tileries` dataset is a good candidate for a pooling specification. Below, for the sake of exposition, we estimate a linearized Cobb-Douglas version of the production function, in order to test a hypothesis of constant returns to scale. It seems appropriate, as a first step, to estimate a pooled specification by OLS:

```
data("Tileries", package = "pder")
til.fm <- log(output) ~ log(labor) + log(machine)
lm.mod <- lm(til.fm, data = Tileries, subset = area == "fayoum")
```

before proceeding to test the restriction $H_0 : \gamma_1 + \gamma_2 = 1$

```
library(car)
lht(lm.mod, "log(labor) + log(machine) = 1")
Linear hypothesis test

Hypothesis:
log(labor)  + log(machine) = 1

Model 1: restricted model
Model 2: log(output) ~ log(labor) + log(machine)

  Res.Df   RSS Df Sum of Sq   F Pr(>F)
1    175 0.602
2    174 0.600  1   0.00104 0.3   0.58
```

Allowing for heteroscedasticity is as easy as passing on `vcovHC` to the `vcov` argument:

```
library(car)
lht(lm.mod, "log(labor) + log(machine) = 1", vcov=vcovHC)
Linear hypothesis test

Hypothesis:
log(labor)  + log(machine) = 1

Model 1: restricted model
Model 2: log(output) ~ log(labor) + log(machine)

Note: Coefficient covariance matrix supplied.

  Res.Df Df    F Pr(>F)
1    175
2    174  1 0.23   0.63
```

The qualitative findings are unchanged, but this is not the point. As the `Note` in the output reminds us, a different covariance estimator has been employed.

Being generic methods, both `lht` and `vcovHC` will select and apply the appropriate particular procedure depending on the object type. Thus, if fed an `lm`, inside `lht.lm` `coef.lm` and `vcovHC.lm` will be applied, with the relevant defaults; if a `plm` is provided instead, `coef.plm` and `vcovHC.plm` will be used.

By default, the most appropriate method for estimating the parameters' covariance in a panel setting is by allowing for clustering. This is what will happen if feeding the `vcovHC` function to the `lht` together with a `plm` object: the `vcovHC` generic will select the `vcovHC.plm` method for doing the actual computing.

Example 1.6 user-supplied covariance, continued – `Tileries` data set
The pooled specification by OLS can be estimated through `plm` as well:

```
plm.mod <- plm(til.fm, data = Tileries, model = "pooling", subset = area == "fayoum")
```

before proceeding to test H_0:

```
library(car)
lht(plm.mod, "log(labor) + log(machine) = 1", vcov = vcovHC)
Linear hypothesis test

Hypothesis:
log(labor)  + log(machine) = 1

Model 1: restricted model
Model 2: log(output) ~ log(labor) + log(machine)

Note: Coefficient covariance matrix supplied.

  Res.Df Df Chisq Pr(>Chisq)
1    167
2    166  1  0.76       0.38
```

Another different covariance has been employed this time, which allows for *clustering* at individual level: an idea that will be explored in Chapter 5. For now it will be sufficient to say that this one, next to heteroscedasticity, allows for error correlation in time within each individual.

Again, constant returns to scale are not rejected; but now our conclusion is valid in a much more general context.

The programmer writing a `lht` method for, say, a hypothetical `mymodel` class will not have to bother about these downstream details because all he needs is for `mymodel` objects to expose `vcov` and `coef` methods and, eventually, to provide alternative covariance estimators, embodied in turn into `vcovXX.mymodel` functions. Then his function will automatically reproduce equation (1.4) in the new context. The **plm** package has been designed to be compliant with this framework and to allow for easy extensions along the lines sketched above.

Next to the issue of designing modular code for easier production and maintenance by re-employing existing functionality in new contexts, object orientation also has important computational advantages in terms of efficiency. As we have seen, object orientation means that the statistical "objects" (the coefficient vector, the covariance) are mapped to computational tools according to types. From the point of view of the developer faced with computational efficiency and accuracy issues, this means that often she is able to exploit the peculiar structure of the problem at hand. Specialized methods (usually written and compiled in C or FORTRAN) are often available, speeding up computations by many orders of magnitude for a specific class of problems.

One simple example is the inversion of block-diagonal symmetric matrices; a typical problem in panel data estimation by GLS, where the estimated error covariance matrix, which is $NT \times NT$, has to be `solved`. An obvious improvement is to exploit the property that the inverse of a block-diagonal matrix is made of the inverses of the individual blocks; nevertheless, defining the error covariance as a `bdsmatrix` object allows to use the fast `solve.bdsmatrix` method from the package by the same name (Therneau, 2014). This solution is used, e.g., for the GGLS estimators described in Chapter 5: a procedure for which computational efficiency is critical, as being statistically appropriate for very large N panels, where on the other hand it becomes computationally problematic.

Another instance where special matrix types greatly extend the feasibility boundaries is in spatial models: here, *sparse* matrices are common, which contain a vast majority of zeros. Simplifying, one could say that sparse matrix algebra methods rely on the additional information on the position of zeros, avoiding both to consume memory for storing them, and to waste resources to compute on them. Sparse matrix methods from the package **spam** (Furrer and Sain, 2010) and from the more general matrix algebra package **Matrix** (Bates and Maechler, 2016) have been extensively employed in the spatial panel methods described in Chapter 10, together with optimizers from **nlme** (Pinheiro et al., 2017) and **MaxLik** (Henningsen and Toomet, 2011) (a discussion is to be found in Millo, 2014, Section 5.2).

On a different but related note, innovation in object types has in turn affected the symbolic descriptions of models: *formulae*, from which model matrices and responses are derived for actual computation. The extension of the `formula` object class into the `Formula` class, which inherits from the former generalizing it to allow multi-part models and multiple responses (Zeileis and Croissant, 2010), is the basis for the consistent specification of a number of estimators based on combining different levels of instrumentation. The consistent and flexible **plm** implementation of the econometric methods described in Chapters 6 and 7 is made possible by the extended functionality of `Formulae`.

This book is on using, rather than developing, panel data methods in R. This short discussion, therefore, cannot but scratch the surface of the wealth of computing infrastructure available to the user who turns toward developing her own methods. We hope to have at least given an intuition and some directions for further inquiry to any user of **plm** and related packages who wants to extend the methods contained herein, leveraging the power of the R environment at large. As Borges put it, "This plan is so vast that each writer's contribution is infinitesimal."

1.6 Notations

This book is necessarily notation-heavy. Moreover, conventions differ across the various subfields of panel data econometrics covered herein. A considerable effort has been made to present formalizations in a consistent way across chapters, although sometimes this can entail a departure from the usual habits.

This section is therefore meant as a reference for the whole book.

1.6.1 General Notation

The probability is denoted by $\boxed{\mathrm{P}}$, the expected value is denoted by $\boxed{\mathrm{E}}$, the variance by $\boxed{\mathrm{V}}$, the trace by $\boxed{\mathrm{tr}}$, the correlation coefficient by $\boxed{\mathrm{cor}}$, and the standard deviation by $\boxed{\sigma}$. A quadratic form is denoted by $\boxed{q}$ and the identity matrix by $\boxed{\mathrm{I}}$. A set of covariates defines two matrices: $\boxed{\mathrm{P}}$, which returns the fitted values when post-multiplied by a vector; and $\boxed{\mathrm{M}}$, which returns the residuals: $\mathrm{P} = X(X^\top X)^{-1}X$ and $\mathrm{M} = \mathrm{I} - \mathrm{P}$. The Cholesky decomposition of a matrix is denoted by $\boxed{C}$, so that:

$$CAC^\top = \mathrm{I}$$

1.6.2 Maximum Likelihood Notations

For models estimated by the maximum likelihood method, the objective function is denoted by $\boxed{\ln L}$, the Jacobian by $\boxed{J}$, the gradient by $\boxed{g}$, the Hessian by $\boxed{H}$ and the information matrix by I. For generic presentations of the log-likelihood method, the generic set of parameters is denoted by $\boxed{\theta}$.

The statistics of the three tests are denoted by $\boxed{\mathrm{LR}}$, $\boxed{\mathrm{LM}}$, and $\boxed{\mathrm{Wald}}$ for, respectively, a likelihood ratio, a Lagrange multiplier, and a Wald test.

1.6.3 Index

A panel is constituted of $\boxed{N}$ individuals denoted by $\boxed{n}$ (when necessary, $\boxed{m}$ is used as an alias for n).

Each individual is observed during $\boxed{T}$ different periods denoted by t (when necessary, s is used as an alias to the t index).

The size of the sample is denoted by $\boxed{O}$, it is equal to $\sum_{n=1}^{N} T_n$, where T_n is the number of time series for individual n. If $T_n = T\ \forall n$ (balanced panel case), we have $O = NT$.

The $\boxed{K}$ covariates are indexed by $\boxed{k}$; note that a column of ones is not consider in this count.

1.6.4 The Two-way Error Component Model

Consider now the two-way error-component model (the more usual one-way individual error component model is obtained as a special case); it writes for an observation:

$$y_{nt} = \alpha + \beta^\top x_{nt} + \epsilon_{nt} = \gamma^\top z_{nt} + \epsilon_{nt}$$

$$\epsilon_{nt} = \eta_n + \mu_t + \nu_{nt}$$

$\boxed{y}$ is the response, $\boxed{\alpha}$ the intercept, $\boxed{x}$ the vector of K covariates with associated coefficients $\boxed{\beta}$. It would be sometimes easier to consider $\boxed{z}$, which is obtained by adding a 1 in the first position of vector x: $z_{nt}^\top = (1, x_{nt}^\top)$, with the vector of associated coefficients $\boxed{\gamma}$ with $\gamma^\top = (\alpha, \gamma^\top)$.

The error of the model $\boxed{\epsilon}$ is the sum of a time-invariant individual effect $\boxed{\eta}$, an individual-invariant time-effect $\boxed{\mu}$, and a residual error $\boxed{\nu}$. Except for some time-series and spatial methods, ν is assumed to be i.i.d..

The variance is denoted by σ^2; we therefore have for the error and its components: σ_ϵ^2, σ_η^2, σ_μ^2 and σ_ν^2.

All estimated values are represented by the "true" value with a hat so that, for example, the estimated coefficients, the residuals, and the estimated variance of the errors are respectively: $\hat{\beta}$, $\hat{\epsilon}$, and $\hat{\sigma}^2_{\epsilon}$.

In matrix form, the same model is written:

$$y = \alpha j + X\beta + \epsilon = Z\gamma + \epsilon$$
$$\epsilon = \mathrm{D}_\eta \eta + \mathrm{D}_\mu \mu + \nu$$

where $\boxed{j}$ is a vector of 1, $\boxed{X}$ and $\boxed{Z}$ the covariate matrices (the latter including a first column of ones, the former without it), η the vector of N individual effects, μ the vector of T time effects, and ν the vector of O residual effects.

$\boxed{\mathrm{D}}$ denotes a matrix of dummy variables; D_η and D_μ are respectively the dummy variable matrices for individuals and periods. In the case of balanced panels, and if the observations are ranked first by individual, then by time series ("the t index changes faster"), these two matrices can be expressed using Kronecker products. Denoting by $\boxed{J} = jj^\top$ a square matrix of ones, we have:

$$\mathrm{D}_\eta = \mathrm{I}_N \otimes J_T$$
$$\mathrm{D}_\mu = J_T \otimes \mathrm{I}_N$$

The covariance matrix of the errors ϵ is denoted by $\boxed{\Omega}$. Some simplifying assumptions lead to:

$$\Omega_\epsilon = \sigma^2_\nu \mathrm{I}_{NT} + \sigma^2_\eta \mathrm{I}_N \otimes J_T + \sigma^2_\mu J_T \otimes \mathrm{I}_N$$

or $\Omega_\epsilon = \sigma^2_\nu \boxed{\Sigma}$, with:

$$\Sigma = \mathrm{I}_{NT} + \frac{\sigma^2_\eta}{\sigma^2_\nu}\mathrm{I}_N \otimes J_T + \frac{\sigma^2_\mu}{\sigma^2_\nu} J_T \otimes \mathrm{I}_N$$

1.6.5 Transformation for the One-way Error Component Model

For the one-way individual error component model, the last term disappears. In this case, we'll denote $\boxed{S}$ the matrix that if post-multiplied by a variable, returns a vector of length O containing the individual sums of the variable, each one being repeated T_N times.

$$S = \mathrm{I}_N \otimes J_T$$

We'll also make use of the matrix $\boxed{\bar{\mathrm{I}}} = \mathrm{I} - \bar{J}$, which post-multiplied by a variable, returns the variable in deviation from its overall mean:

$$\bar{\mathrm{I}} = \mathrm{I} - \bar{J}$$

In the case of balanced panels, the *between* and *within* matrices, respectively denoted by $\boxed{B}$ and $\boxed{W}$, can be defined:

$$B = \frac{1}{T} S = \frac{1}{T} \mathrm{I}_N \otimes J_T$$
$$W = \mathrm{I} - B = \mathrm{I}_{NT} - \frac{1}{T} \mathrm{I}_N \otimes J_T$$

Denoting $\boxed{\sigma^2_\iota} = \sigma^2_\nu + T\sigma^2_\eta$, the covariance matrix of the error can be written:

$$\Omega_\epsilon = \sigma^2_\nu \left(W + \frac{\sigma^2_\iota}{\sigma^2_\nu} B \right) = \sigma^2_\nu \left(W + \frac{1}{\phi^2} B \right)$$

with $\boxed{\phi} = \frac{\sigma_\nu}{\sigma_i}$. We'll also denote $\boxed{\theta} = 1 - \phi$ the fraction of the individual mean that is subtracted in the GLS model.

Two other transformation matrices are used: $\boxed{D}$ and $\boxed{O}$ perform respectively the first-difference and the orthogonal deviation of a vector.

1.6.6 Transformation for the Two-ways Error Component Model

Back to the two-ways error component model, we now have two between matrices:

$$B_\eta = \mathrm{I}_N \otimes J_T / T$$
$$B_\mu = J_T \otimes \mathrm{I}_N / N$$

The *within* matrix is, denoting by $\boxed{\bar{J}} = J_{NT}/NT$ the matrix that post multiplied by a vector representing a variable returns its overall mean repeated NT times:

$$W = \mathrm{I} - B_\eta - B_\mu + \bar{J}$$

Denote further $\bar{B}_\eta = B_\eta - \bar{J}$ and $\bar{B}_\mu = B_\mu - \bar{J}$. The covariance matrix of the errors then writes:

$$\Omega_\epsilon = \sigma_\nu^2 \left(W + \frac{1}{\phi_\eta^2}\bar{B}_\eta + \frac{1}{\phi_\mu^2}\bar{B}_\eta + \frac{1}{\phi_2^2}\bar{J} \right)$$

with:

$$\phi_\eta^2 = \frac{\sigma_\nu}{\sqrt{\sigma_\nu^2 + T\sigma_\eta^2}}$$
$$\phi_\mu^2 = \frac{\sigma_\nu}{\sqrt{\sigma_\nu^2 + N\sigma_\mu^2}}$$
$$\phi_2^2 = \frac{\sigma_\nu}{\sqrt{\sigma_\nu^2 + T\sigma_\eta^2 + N\sigma_\mu^2}}$$

as for the one-way individual error component model, $\theta_i = 1 - \phi_i$ for $i = \eta, \mu, 2$

1.6.7 Groups and Nested Models

The group effect is denoted by $\boxed{\lambda}$, the $\boxed{G}$ groups are indexed by $\boxed{g}$.

1.6.8 Instrumental Variables

The matrix of instruments is denoted by $\boxed{L}$, the number of instruments by $\boxed{M}$.

1.6.9 Systems of Equations

We consider a system of $\boxed{L}$ equations indexed by $\boxed{l}$, aliased by $\boxed{m}$ when necessary. $\boxed{\Xi}$ is the matrix of dimension $O \times L$ where each column contains the error vector for one of the equations. Its covariance matrix is denoted by $\boxed{\Sigma}$:

$$\Sigma = \mathrm{E}(\Xi^\top \Xi)$$

its elements being denoted by $\boxed{\sigma}$.

1.6.10 Time Series

The most general time-series model considered in the book is defined by the following equations:

$$
\begin{aligned}
y_{nt} &= \rho y_{t-1} + \gamma^{\top} z_{nt} + \epsilon_{nt} \\
\epsilon_{nt} &= \eta_n + \nu_{nt} \\
\nu_{nt} &= \psi \nu_{nt-1} + \zeta_{nt}
\end{aligned}
$$

where $\boxed{\rho}$ is the auto-regressive coefficient of the AR(1) model, $\boxed{\psi}$ the coefficient of the AR(1) process for ν, and $\boxed{\zeta}$ are iid errors.

When lags are needed, they'll be denoted by $\boxed{l}$ and the number of lags is denoted by $\boxed{L}$. It is sometimes useful to know the beginning of the process; this will be denoted by $-\boxed{S}$, 1 being the first available observation.

1.6.11 Limited Dependent and Count Variables

For the ordered model, the $\boxed{J}$ levels of the dependent variables, defined by the thresholds $\boxed{\mu}$, are denoted by $\boxed{j}$.

For Poisson models, the Poisson parameter is denoted by $\boxed{\theta}$. In the cross-series case, it is equal to:

$$\theta_n = \eta_n \lambda_{nt}$$

where $\boxed{\lambda}$ is a linear combination of the covariates. $\boxed{\Lambda}$ and $\boxed{Y}$ are respectively the sums of λ and of the response for one individual. For the Negbin model, $\boxed{\delta}$ is the parameter of the distribution and $\boxed{\nu}$ the parameter that links the expected value and the variance of the response.

1.6.12 Spatial Panels

The proximity matrix is denoted by $\boxed{W}$, its elements by $\boxed{w}$. It can be constructed either on binary neighborhood or on the distance $\boxed{d}$ between two individuals. The most general model considered in this book is described by the following equations:

$$
\begin{aligned}
y &= \lambda(\mathrm{I}_T \otimes W)y + Z\gamma + \epsilon \\
\epsilon &= (j_T \otimes \eta) + \nu \\
\nu &= \rho(\mathrm{I}_T \otimes W)\nu + \zeta \\
\zeta_t &= \psi \zeta_{t-1} + \xi_t, \quad t = 1, \ldots, T
\end{aligned}
$$

The first equation defines a SAR model, and the auto-regressive spatial coefficient for the response is denoted by $\boxed{\lambda}$. The third equation defines a SEM model: it indicates that the non-individual part of the error of the model is also spatially auto-correlated, the auto-regressive spatial coefficient being denoted by $\boxed{\rho}$. Finally, the last equation indicates that the residual component of the model is serially auto-correlated, with an AR(1) coefficient denoted by $\boxed{\psi}$; and $\boxed{\xi}$ defines i.i.d.errors.

To simplify the notation, we'll also define matrices $\boxed{A} = \mathrm{I} - \lambda W$ and, $\boxed{B} = \mathrm{I} - \rho W$ which return respectively the spatial filter of the response and the error.

2

The Error Component Model

The error component model is relevant when the slopes, i.e., the marginal effects of the covariates on the response, are the same for all the individuals, the intercepts being *a priori* different. Note that for some authors, the error component model is a byword for the "random-effects model" as opposed to the "fixed-effects model." These two estimators will be analyzed in this chapter as two different ways to consider the individual component of the error terms for the same error component model (assuming no correlation and correlation with the regressors, respectively).

This is the landmark model of panel data econometrics, and this chapter presents the main results about it.

2.1 Notations and Hypotheses

2.1.1 Notations

For the observation of individual n at period t, we can write the model to be estimated, denoting by y_{nt} the response, x_{nt} the vector of K covariates, ϵ_{nt} the error, α the intercept, and β the vector of parameters associated to the covariates:

$$y_{nt} = \alpha + x_{nt}^{\top}\beta + \epsilon_{nt} \tag{2.1}$$

It'll be sometimes easier to store the intercept and the slopes in the same vector of coefficients. Denoting by $\gamma^{\top} = (\alpha, \beta^{\top})$ this vector and $z_{nt}^{\top} = (1, x_{nt}^{\top})$ the associated vector of covariates, the model can then be written:

$$y_{nt} = z_{nt}^{\top}\gamma + \epsilon_{nt} \tag{2.2}$$

For the error component model, the error is the sum of two effects:

- the first, η_n is the individual effect for individual n,
- the second, ν_{nt} is the residual effect, also called the idiosyncratic effect.

$$\epsilon_{nt} = \eta_n + \nu_{nt} \tag{2.3}$$

For the whole sample, we'll denote by y the vector containing the response and X the matrix of covariates, storing the observations ordered by individual first and then by period. We'll suppose from now that the panel is *balanced*, which means that we have the same number of observations (T) for all the individuals (N). In this case, y is a vector of length NT and X a matrix of dimension $NT \times K$.

Panel Data Econometrics with R, First Edition. Yves Croissant and Giovanni Millo.

Companion website: www.wiley.com/go/croissant/data-econometrics-with-R

$$y = \begin{pmatrix} y_{11} \\ y_{12} \\ \vdots \\ y_{1T} \\ y_{21} \\ y_{22} \\ \vdots \\ y_{2T} \\ \vdots \\ y_{N1} \\ y_{N2} \\ \vdots \\ y_{NT} \end{pmatrix} \text{ and } X = \begin{pmatrix} x_{11}^1 & x_{11}^2 & \dots & x_{11}^K \\ x_{12}^1 & x_{12}^2 & \dots & x_{12}^K \\ \vdots & \vdots & \ddots & \vdots \\ x_{1T}^1 & x_{1T}^2 & \dots & x_{1T}^K \\ x_{21}^1 & x_{21}^2 & \dots & x_{21}^K \\ x_{22}^1 & x_{22}^2 & \dots & x_{22}^K \\ \vdots & \vdots & \ddots & \vdots \\ x_{2T}^1 & x_{2T}^2 & \dots & x_{2T}^K \\ \vdots & \vdots & \ddots & \vdots \\ x_{N1}^1 & x_{N1}^2 & \dots & x_{N1}^K \\ x_{N2}^1 & x_{N2}^2 & \dots & x_{N2}^K \\ \vdots & \vdots & \ddots & \vdots \\ x_{NT}^1 & x_{NT}^2 & \dots & x_{NT}^K \end{pmatrix}$$

Denoting by j a vector of ones of length NT, we get:

$$y = \alpha j + X\beta + \epsilon \tag{2.4}$$

When we want to use the extended vector of coefficients, we denote $Z = (j, X)$, and the model to be estimated is:

$$y = Z\gamma + \epsilon \tag{2.5}$$

2.1.2 Some Useful Transformations

Panel data econometricians usually break the total variation up into the sum of intra-individual and inter-individual variations. These two variations can easily be obtained by transforming the data using different transformation matrices, which can be written using Kronecker products.

The Kronecker product of 2 matrices, denoted $A \otimes B$, is the matrix obtained by multiplying each element of A by B.

I_k denotes the identity matrix of dimension k, j_l is a vector of ones of length l and $J_l = j_l \times j_l^\top$ is a matrix of 1 of dimension $l \times l$.

The inter-individual (or *between*) transformation is obtained by using a transformation matrix denoted by B, which is defined by:

$$B = \mathrm{I}_N \otimes J_T / T$$

For example, we have, for $N = 2$ and $T = 3$:

$$B = \begin{pmatrix} 1 & 0 \\ 0 & 1 \end{pmatrix} \otimes \left[\begin{pmatrix} 1 \\ 1 \\ 1 \end{pmatrix} \begin{pmatrix} 1 & 1 & 1 \end{pmatrix} / 3 \right]$$

$$= \begin{pmatrix} 1 & 0 \\ 0 & 1 \end{pmatrix} \otimes \begin{pmatrix} 1/3 & 1/3 & 1/3 \\ 1/3 & 1/3 & 1/3 \\ 1/3 & 1/3 & 1/3 \end{pmatrix}$$

$$= \begin{pmatrix} 1/3 & 1/3 & 1/3 & 0 & 0 & 0 \\ 1/3 & 1/3 & 1/3 & 0 & 0 & 0 \\ 1/3 & 1/3 & 1/3 & 0 & 0 & 0 \\ 0 & 0 & 0 & 1/3 & 1/3 & 1/3 \\ 0 & 0 & 0 & 1/3 & 1/3 & 1/3 \\ 0 & 0 & 0 & 1/3 & 1/3 & 1/3 \end{pmatrix}$$

We then have:

$$(Bx)^\top = (\bar{x}_1, \bar{x}_1, \dots, \bar{x}_1, \bar{x}_2, \bar{x}_2, \dots, \bar{x}_2, \dots, \bar{x}_{N.}, \bar{x}_{N.}, \dots, \bar{x}_{N.})$$

To get the intra-individual (or *within*) transformation, we'll use a transformation matrix W defined as:

$$W = \mathrm{I}_{NT} - \mathrm{I}_N \otimes J_T/T = \mathrm{I}_{NT} - B$$

These two matrices have very important properties:

- they are symmetric, so we then have $B^\top = B$ and $W^\top = W$,
- they are idempotent, which means that $W \times W = W$ and $B \times B = B$. For example, for the *between* transformation, if we apply it twice to x, we obtain: $(B \times B) \times z = B \times (B \times z)$. One computes the individual means of a vector, which already contains individual means; the vector is, therefore, unchanged; we then have $(B \times B) \times z = B \times z$, and the same reasoning applies to W,
- they perform a decomposition of a vector, which means that $B \times z + W \times z = z$, as $W = \mathrm{I} - B$ and therefore $W + B = \mathrm{I}$,
- they are orthogonal: $W^\top B = 0$. Indeed, as the two matrices are symmetric and using the result that $W = \mathrm{I} - B$, we have: $W^\top B = W \times B = (\mathrm{I} - B) \times B = B - B \times B = B - B = 0$. $W(Bz)$ consist in taking the deviations from individual means of the individual means and is therefore equal to 0 irrespective of z.

W and B therefore perform an *orthogonal decomposition* of a vector z; this means that pre-multiplying z by each of the two matrices, we obtain two vectors that sum to z and for which the inner product is 0.

2.1.3 Hypotheses Concerning the Errors

ϵ is the sum of a vector ν of length NT containing the idiosyncratic part of the error and of the individual effect η, which is a vector of length N for which each element is repeated T times. This can be written in matrix form:

$$\epsilon = (\mathrm{I}_N \otimes j_T)\eta + \nu \tag{2.6}$$

The estimated model will be defined by estimated parameters $\hat{\gamma}^\top = (\hat{\alpha}, \hat{\beta}^\top)$ and by a vector of residuals $\hat{\epsilon}$.

$$y = \hat{\alpha} j + X\hat{\beta} + \hat{\epsilon} \tag{2.7}$$

$$y = Z\hat{\gamma} + \hat{\epsilon} \tag{2.8}$$

Subtracting (2.5) from (2.8) enables to write the residuals as a function of the errors:

$$\hat{\epsilon} = \epsilon - Z(\hat{\gamma} - \gamma) \tag{2.9}$$

To get a similar expression in terms of X and β, we use (2.4) and (2.7):

$$\hat{\epsilon} = \epsilon - (\hat{\alpha} - \alpha)j - X(\hat{\beta} - \beta)$$

The mean of this expression is, denoting $\bar{j} = j/(N \times T)$:

$$\bar{j}^\top \hat{\epsilon} = \bar{j}^\top \epsilon - (\hat{\alpha} - \alpha) - \bar{j}^\top X(\hat{\beta} - \beta)$$

In a linear model with an intercept, $\bar{j}^\top \hat{\epsilon}$, which is the average of the residuals, is 0. Using the two previous equations, we get:

$$\hat{\epsilon} = (\mathrm{I} - \bar{J})(\epsilon - X(\hat{\beta} - \beta)) = \bar{\mathrm{I}}(\epsilon - X(\hat{\beta} - \beta)) \tag{2.10}$$

with $\bar{J}_{NT} = jj^\top/(NT)$ a matrix that post-multiplied by a vector returns a vector of the same length containing the overall mean. $\bar{\mathrm{I}} = \mathrm{I} - \bar{J}$, post-multiplied by a vector returns the vector in deviations from the overall mean.

The expressions (2.9 and 2.10) will be used all along this chapter to analyze the properties of the estimators.

The following hypotheses are made concerning the errors:

- the expected values of the two components of the error are supposed to be 0; anyway, their means can't be identified if there is an intercept in the model,
- the individual effects η_n are homoscedastic and mutually uncorrelated,
- the idiosyncratic part of the error ν_{nt} is also homoscedastic and uncorrelated,
- the two components of the errors are uncorrelated.

In this case, the covariance matrix of the errors depends only on the variance of the two components of the errors, i.e., the two parameters σ_ν^2 and σ_η^2. Concerning the variance and covariances of the errors, we then have:

- for the variance of one error: $\mathrm{E}(\epsilon_{nt}^2) = \sigma_\eta^2 + \sigma_\nu^2$,
- for the covariance of two errors of the same individual for two different periods: $\mathrm{E}(\epsilon_{nt}\epsilon_{ns}) = \sigma_\eta^2$,
- for the covariance of two errors of two different individuals (belonging to the same period or not): $\mathrm{E}(\epsilon_{nt}\epsilon_{mt}) = \mathrm{E}(\epsilon_{nt}\epsilon_{ms}) = 0$.

For a given individual n, the covariance matrix of the vector of errors for this individual $\epsilon_n^\top = (\epsilon_{n1}, \epsilon_{n2}, \dots, \epsilon_{nt})$ is:

$$\Omega_{nn} = \mathrm{E}(\epsilon_n \epsilon_n^\top) = \sigma_\nu^2 \mathrm{I}_T + \sigma_\eta^2 J_T \tag{2.11}$$

For the whole sample, we have $\epsilon^\top = (\epsilon_1^\top, \epsilon_2^\top, \dots, \epsilon_N^\top)$, and the covariance matrix is a square matrix of dimension NT that contains submatrices $\mathrm{E}(\epsilon_n \epsilon_m)$. For $n = m$, this submatrix is given by (2.11); for $n \neq m$, this is a 0 matrix given the hypothesis of no correlation between the errors of two different individuals. The covariance matrix of the errors Ω is then a block-diagonal matrix, the N blocks being the matrix given by the equation (2.11). This matrix can then be expressed as a Kronecker product:

$$\Omega = \mathrm{I}_N \otimes (\sigma_\nu^2 \mathrm{I}_T + \sigma_\eta^2 J_T) = \sigma_\nu^2 \mathrm{I}_{NT} + \sigma_\eta^2 (\mathrm{I}_N \otimes J_T)$$

This matrix can also be usefully expressed in terms of the two transformation matrices *within* and *between* described in subsection 2.1.2. In fact, $B = \frac{1}{T}\mathrm{I}_N \otimes J_T$ and $W = \mathrm{I} - B$. Introducing these two matrices in the expression of Ω, we get:

$$\Omega = \sigma_\nu^2 (B + W) + T\sigma_\eta^2 B$$

which finally implies, denoting $\sigma_\iota^2 = \sigma_\nu^2 + T\sigma_\eta^2$:

$$\Omega = \sigma_\nu^2 W + \sigma_\iota^2 B \tag{2.12}$$

Finally, all along this chapter, we'll suppose that both components of the errors are uncorrelated with the covariates: $\mathrm{E}(\eta \mid x) = \mathrm{E}(\nu \mid x) = 0$.

2.2 Ordinary Least Squares Estimators

The variability in a panel has two components:

- the *between* or inter-individual variability, which is the variability of panel's variables measured in individual means, which is $\bar{z}_n$ or, in matrix form, Bz,
- the *within* or intra-individual variability, which is the variability of panel's variables measured in deviation from individual means, which is $z_{nt} - \bar{z}_n$ or, in matrix form $Wz = z - Bz$.

Three estimations by ordinary least squares can then be performed: the first one on raw data, the second one on the individual means of the data (*between* model), and the last one on the deviations from individual means (*within* model).

2.2.1 Ordinary Least Squares on the Raw Data: The *Pooling* Model

The model to be estimated is $y = \alpha j + X\beta\epsilon = Z\gamma + \epsilon$. Using the second formulation, the sum of squares residuals can be written:

$$(y^\top - \gamma^\top Z^\top)(y - Z\gamma)$$

and the first-order conditions for a minimum are (up to the -2 multiplicative factor):

$$Z^\top \hat{\epsilon} = 0 \tag{2.13}$$

The first column of Z is a vector of ones associated to α, which is the first element of γ. Therefore, dividing the first element of this vector by the number of observations leads to:

$$\bar{y} = \hat{\alpha} + \bar{x}^\top \hat{\beta} \tag{2.14}$$

This is the well-known result that the mean of the sample, i.e., $(\bar{x}, \bar{y})$ is on the regression line of the ordinary least squares estimator. The K other first-order conditions imply that $\sum_n \sum_t \hat{\epsilon}_{nt} x_{knt} = 0$, which can be rewritten, the average residual $\bar{\hat{\epsilon}}$ being equal to 0:

$$\frac{\sum_n \sum_t (\hat{\epsilon}_{nt} - \bar{\hat{\epsilon}})(x_{knt} - \bar{x}_k)}{N \times T} = 0 \tag{2.15}$$

which means that the sample covariances between the residuals and the covariates are 0. Solving (2.13), we get the ordinary least squares estimator for the whole vector of coefficients:

$$\hat{\gamma}_{\mathrm{OLS}} = (Z^\top Z)^{-1} Z^\top y \tag{2.16}$$

Substituting y by $Z\gamma + \epsilon$ in (2.16),

$$\hat{\gamma}_{\mathrm{OLS}} - \gamma = (Z^\top Z)^{-1} Z^\top \epsilon \tag{2.17}$$

To get the estimator of the slopes, one splits Z in (j, X) and $\hat{\gamma}^\top$ in $(\hat{\alpha}, \hat{\beta}^\top)$:

$$\begin{pmatrix} \hat{\alpha} \\ \hat{\beta} \end{pmatrix} = \begin{pmatrix} NT & j^\top X \\ X^\top j & X^\top X \end{pmatrix}^{-1} \begin{pmatrix} j^\top y \\ X^\top y \end{pmatrix}$$

The formula for the inverse of a partitioned matrix is given by:

$$\begin{pmatrix} A_{11} & A_{12} \\ A_{21} & A_{22} \end{pmatrix}^{-1} = \begin{pmatrix} A_{11}^{-1}(I + A_{12}F_2A_{21}A_{11}^{-1}) & -A_{11}^{-1}A_{12}F_2 \\ -F_2A_{21}A_{11}^{-1} & F_2 \end{pmatrix} \tag{2.18}$$

with $F_2 = (A_{22} - A_{21}A_{11}^{-1}A_{12})^{-1}$. The upper left block may also be written: $F_1 = (A_{11} - A_{12}A_{22}^{-1}A_{21})^{-1}$

We have here:

$$(Z^\top Z)^{-1} = \begin{pmatrix} 1/NT + j^\top XFX^\top j/(NT)^2 & -j^\top XF^\top/NT \\ -FX^\top j/NT & F \end{pmatrix}$$

with $F = (X^\top(\mathrm{I} - \bar{J})X)^{-1}$. $\bar{J}z$ returns a vector of length NT for which all the elements are the vector mean $\bar{z}$. One can easily check that this matrix is idempotent. We then have:

$$\hat{\beta} = (X^\top(\mathrm{I} - \bar{J})X)^{-1}X^\top(\mathrm{I} - \bar{J})y \tag{2.19}$$

which is a formula similar to (2.16), but with variables pre-multiplied by $\mathrm{I} - \bar{J}$, this transformation removing the overall mean of every variable. For the intercept $\hat{\alpha}$, we find the same expression as (2.14). In order to analyze the characteristics of the OLS estimator, we substitute in (2.19) y by $\alpha j + X\beta + \epsilon$:

$$\hat{\beta} = \beta + (X^\top(\mathrm{I} - \bar{J})X)^{-1}X^\top(\mathrm{I} - \bar{J})\epsilon$$

The estimator is then unbiased ($\mathrm{E}(\hat{\beta}) = \beta$) if $\mathrm{E}(X^\top(\mathrm{I} - \bar{J})\epsilon) = 0$, i.e., if the theoretical covariances between the covariates and the errors are all 0. This result is directly linked with expression (2.13), which indicates that the OLS estimator is computed so that empirical covariances between the residuals and the covariates are all 0. The estimator is consistent if: $\mathrm{plim}\ \hat{\beta} = \beta$. This expression is:

$$\mathrm{plim}\ \hat{\beta} = \beta + \mathrm{plim}\left(\frac{1}{NT}X^\top(\mathrm{I} - \bar{J})X\right)^{-1} \mathrm{plim}\frac{1}{NT}X^\top(\mathrm{I} - \bar{J})\epsilon$$

The first term is the population covariance matrix of the covariates and the second one the population covariance vector of the covariates and the errors. The estimator is therefore consistent if the covariance matrix of the covariates exists, is not 0, and if the covariances between the covariates and the errors are all 0. The variance of the OLS estimator is given by:

$$\mathrm{V}(\hat{\gamma}_{\text{OLS}}) = \mathrm{E}((\hat{\gamma}_{\text{OLS}} - \gamma)(\hat{\gamma}_{\text{OLS}} - \gamma)^\top) = (Z^\top Z)^{-1}Z^\top\Omega Z(Z^\top Z)^{-1} \tag{2.20}$$

Note that for the error component model, the covariance matrix of the errors Ω doesn't reduce to a scalar times the identity matrix because of the correlation induced by the individual effects. Therefore, the variance of the OLS estimator doesn't reduce to $\mathrm{V}(\hat{\gamma}_{\text{OLS}}) = \sigma^2(Z^\top Z)^{-1}$, and using this expression in tests will lead to biased inference.

In conclusion, the OLS estimator, even if it is unbiased and consistent, has two limitations:

- the first one is that the usual estimator of the variance is not correct and should be replaced by a more complex expression,
- the second is that, in this context, OLS is not the best linear unbiased estimator, which means that there exist other linear unbiased estimators that are more efficient.

2.2.2 The *between* Estimator

The *between* estimator is the OLS estimator applied to the model pre-multiplied by B, i.e., the model in individual means.

$$By = BZ\gamma + B\epsilon = \alpha j + BX\beta + B\epsilon$$

Note that the items of the model that don't exhibit intra-individual variations are unaffected by this transformation. This is the case of the column of 1 associated to the intercept, of the matrix $(I_N \otimes j_T)$ associated to the individual effects and also of some covariates with no intra-individual as, for exemple, the gender in a sample of individuals. Note also that the $N \times T$ observations of this model are in fact N distinct observations of individual means repeated T times. Using as in the case of the OLS estimator, the formula of the inverse of a partitioned matrix, the *between* estimator is:

$$\hat{\beta}_B = (X^\top(B - \bar{J})X)^{-1}X^\top(B - \bar{J})y = (X^\top\bar{B}X)^{-1}X^\top\bar{B}y \tag{2.21}$$

$\bar{B} = B - \bar{J}$ is a matrix that transforms a variable in its individual means in deviation from the overall mean. The variance of $\hat{\beta}$ is obtained by replacing y by $\alpha j + X\beta + \epsilon$:

$$\hat{\beta}_B - \beta = (X^\top\bar{B}X)^{-1}X^\top\bar{B}\epsilon$$
$$V(\hat{\beta}_B) = (X^\top\bar{B}X)^{-1}X^\top\bar{B}\Omega\bar{B}X(X^\top\bar{B}X)^{-1}$$

The expression of Ω given by (2.12) implies that $\bar{B}\Omega = \sigma_\iota^2(B - \bar{J})$. Consequently, the expression of the variance of the *between* estimator is simply:

$$V(\hat{\beta}_B) = \sigma_\iota^2(X^\top\bar{B}X)^{-1} \tag{2.22}$$

For the full vector of the coefficients (including the intercept α), the *between* estimator and its variance are:

$$\hat{\gamma}_B = (Z^\top BZ)^{-1}Z^\top By \tag{2.23}$$
$$V(\hat{\gamma}_B) = \sigma_\iota^2(Z^\top BZ)^{-1} \tag{2.24}$$

To estimate σ_ι^2, we use the deviance of the *between* model: $\hat{q}_B = \hat{\epsilon}_B^\top B\hat{\epsilon}_B$. Using (2.23) and (2.9):

$$\hat{\epsilon}_B = (I - Z(Z^\top BZ)^{-1}Z^\top B)\epsilon$$
$$B\hat{\epsilon}_B = (B - BZ(Z^\top BZ)^{-1}Z^\top B)\epsilon = M_B\epsilon$$

The M_B matrix is idempotent, and its trace is, using the property that the trace is invariant under cyclical permutations: $\mathrm{tr}(M_B) = \mathrm{tr}(B) - \mathrm{tr}(I_{K+1}) = N - K - 1$. We then have $\hat{q}_B = \epsilon^\top M_B\epsilon$ and $E(\hat{q}_B) = E(\mathrm{tr}(\epsilon^\top M_B\epsilon)) = E(\mathrm{tr}(M_B\epsilon\epsilon^\top)) = \mathrm{tr}(M_B\Omega)) = \sigma_\iota^2\mathrm{tr}(M_B)$. The unbiased estimator of σ_ι^2 is then $\hat{\sigma}_\iota^2 = \hat{q}_B/(N - K - 1)$. The one returned by an OLS program is: $\hat{q}_B/(NT - K - 1)$ and the covariance matrix of the coefficients should then be multiplied by $(NT - K - 1)/(N - K - 1)$.

2.2.3 The *within* Estimator

The *within* estimator is obtained by applying the OLS estimator to the model pre-multiplied by the W matrix.

$$Wy = W(\alpha j + X\beta + \epsilon) = WX\beta + Wv$$

The *within* transformation removes the vector of 1 associated to the intercept and the matrix associated to the vector of individual effects. It also removes covariates that don't exhibit intra-individual variation. Applying OLS to the transformed model leads to the *within* estimator:

$$\hat{\beta}_W = (X^\top WX)^{-1}X^\top Wy \tag{2.25}$$

The variance of $\hat{\beta}_W$ is:

$$V(\hat{\beta}_W) = (X^\top WX)^{-1}X^\top W\Omega WX(X^\top WX)^{-1}$$

$W\Omega = W(\sigma_\nu^2 W + \sigma_\iota^2 B) = \sigma_\nu W$. The *within* transformation therefore induces a correlation among the errors of the model. The variance of the *within* estimator reduces to:

$$\mathrm{V}(\hat{\beta}_{\mathrm{W}}) = \sigma_\nu^2 (X^\top W X)^{-1} \tag{2.26}$$

we then have, in spite of this correlation, the standard expression of the variance. In order to estimate σ_ν^2, one uses the deviance of the *within* estimator: $\hat{q}_{\mathrm{W}} = \hat{\epsilon}_{\mathrm{W}}^\top W \hat{\epsilon}_{\mathrm{W}}$. Using (2.25) and (2.10):

$$\hat{\epsilon}_{\mathrm{W}} = (\mathrm{I} - X(X^\top WX)^{-1}X^\top W)\epsilon$$

$$W\hat{\epsilon}_{\mathrm{W}} = (W - WX(X^\top WX)^{-1}X^\top W)\epsilon = \mathrm{M}_{\mathrm{W}}\epsilon$$

The matrix M_{W} is idempotent and its trace is $\mathrm{tr}(\mathrm{M}_{\mathrm{W}}) = \mathrm{tr}(W) - \mathrm{tr}(\mathrm{I}_K) = NT - N - K$. We then have $\mathrm{E}(\hat{q}_{\mathrm{W}}) = \mathrm{E}(\mathrm{tr}(\epsilon^\top \mathrm{M}_{\mathrm{W}}\epsilon)) = \mathrm{E}(\mathrm{tr}(\mathrm{M}_{\mathrm{W}}\epsilon\epsilon^\top)) = \mathrm{tr}(\mathrm{M}_{\mathrm{W}}\Omega)) = \sigma_\nu^2 \mathrm{tr}(\mathrm{M}_{\mathrm{W}})$. The unbiased estimator of σ_ν^2 is then $\hat{\sigma}_\nu^2 = \hat{q}_{\mathrm{W}}/(NT - N - K)$, and the one returned by an OLS program is: $\hat{q}_{\mathrm{W}}/(NT - K - 1)$. The covariance matrix of the coefficients should then be multiplied by: $(NT - K - 1)/(NT - N - K)$.

The *within* model is also called the "fixed-effects model" or the least-squares dummy variable model, because it can be obtained as a linear model in which the individual effects are estimated and then taken as fixed parameters. This model can be written:

$$y = X\beta + (\mathrm{I}_N \otimes j_T)\eta + \nu$$

where η is now a vector of parameters to be estimated. There are therefore $N + K$ parameters to estimate in this model.[1] The estimation of this model is computationally feasible if N is not too large. In a micro panel of large size, the estimation becomes problematic.

The equivalence between both models may be established using the Frisch-Waugh theorem or using the formula of the inverse of a partitioned matrix. The Frisch-Waugh theorem states that it is equivalent to regress y on a set of covariates X_1, X_2 or to regress the residuals of y from a regression on X_2 on the residuals of X_1 on a regression on X_2. The application of the Frisch-Waugh theorem in this context consists in regressing each variable with respect to $X_2 = \mathrm{I}_N \otimes j_T$ and getting the residuals. Here, for each variable, the residual is $z_{nt} - \hat{\eta}_n$. The first-order condition of the sum of squared residuals minimization is $X_2^\top \hat{\epsilon} = 0$. X_2 being a matrix which selects the individuals, we finally get for every individual, denoting $\bar{z}_{n.} = \frac{\sum_{t=1}^T z_{nt}}{T}$:

$$\sum_{t=1}^T (z_{nt} - \hat{\eta}_n) = \sum_{t=1}^T z_{nt} - T\hat{\eta}_n = 0$$

Consequently, we have $\hat{\eta}_n = \bar{z}_{n.}$ and the residuals are the deviations of the variable from its individual means. Therefore, the Frisch-Waugh theorem implies that the fixed effect model can be estimated by applying the OLS estimator to the model transformed in deviations from the individual means, i.e., by regressing Wy on WX.

With the *within* coefficients in hand, specific intercepts for every individual in the sample $\alpha + \eta_n$ can then be computed:

$$\hat{\alpha}_n = \bar{y}_{n.} - \bar{x}_{n.}^\top \hat{\beta}$$

where $\bar{z}_{n.}$ is the vector of individual means of z.

If one wants to define individual effects with 0 mean in the sample, a general intercept can be computed: $\hat{\alpha} = \bar{\bar{y}} - \bar{\bar{x}}^\top \hat{\beta}$, $\bar{\bar{z}}$ being the overall mean of z. We then have for every individual in the sample $\hat{\eta}_n = \hat{\alpha}_n - \hat{\alpha} = (\bar{y}_{n.} - \bar{\bar{y}}) - (\bar{x}_{n.} - \bar{\bar{x}})^\top \hat{\beta}$

1 The N individual effects and the intercept α can't both be identified. The choice made here consists in setting α to 0.

Example 2.1 within estimator – `TobinQ` data set
To illustrate the estimation of the estimators seen in this chapter, we use the `TobinQ` dataset of the **pder** package. These data concern 188 American firms for 35 years (from 1951 to 1985).

```
data("TobinQ", package = "pder")
```

Schaller (1990) wishes to test Tobin (1969)'s theory of investment. In this model, the main variable that explains investment is the ratio between the value of the firm and the replacement cost of its physical capital, this ratio being called "Tobin Q". If the financial market is perfect, the value of the firm equals the actual value of its future profits. If the Tobin Q is greater than 1, this means that the profitability of investment is greater than its cost and so that the investment is valuable. The response is therefore the rate of investment (investment divided by the capital stock) and the covariate is Tobin Q.

The **plm** package provides the `plm` function to estimate linear models on panel data. Its main arguments are:

- `formula`, the symbolic description of the model,
- `data`, the `data.frame`, which can be either an ordinary `data.frame` or a `pdata.frame`; in the first case, the `index` may be added to indicate the individual and time index,
- `model`, the estimator one wants to compute: `'within'`, `'between'`, `'pooling'` (which is the OLS estimator) and `'random'` (which is the GLS estimator that will be presented in the next section).

We first create a `pdata.frame` using the `pdata.frame` function. This is done indicating in the `index`:

- a character vector of length two indicating the individual and time index,
- a character vector of length one indicating the individual index (in this case, it is assumed that there is no time index in the data),
- an integer indicating the number of periods (only for a balanced panel with observations first ordered by individuals and then by period),
- `NULL`, the default: in this case, it is assumed that the first two columns of the `data.frame` contain the individual and the time index.

These different possibilities are illustrated below, the first two columns of `TobinQ` containing the individual and the time index.

```
pTobinQ <- pdata.frame(TobinQ)
pTobinQa <- pdata.frame(TobinQ, index = 188)
pTobinQb <- pdata.frame(TobinQ, index = c('cusip'))
pTobinQc <- pdata.frame(TobinQ, index = c('cusip', 'year'))
```

The `pdim` function can be used to inspect the individual and time dimensions of the data. It has a method for `pdata.frame` objects (without any further argument) and for `data.frame`. In the latter case, the `index` argument can be set; if not, it is once more assumed that the first two columns of the `data.frame` contain the individual and the time index.

```
pdim(pTobinQ)
Balanced Panel: n = 188, T = 35, N = 6580
```

```
pdim(TobinQ, index = 'cusip')
pdim(TobinQ)
```

A `pdata.frame` has an `index` attribute, which is a `data.frame` that contains the index. It can be extracted using the `index` function:

```
head(index(pTobinQ))
  cusip year
2  2824 1951
3  2824 1952
4  2824 1953
5  2824 1954
6  2824 1955
7  2824 1956
```

We then estimate the three models we have described:

```
Qeq <- ikn ~ qn
Q.pooling <- plm(Qeq, pTobinQ, model = "pooling")
Q.within <- update(Q.pooling, model = "within")
Q.between <- update(Q.pooling, model = "between")
```

Either simple or extended printing of the results is obtained as usual with R applying the `print.plm` or `summary.plm` methods to the object containing the fitted model. For example, for the *within* estimator, we get:

```
Q.within

Model Formula: ikn ~ qn

Coefficients:
     qn
0.00379
summary(Q.within)
Oneway (individual) effect Within Model

Call:
plm(formula = Qeq, data = pTobinQ, model = "within")

Balanced Panel: n = 188, T = 35, N = 6580

Residuals:
    Min.  1st Qu.    Median  3rd Qu.      Max.
-0.21631 -0.04525 -0.00849  0.03365   0.61844

Coefficients:
   Estimate Std. Error t-value Pr(>|t|)
qn 0.003792   0.000173      22   <2e-16 ***
---
```

```
Signif. codes:
0 '***' 0.001 '**' 0.01 '*' 0.05 '.' 0.1 ' ' 1

Total Sum of Squares:    36.7
Residual Sum of Squares: 34.1
R-Squared:        0.0702
Adj. R-Squared: 0.0428
F-statistic: 482.412 on 1 and 6391 DF, p-value: <2e-16
```

For the *within* estimator, the `fixef.plm` method computes the individual effects. Three flavors of fixed effects may be obtained depending on the value of the `type` argument:

- `'level'`, the default value, returns the individual intercepts, i.e., $\hat{\alpha} + \hat{\eta}_n$,
- `'dfirst'` returns the individual effects in deviations from the first individual; $\hat{\alpha}$ is in this case the intercept for the first individual,
- `'dmean'` returns the individual effects in deviations from their mean; in this case, $\hat{\alpha}$ is the average of the individual intercepts.

```
head(fixef(Q.within))
  2824   6284   9158  13716  17372  19411
0.1453 0.1281 0.2581 0.1100 0.1267 0.1695
head(fixef(Q.within, type = "dfirst"))
    6284      9158     13716     17372     19411     19519
-0.01723  0.11279 -0.03528 -0.01856  0.02420 -0.01038
head(fixef(Q.within, type = "dmean"))
     2824      6284      9158     13716     17372     19411
-0.014213 -0.031448  0.098581 -0.049492 -0.032778  0.009986
```

We then illustrate the equivalence of the *within* estimator and the least-squares dummy variables estimator. For this later estimator, we use the `lm` function with the `cusip` variable used as a covariate, as it is the individual index. The default behavior of `lm` is to remove the first level of the factor. The fixed effects are then equal to those obtained using the `fixef.plm` function with the argument `type` equal to `'dfirst'`.

```
head(coef(lm(ikn ~ qn + factor(cusip), pTobinQ)))
        (Intercept)                   qn  factor(cusip)6284
           0.145290             0.003792          -0.017235
 factor(cusip)9158 factor(cusip)13716 factor(cusip)17372
           0.112794            -0.035279          -0.018564
```

2.3 The Generalized Least Squares Estimator

The *within* estimator is a regression on data that have been transformed so that the individual effects vanish (they are, so to say, "transformed out"), while the least squares dummy variables considers the individual effects as parameters to be estimated (they are "estimated out"); both give identical estimates of the slopes. On the contrary, the GLS estimator considers the individual effects as random draws from a specific distribution and seeks to estimate the parameters of this distribution in order to obtain efficient estimators of the slopes.

2.3.1 Presentation of the GLS Estimator

When the errors are not correlated with the covariates but are characterized by a non-scalar covariance matrix Ω, the efficient estimator is the generalized least squares estimator:

$$\hat{\gamma}_{\mathbf{GLS}} = (Z^\top\Omega^{-1}Z)^{-1}(Z^\top\Omega^{-1}y) \tag{2.27}$$

In order to compute the variance of $\hat{\gamma}_{\mathbf{GLS}}$, we substitute as previously y by $Z\gamma + \epsilon$. We then have:

$$\hat{\gamma}_{\mathbf{GLS}} - \gamma = (Z^\top\Omega^{-1}Z)^{-1}Z^\top\Omega^{-1}\epsilon$$

Using a reasoning similar to (2.20), we obtain the variance of the estimator:

$$\begin{aligned} \mathrm{V}(\hat{\gamma}_{\mathbf{GLS}}) &= (X^\top\Omega^{-1}X)^{-1}X^\top\Omega^{-1}\mathrm{E}(\epsilon\epsilon^\top)\Omega^{-1}X(X^\top\Omega^{-1}X)^{-1} \\ &= (X^\top\Omega^{-1}X)^{-1} \end{aligned} \tag{2.28}$$

The hypothesis we have made concerning the errors implies that the covariance matrix of the errors is given by (2.12): $\Omega = \sigma_\nu^2 W + (T\sigma_\eta^2 + \sigma_\nu^2)B$, which is a linear combination of two idempotent and orthogonal matrices. Ω depends only on two parameters: the variances of the two components of the error terms (σ_ν^2 and σ_η^2). We have shown, in subsection 2.1.2, that these two matrices are idempotent ($B \times B = B$ and $W \times W = W$) and orthogonal ($B \times W = 0$). The expression of powers of Ω is then particularly simple:

$$\Omega^\upsilon = \sigma_\iota^{2\upsilon} B + \sigma_\nu^{2\upsilon} W \tag{2.29}$$

which can be easily checked, for example for $\upsilon = 2$. This result can also be extended to negative integers and to rationals; we then have, for $\upsilon = -1$:

$$\Omega^{-1} = \frac{1}{\sigma_\iota^2}B + \frac{1}{\sigma_\nu^2}W$$

and the GLS estimator of the random error model and its variance are then:

$$\hat{\gamma}_{\mathbf{GLS}} = \left(\frac{1}{\sigma_\nu^2}Z^\top WZ + \frac{1}{\sigma_\iota^2}Z^\top BZ\right)^{-1}\left(\frac{1}{\sigma_\nu^2}Z^\top Wy + \frac{1}{\sigma_\iota^2}Z^\top By\right) \tag{2.30}$$

$$\mathrm{V}(\hat{\gamma}_{\mathbf{GLS}}) = \left(\frac{1}{\sigma_\nu^2}Z^\top WZ + \frac{1}{\sigma_\iota^2}Z^\top BZ\right)^{-1} \tag{2.31}$$

For the vector of slopes, we obtain:

$$\hat{\beta}_{\mathbf{GLS}} = \left(\frac{1}{\sigma_\nu^2}X^\top WX + \frac{1}{\sigma_\iota^2}X^\top \bar{B}X\right)^{-1}\left(\frac{1}{\sigma_\nu^2}X^\top Wy + \frac{1}{\sigma_\iota^2}X^\top \bar{B}y\right) \tag{2.32}$$

$$\mathrm{V}(\hat{\beta}_{\mathbf{GLS}}) = \left(\frac{1}{\sigma_\nu^2}X^\top WX + \frac{1}{\sigma_\iota^2}X^\top \bar{B}X\right)^{-1} \tag{2.33}$$

This estimator is called the random effects model, as opposed to the fixed effects model. This results from the fact that, as observed, in this case, the individual effects are considered as random deviates, the parameters of whose distribution we seek to estimate.

The dimension of the matrix Ω is given by the size of the sample. If the sample is large, it is therefore not practical to compute the estimator according to the matrix formula (2.27). A more efficient way is to apply OLS on suitably pre-transformed data. To this end, one has to compute the C matrix such that: $C^\top C = \Omega^{-1}$ and then use this matrix to transform all the variables of the model. Denoting $\tilde{y} = Cy$ and $\tilde{Z} = CZ$ the transformed variables, the estimation by **OLS** on transformed variables gives:

$$\hat{\gamma} = (\tilde{Z}^\top \tilde{Z})^{-1} \tilde{Z}^\top \tilde{y} = (Z^\top C^\top C Z)^{-1} Z^\top C^\top C y = (Z^\top \Omega^{-1} Z)^{-1} Z^\top \Omega^{-1} y$$

which is the **GLS** given by (2.30). The expression of the matrix C is obtained using equation (2.29) for $v = -0.5$:

$$C = \Omega^{-0.5} = \frac{1}{\sigma_\iota} B + \frac{1}{\sigma_\nu} W$$

This transformation consists in a linear combination of the *between* and *within* transformations with weights depending on the variances of the two error components. In fact, pre-multiplying the variables by $\sigma_\nu \Omega^{-0.5}$ (which is equivalent to premultiplication by $\Omega^{-0.5}$ and simplifies notation), the weights become respectively $\frac{\sigma_\nu}{\sigma_\iota}$ and 1. The transformed variable is therefore:

$$\tilde{z}_{nt} = \frac{\sigma_\nu}{\sigma_\iota} \bar{z}_{n.} + (z_{nt} - \bar{z}_{n.}) = z_{nt} - \left(1 - \frac{\sigma_\nu}{\sigma_\iota}\right) \bar{z}_{n.} = z_{nt} - \theta \bar{z}_{n.}$$

with, denoting $\phi = \frac{\sigma_\nu}{\sigma_\iota}$:

$$\theta = 1 - \phi = 1 - \sqrt{\frac{\sigma_\nu^2}{T\sigma_\eta^2 + \sigma_\nu^2}} = 1 - \frac{1}{\sqrt{1 + T\frac{\sigma_\eta^2}{\sigma_\nu^2}}}$$

As will be explained in detail below, the importance of the individual effects in the composite errors, measured by their share of the total variance, determines how close the estimator will be to either the *within* or the pooled **OLS**, which are obtained as special cases, respectively, when the variance of the individual effects σ_η dominates ($\theta \rightarrow 1$) or vanishes ($\theta \rightarrow 0$).

2.3.2 Estimation of the Variances of the Components of the Error

In order to make operational the estimator, residuals from consistent estimators are used to estimate the unknown parameters σ_ν and σ_η (and hence σ_ι). The estimator obtained is then called the feasible generalized least squares estimator.

Consider the errors of the model ϵ_{nt}, their individual mean $\bar{\epsilon}_{n.}$ and their deviations from these individual means $\epsilon_{nt} - \bar{\epsilon}_{n.}$. By hypothesis, we have: $\mathrm{V}(\epsilon_{nt}) = \sigma_\nu^2 + \sigma_\eta^2$. For the individual means, we get:

$$\bar{\epsilon}_{n.} = \frac{1}{T} \sum_{t=1}^{T} \epsilon_{nt} = \eta_n + \frac{1}{T} \sum_{t=1}^{T} \nu_{nt}$$

$$\mathrm{V}(\bar{\epsilon}_{n.}) = \sigma_\eta^2 + \frac{1}{T} \sigma_\nu^2 = \sigma_\iota^2 / T$$

The variance of the deviation from the individual means is easily obtained by isolating terms in ϵ_{nt}:

$$\epsilon_{nt} - \bar{\epsilon}_{n.} = \epsilon_{nt} - \frac{1}{T} \sum_{t=1}^{T} \epsilon_{nt} = \left(1 - \frac{1}{T}\right) \epsilon_{nt} - \frac{1}{T} \sum_{s \neq t} \epsilon_{st}$$

the sum then contains $T - 1$ terms. The variance is:

$$\mathrm{V}(\epsilon_{nt} - \bar{\epsilon}_{n.}) = \left(1 - \frac{1}{T}\right)^2 \sigma_\nu^2 + \frac{1}{T^2} (T-1) \sigma_\nu^2$$

which finally leads to:

$$\mathrm{V}(\epsilon_{nt} - \bar{\epsilon}_{n.}) = \frac{T-1}{T}\sigma_\nu^2$$

If ϵ were known, natural estimators of these two variances σ_ι^2 et σ_ν^2 would be:

$$\hat{\sigma}_\iota^2 = T\frac{\sum_{n=1}^{N}\bar{\epsilon}_{n.}^2}{N} = T\frac{\sum_{n=1}^{N}\sum_{t=1}^{T}\bar{\epsilon}_{n.}^2}{NT} = T\frac{\epsilon^\top B\epsilon}{NT} = \frac{\epsilon^\top B\epsilon}{N} \tag{2.34}$$

$$\begin{aligned}\hat{\sigma}_\nu^2 &= \frac{T}{T-1}\frac{\sum_{n=1}^{N}\sum_{t=1}^{T}(\epsilon_{nt}-\bar{\epsilon}_{n.})^2}{NT} = \frac{\sum_{n=1}^{N}\sum_{t=1}^{T}(\epsilon_{nt}-\bar{\epsilon}_{n.})^2}{N(T-1)} \\ &= \frac{\epsilon^\top W\epsilon}{N(T-1)}\end{aligned} \tag{2.35}$$

i.e., estimators based on the norm of the errors transformed using the *between* and *within* matrices. Of course, the errors are unknown, but consistent estimation of the variances may be obtained by substituting the errors by residuals obtained from a consistent estimation of the model. Among the numerous estimators available, the one proposed by Wallace and Hussain (1969) is particularly simple as it consists on using the **OLS** residuals to write the sample counterpart of equations (2.34) and (2.35)

$$\hat{\sigma}_\iota^2 = \frac{\hat{\epsilon}_{\text{OLS}}^\top B\hat{\epsilon}_{\text{OLS}}}{N}$$

$$\hat{\sigma}_\nu^2 = \frac{\hat{\epsilon}_{\text{OLS}}^\top W\hat{\epsilon}_{\text{OLS}}}{N(T-1)}$$

The estimated variance of the individual effects can then be obtained:

$$\hat{\sigma}_\eta^2 = \frac{\hat{\sigma}_\iota^2 - \hat{\sigma}_\nu^2}{T}$$

The estimator of Amemiya (1971) is based on the estimation of the *within* model. We first compute the overall intercept

$$\hat{\alpha} = \bar{\bar{y}} - \hat{\beta}_{\text{W}}\bar{\bar{x}}$$

and then compute the residuals $\hat{\epsilon}_{\text{W}}$:

$$\hat{\epsilon}_{\text{W}} = y - \hat{\alpha}j - \hat{\beta}_{\text{W}}X$$

These residuals are then used to compute the two quadratic form.

$$\hat{\sigma}_\iota^2 = \frac{\hat{\epsilon}_{\text{W}}^\top B\hat{\epsilon}_{\text{W}}}{N}$$

$$\hat{\sigma}_\nu^2 = \frac{\hat{\epsilon}_{\text{W}}^\top W\hat{\epsilon}_{\text{W}}}{N(T-1)}$$

Note that the later is just the deviance of the *within* estimation divided by $N \times (T-1)$. Note also that the variance of the individual effect is overestimated if the model contains some time-invariant variables which disappear with the within transformation.

In this case, Hausman and Taylor (1981) proposed the following adjustment: $\hat{\epsilon}_{\text{W}}$ are regressed on all the time-invariant variables in the model and the residuals of this regression $\hat{\epsilon}_{\text{HT}}$ are substituted with $\hat{\epsilon}_{W}$ in the computation of the quadratic forms. This will reduce the estimate of $\hat{\sigma}_{\iota}^{2}$ and leave unchanged the estimate of $\hat{\sigma}_{\nu}^{2}$, so that the estimate of $\hat{\sigma}_{\eta}$ will also decrease.

For the Swamy and Arora (1972) estimator, the *within* and the *between* models are estimated. The residuals of the *between* model are used for the first quadratic form and those of the *within* model for the second one.

$$\hat{\sigma}_{\iota}^{2} = \frac{\hat{\epsilon}_{B}^{\top} B \hat{\epsilon}_{B}}{N-K-1}$$

$$\hat{\sigma}_{\nu}^{2} = \frac{\hat{\epsilon}_{W}^{\top} W \hat{\epsilon}_{W}}{N(T-1)-K}$$

Note that Swamy and Arora (1972) use the degrees of freedom of both regressions for the estimation of the variances, i.e., K is deduced from the number of observations. Note also that $B\hat{\epsilon}_{B}$ and $W\hat{\epsilon}_{W}$ are the residuals of the *between* and *within* regressions computed on the transformed data, so that the numerators of the two quadratic forms are the deviances of the two regressions.

For all these estimators, σ_{η}^{2} is not directly estimated but obtained by subtracting $\hat{\sigma}_{\nu}^{2}$ from $\hat{\sigma}_{\iota}^{2}$. In small samples, it can therefore be negative, and in this case it is set to 0.

On the contrary, for the Nerlove (1971) estimator, σ_{η}^{2} is estimated by computing the empirical variance of the fixed effects of the *within* model, as the estimate of σ_{ν} is obtained by dividing the quadratic form of the within residuals by the number of observations.

$$\hat{\eta}_{n} = \bar{y}_{n.} - \hat{\beta}_{W}\bar{x}_{n.}$$

$$\hat{\sigma}_{\eta}^{2} = \sum_{n=1}^{N} (\hat{\eta}_{n} - \bar{\hat{\eta}})^{2}/(N-1)$$

$$\hat{\sigma}_{\nu}^{2} = \frac{\hat{\epsilon}_{W}^{\top} W \hat{\epsilon}_{W}}{N \times T}$$

Example 2.2 random effects model – `TobinQ` data set

The random effects model is obtained by setting `model` to `'random'`. Specific arguments indicate how the variances are estimated.

- `random.method` is one of `'walhus'` for Wallace and Hussain (1969), `'swar'` for Swamy and Arora (1972), `amemiya` for Amemiya (1971), `'ht'` for Hausman and Taylor (1981) and `'nerlove'` for Nerlove (1971).
- `random.models` is an alternative to the `random.methods` argument : it is a character vector of length 1 or 2 that indicates which preliminary estimations are performed in order to estimate the variances; for example, `c("within", "between")` use the *within* residuals to estimate σ_{ν} and the *between* residuals to estimate σ_{ι}, `c("pooling")` or `("pooling", "pooling")` use the pooling residuals for the estimation of both variances,
- `random.dfcor` is a numeric vector of length 2; it indicates what is the denominator of the two quadratic forms. If :
  - 0 the number of observations is used (NT, N),
  - 1, the numerators of the theoretical formulas are used $(N(T-1), N)$
  - 2, the number of estimated parameters are deduced $(N(T-1)-K, N-K-1)$.

The following two commands estimate the same Swamy and Arora (1972) model :

```
Q.swar <- plm(Qeq, pTobinQ, model = "random", random.method = "swar")
Q.swar2 <- plm(Qeq, pTobinQ, model = "random",
               random.models = c("within", "between"),
               random.dfcor = c(2, 2))
summary(Q.swar)
Oneway (individual) effect Random Effect Model
   (Swamy-Arora's transformation)

Call:
plm(formula = Qeq, data = pTobinQ, model = "random", random.method = "swar")

Balanced Panel: n = 188, T = 35, N = 6580

Effects:
                  var std.dev share
idiosyncratic 0.00533 0.07303  0.73
individual    0.00202 0.04493  0.27
theta: 0.735

Residuals:
   Min. 1st Qu.  Median 3rd Qu.    Max.
-0.2330 -0.0475 -0.0103  0.0336  0.6211

Coefficients:
            Estimate Std. Error t-value Pr(>|t|)
(Intercept) 0.159327   0.003425    46.5   <2e-16 ***
qn          0.003862   0.000168    22.9   <2e-16 ***
---
Signif. codes:
0 '***' 0.001 '**' 0.01 '*' 0.05 '.' 0.1 ' ' 1

Total Sum of Squares:    37.9
Residual Sum of Squares: 35.1
R-Squared:      0.0742
Adj. R-Squared: 0.074
F-statistic: 526.854 on 1 and 6578 DF, p-value: <2e-16
```

The results indicate that the part in the variance of the individual effect is about one fourth. The parameter called θ is the part of the individual mean that is removed from each variable for the **GLS** estimator. It can be written as $1 - \frac{1}{\sqrt{1+T\sigma_\eta^2/\sigma_\nu^2}}$ and is here equal to 73%. This high value is due to the large time dimension of this panel ($T = 35$). This implies that the **GLS** estimator is closer to the *within* estimator ($\theta = 1$) than to the **OLS** estimator ($\theta = 0$).

The part of the result that deals with the estimation of the two components of the error may also be obtained by applying the `ercomp` function either to the **GLS** fitted model or using a `formula` – `data` interface:

```
ercomp(Qeq, pTobinQ)
ercomp(Q.swar)
```

We then compare the results obtained with the 4 estimation methods we've presented:

```
Q.walhus <- update(Q.swar, random.method = "swar")
Q.amemiya <- update(Q.swar, random.method = "amemiya")
Q.nerlove <- update(Q.swar, random.method = "nerlove")
Q.models <- list(swar = Q.swar, walhus = Q.walhus,
                 amemiya = Q.amemiya, nerlove = Q.nerlove)
sapply(Q.models, function(x) ercomp(x)$theta)
   swar.id  walhus.id amemiya.id nerlove.id
    0.7351     0.7351     0.7361     0.7489
sapply(Q.models, coef)
                 swar   walhus  amemiya  nerlove
(Intercept) 0.159327 0.159327 0.159328 0.159344
qn          0.003862 0.003862 0.003862 0.003855
```

The first `sapply` command extracts from the ercomp object the `theta` element, indicating the proportion of the individual mean that is removed from the variables. These are very close to each other, and consequently, the estimated coefficients for the 4 models are almost identical.

2.4 Comparison of the Estimators

We have four different estimators of the same model : the *between* and the *within* estimators use only one source of the variance of the sample, while the OLS and the GLS estimators use both.

Note first that, if the hypothesis that the errors and the covariates are uncorrelated is true, all these models are unbiased and consistent, which means that they should give similar results, at least in large samples.

We'll first analyze the relations between these estimators; we'll then compare their variances; and finally we'll analyze in which circumstances we should use fixed or random effects.

2.4.1 Relations between the Estimators

We can expect the OLS and GLS estimators to give intermediate results between the *within* and the *between* estimators as they use both sources of variance. From equation (2.32), the GLS estimator can be written :

$$\hat{\beta}_{\text{GLS}} = (X^\top WX + \phi^2 X^\top \bar{B}X)^{-1}(X^\top Wy + \phi^2 X^\top \bar{B}y)$$

Using (2.21) and (2.25), $\hat{\beta}_{\text{GLS}}$ can then be expressed as a weighted average of the *within* and the *between* estimators.

$$\hat{\beta}_{\text{GLS}} = (X^\top WX + \phi^2 X^\top \bar{B}X)^{-1}(X^\top WX\hat{\beta}_{\text{W}} + \phi^2 X^\top \bar{B}X\hat{\beta}_{\text{B}})$$

A similar result applies to the OLS estimator which is the GLS estimator for $\phi = 1$.

$$\hat{\beta}_{\text{OLS}} = (X^\top WX + X^\top \bar{B}X)^{-1}(X^\top WX\hat{\beta}_{\text{W}} + X^\top \bar{B}X\hat{\beta}_{\text{B}})$$

For the OLS estimator, the weights are very intuitive because they are just the shares of the intra- and the inter-individual variances of the covariates. For the GLS estimator, the weights depend not only on the shares of the variance of the covariates but also on the variance of the errors, which determines the ϕ parameter. The **GLS** estimator will always give less weight to the *between* variation, as ϕ is lower than 1. It leads to two special cases :

- $\phi \rightarrow 0$; this means that σ_ν is "small" compared to σ_η. In this case, the **GLS** estimator converges to the *within* estimator,
- $\phi \rightarrow 1$; this means that σ_ν is "large" compared to σ_η. In this case, the **GLS** estimator converges to the **OLS** estimator.

The relation between the estimators can also be illustrated by the fact that the **OLS** and the **GLS** can be obtained by stacking the *within* and *between* transformations of the model:[2]

$$\begin{pmatrix} Wy \\ By \end{pmatrix} = \begin{pmatrix} WZ \\ BZ \end{pmatrix} \gamma + \begin{pmatrix} W\epsilon \\ B\epsilon \end{pmatrix} \tag{2.36}$$

The matrix of covariance of the errors of this stacked model is :

$$\begin{pmatrix} \sigma_\nu^2 W & 0 \\ 0 & \sigma_\iota^2 B \end{pmatrix} \tag{2.37}$$

Applying **OLS** to (2.36), we get;

$$(Z^\top WZ + Z^\top BZ)^{-1}(Z^\top Wy + Z^\top By) = (Z^\top Z)^{-1} Z^\top y$$

which is the **OLS** estimator,
while applying **GLS** to (2.36) yields the **GLS** estimator of equation (2.30).

$$(Z^\top WZ + \phi^2 Z^\top BZ)^{-1}(Z^\top Wy + \phi^2 Z^\top By)$$

2.4.2 Comparison of the Variances

From equation (2.33), the variance of the **GLS** estimator can be written :

$$\mathrm{V}(\hat{\beta}_{\mathrm{GLS}}) = \sigma_\nu^2 (X^\top WX + \phi^2 X^\top \bar{B}X)^{-1} \tag{2.38}$$

The variance of the *within* estimator being : $\sigma_\nu^2(X^\top WX)^{-1}$, $\mathrm{V}(\hat{\beta}_{\mathrm{W}}) - \mathrm{V}(\hat{\beta}_{\mathrm{GLS}})$ is a positive definite matrix, and the **GLS** estimator is therefore more efficient than the *within* estimator. Similarly, equation (2.22) shows that the variance of the *between* may be written $\sigma_\nu^2(\phi^2 X^\top \bar{B}X)^{-1}$ and therefore $\mathrm{V}(\hat{\beta}_{\mathrm{B}}) - \mathrm{V}(\hat{\beta}_{\mathrm{GLS}})$ is also a positive definite matrix.

2.4.3 Fixed *vs* Random Effects

The individual effects are not fixed or random by nature. Within the same framework (the individual effects model), they are treated as either a vector of constant parameters or the realization of random deviates for the purpose of estimation, depending on their probabilistic structure and, in particular, on their correlation with the explanatory variables.

In a micro-panel, the random effects approach is appealing, as we work on a sample with numerous individuals who are randomly drawn from a very large population. There is no interest in estimating the individual effects, and the random effect approach is more appropriate, given the way the sample was obtained.

2 See Baltagi (2013).

On the contrary, in a macro-panel, the sample is fixed or quasi-fixed and almost exhaustive (think of the countries of the world or the large enterprises of a country). In this case, the estimation of the individual effects may be an interesting result, and the fixed effects approach seems relevant.

Anyway, the main argument that leads to choose one of the two approaches is the possibility of correlation between some covariates and the individual effects. If we maintain the hypothesis that the idiosyncratic error is uncorrelated with the covariates ($E(X^\top \nu) = 0$), two situations can occur :

- $E(X^\top \eta) = 0$: the individual effects are not correlated; in this case, both models are consistent, but the random effects estimator is more efficient that the fixed effects model,
- $E(X^\top \eta) \neq 0$: the individual effects are correlated; in this case, only the fixed effects method gives consistent estimates as, with the *within* transformation, the individual effects vanish.

Example 2.3 comparison of the estimators – `TobinQ` data set
The following command extracts the coefficient of `qn` and its standard deviation for the four estimators (we consider only the Swamy and Arora (1972) method for the GLS estimator, as all the random effects models give very similar results).

```
sapply(list(pooling = Q.pooling, within = Q.within,
            between = Q.between, swar = Q.swar),
       function(x) coef(summary(x))["qn", c("Estimate", "Std. Error")])
             pooling    within   between      swar
Estimate   0.0043920 0.0037919 0.0051847 0.0038622
Std. Error 0.0001529 0.0001726 0.0007491 0.0001683
```

The OLS and GLS estimators are in the interval defined by the *within* and *between* estimators, and the GLS estimator is closer to the *within* estimator than OLS.

Looking at the standard deviations, OLS seems to be the most efficient model, but remember that the standard formula for computing the variance of the OLS estimator is biased if individual effects are present. The standard deviation for the GLS estimator (1.683E-04) is slightly lower than for the *within* estimator (1.726E-04) and much lower than for the *between* estimator (7.491E-04).

The formal relation between the different estimators is then illustrated by computing the shares of the variances for the covariate `qn`. For this purpose, we'll extract this series from the `padata.frame`, which is not, as for `data.frame`, a numeric vector, but a `pseries` object, which inherits from the `pdata.frame` it has been extracted from the `index` attribute. The `summary.psries` method applied to this object indicate the variance structure of the series:

```
summary(pTobinQ$qn)
total sum of squares: 314300
     id    time
0.43081 0.09393
```

We can use the `Within` and the `Between` function with this series in order to compute its *within* and the *between* transformations, and then the weights of the *within* and the *between* estimators in the OLS estimator.

```
SxxW <- sum(Within(pTobinQ$qn) ^ 2)
SxxB <- sum((Between(pTobinQ$qn) - mean(pTobinQ$qn)) ^ 2)
SxxTot <- sum( (pTobinQ$qn - mean(pTobinQ$qn)) ^ 2)
pondW <- SxxW / SxxTot
pondW
[1] 0.5692
pondW * coef(Q.within)[["qn"]] +
  (1 - pondW) * coef(Q.between)[["qn"]]
[1] 0.004392
```

The weight of the *within* model is 57%. The **OLS** estimator (0.0044) is then about half way between the *between* estimator (0.0052) and the *within* estimator (0.0038). To get the **GLS** estimator, we first estimate the parameter ϕ using the residuals of the *within* and the *between* estimators:

```
T <- 35
N <- 188
smxt2 <- deviance(Q.between) * T / (N - 2)
sidios2 <- deviance(Q.within) / (N * (T - 1) - 1)
phi <- sqrt(sidios2 / smxt2)
```

The weights for the *within* and the *between* estimators and the **GLS** estimator are then computed:

```
pondW <- SxxW / (SxxW + phi^2 * SxxB)
pondW
[1] 0.9496
pondW * coef(Q.within)[["qn"]] +
  (1 - pondW) * coef(Q.between)[["qn"]]
[1] 0.003862
```

The weight of the *within* estimator (0.95) is much larger for the **GLS** estimator than for the **OLS** estimator. This is mainly due to the fact that T is large (35 years). The **GLS** estimator (0.039) is therefore very close to the *within* estimator (0.0038).

2.4.4 Some Simple Linear Model Examples

Even if they are of limited practical interest, given that relevant econometric models usually contain several covariates, simple linear models have a great pedagogical value, as they enable the graphical representation of the sample and estimators using regression lines. They are for this reason very useful to illustrate the relationship between the estimators. We'll use successively four data sets.

Example 2.4 simple linear model – `ForeignTrade` data set
The first one, called `ForeignTrade`, has been used by Kinal and Lahiri (1993) to construct a full model of external exchange for developing countries, which will be presented in details in chapter 6. For now, we'll simply analyze the link between the imports (`imports`) and the national product (gnp). Both variables are measured in log and per capita.

The following commands create a `pdata.frame`, extract the covariate and apply to it the `summary.pdata.frame` method, which computes the decomposition of its variance.

We then use the `ercomp` function in order to compute the variances of the error components. Finally, to estimate all the models, we first create a vector containing the names of the models, and we then use the `sapply` function in order to extract the coefficient from these fitted models.

```
data("ForeignTrade", package = "pder")
FT <- pdata.frame(ForeignTrade)
summary(FT$gnp)
total sum of squares: 4111
      id     time
0.982480 0.007638
ercomp(imports ~ gnp, FT)
                  var std.dev share
idiosyncratic 0.0863  0.2938  0.07
individual    1.0779  1.0382  0.93
theta: 0.942
models <- c("within", "random", "pooling", "between")
sapply(models, function(x) coef(plm(imports ~ gnp, FT, model = x))["gnp"])
 within.gnp  random.gnp pooling.gnp between.gnp
    0.90236     0.76816     0.06366     0.04871
```

For this model, the variance of the covariate and of the error is almost only due to the inter-individual variation (respectively 98 and 93%). In this case, the GLS estimator consists in removing 94% of the individual mean and is therefore almost identical to the *within*

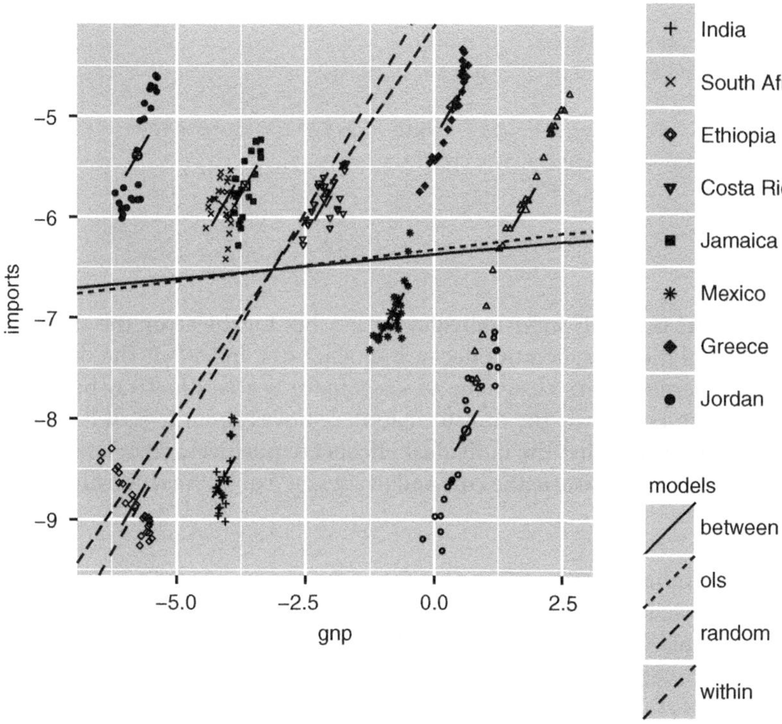

Figure 2.1 Imports in terms of the national product for the ForeignTrade data.

model. Concerning the **OLS** estimator, which takes into account almost all the inter-individual variation, it is very close to the *between* estimator. Finally, the first two models give results that are very different from the last two models and return a much higher elasticity. The Figure 2.1 indicates that there is a strong negative correlation between the individual effects and the covariate. In this case, the estimators that do not control for the individual effects are biased downward. This is the case for the **OLS** and the *between* estimators, and to a much lesser extent for the **GLS** estimator, which uses only a very small part of the inter-individual variation.

Example 2.5 simple linear model – `TurkishBanks` data set

The `TurkishBanks` data were used by El-Gamal and Inanoglu (2005) to analyze production costs of banks. The only covariate is the production, and both variables are in logs. Computing as before, we get:

```
data("TurkishBanks", package = "pder")
TurkishBanks <- na.omit(TurkishBanks)
TB <- pdata.frame(TurkishBanks)
summary(log(TB$output))
total sum of squares: 2692
      id     time
0.84730 0.01255
ercomp(log(cost) ~ log(output), TB)
                 var std.dev share
idiosyncratic 0.329   0.574   0.6
individual    0.216   0.464   0.4
theta:
   Min. 1st Qu.  Median    Mean 3rd Qu.    Max.
  0.619   0.651   0.651   0.647   0.651   0.651
sapply(models, function(x)
       coef(plm(log(cost) ~ log(output), TB, model = x))["log(output)"])
 within.log(output)  random.log(output) pooling.log(output)
             0.5064              0.6471              0.8007
between.log(output)
             0.8531
```

The variation of the covariate is mainly inter-individual (85%), but for the error, the share of the individual effect and that of the idiosyncratic effect are similar (40% and 60%). The **OLS** and the *between* estimators are therefore very close. The **GLS** estimator is about halfway between the **OLS** and the *within* estimators because the transformation removes about 65% of the individual mean. The Figure 2.2 indicates that the individual effects are positively correlated with the covariate, and consequently, the *between*, the **OLS** and in a lesser extent the **GLS** estimators are upward-biased.

Example 2.6 simple linear model – `TexasElectr` data set

The `TexasElectr` data are used by Kumbhakar (1996) and Horrace and Schmidt (1996) and concern the production cost of electric firms in Texas. We first define the cost as being the sum of labor expense `explab`, capital expense `expcap`, and fuel expense `exfuel`. The same computations are then done as above.

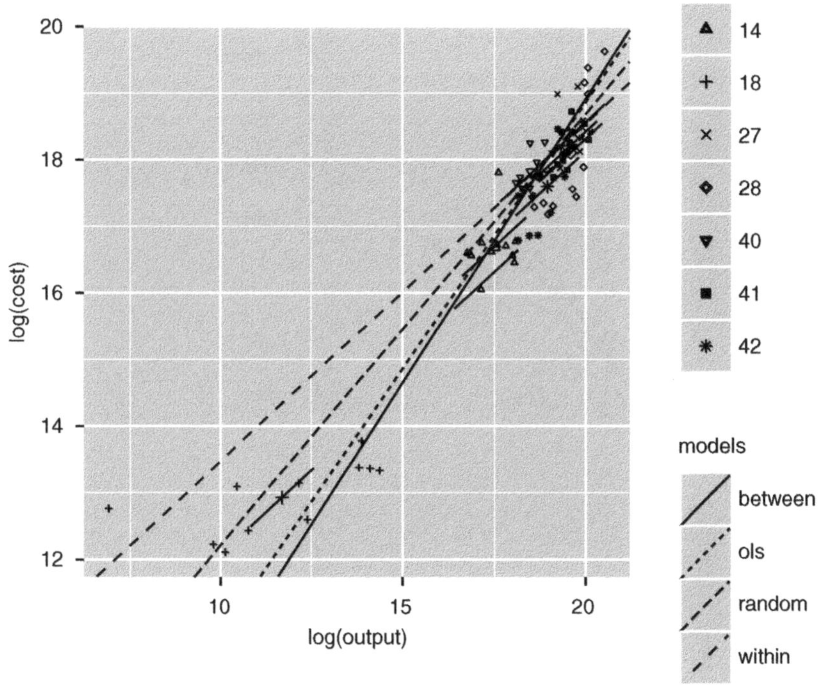

Figure 2.2 Cost in terms of output for the TurkishBanks data.

```
data("TexasElectr", package = "pder")
TexasElectr$cost <- with(TexasElectr, explab + expfuel + expcap)
TE <- pdata.frame(TexasElectr)
summary(log(TE$output))
total sum of squares: 113.5
    id   time
0.8234 0.1685
ercomp(log(cost) ~ log(output), TE)
                  var std.dev share
idiosyncratic 0.10681 0.32681  0.99
individual    0.00109 0.03299  0.01
theta: 0.0808
sapply(models, function(x)
       coef(plm(log(cost) ~ log(output), TE, model = x))["log(output)"])
 within.log(output)  random.log(output) pooling.log(output)
             2.6325              1.2260              1.1804
between.log(output)
             0.8689
```

The variation of the covariate is mainly inter-individual (82%); yet this is not the case for the error, for which the idiosyncratic share is very important: therefore, only a very small part of the individual mean is removed while applying the GLS estimator. The GLS and OLS estimators are therefore almost equal. The *within* estimator is much higher because the individual effects and the covariate are negatively correlated (see Figure 2.3).

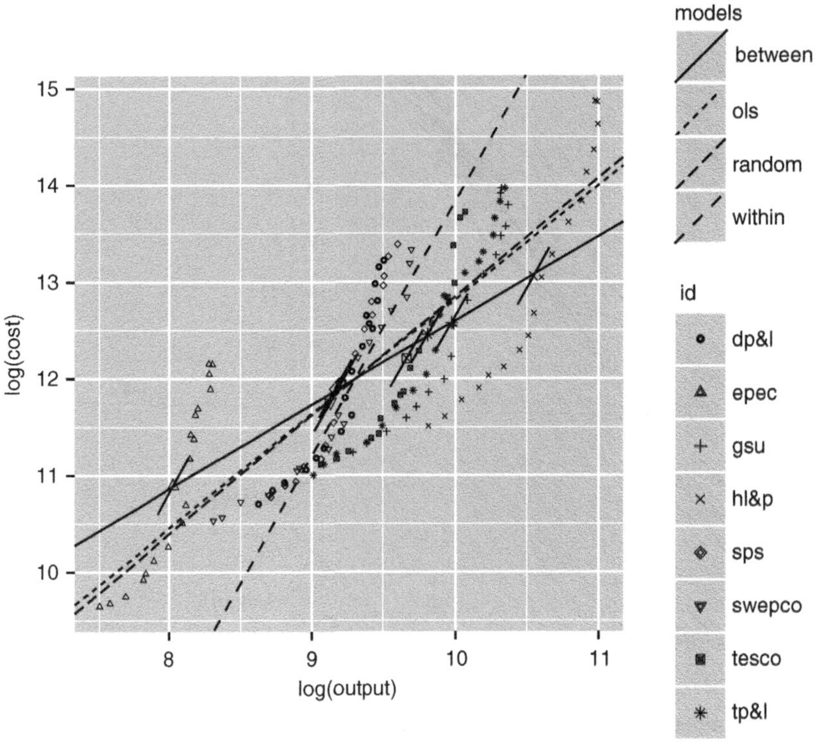

Figure 2.3 Cost and output for the TexasElectr data set.

Example 2.7 simple linear model – `DemocracyIncome25` data set
The last dataset used is `DemocracyIncome25` used by Acemoglu, Johnson, Robinson, and Yared (2008). This dataset deals with 25 countries, observed over 7 25-year periods between 1850 and 2000. The authors analyze the dynamic causal relationship between wealth and democracy. Their analysis will be reproduced in detail in chapter 7. For now, we'll simply analyze the relationship between democracy (`democracy`) and wealth (`income`) lagged one period.

```
data("DemocracyIncome25", package = "pder")
DI <- pdata.frame(DemocracyIncome25)
summary(lag(DI$income))
total sum of squares: 135
    id   time
0.4298 0.4891
ercomp(democracy ~ lag(income), DI)
                 var std.dev share
idiosyncratic 0.0586  0.2422  0.79
individual    0.0155  0.1243  0.21
theta: 0.378
sapply(models, function(x)
       coef(plm(democracy ~ lag(income), DI, model = x))["lag(income)"])
 within.lag(income)  random.lag(income) pooling.lag(income)
             0.1870              0.2101              0.2309
between.lag(income)
             0.2892
```

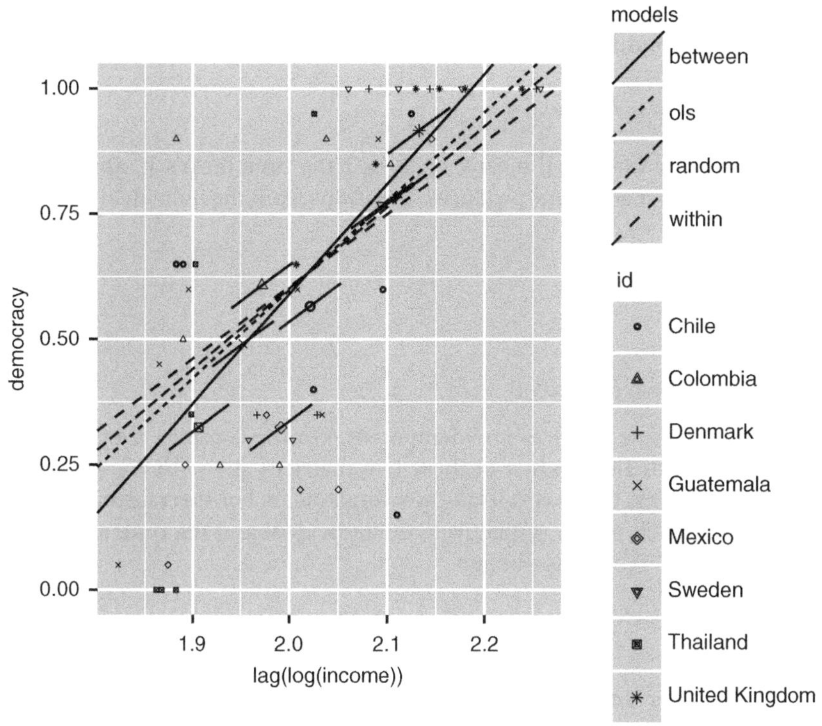

Figure 2.4 Democracy and lagged income for the data DemocracyIncome25.

The share of the inter-individual variation for the covariate and for the error are rather weak (43 and 21%). 41% of the individual mean is removed from the variables in order to compute the GLS estimator. Finally, Figure 2.4 shows that there is no obvious correlation between the individual effects and the covariate; consequently, the 4 estimators are rather close to each other.

2.5 The Two-ways Error Components Model

The two-ways error component is obtained by adding a time-invariant effect μ_t to the model.

$$y_{nt} = \alpha + \beta x_{nt} + \eta_n + \mu_t + \nu_{nt}$$

2.5.1 Error Components in the Two-ways Model

We make for the time effects the same hypotheses that we made for the individual effects:

- μ has a zero mean and is homoscedastic, its variance is denoted by σ_μ^2,
- the time effects are mutually uncorrelated, $\mathrm{E}(\mu_t \mu_s) = 0 \; \forall t \neq s$,
- the time effects are uncorrelated with the individual effects and the idiosyncratic terms.

With these hypotheses, the covariance matrix of the errors becomes:

$$\Omega = \sigma_\nu^2 \mathrm{I}_{NT} + \sigma_\eta^2 \mathrm{I}_N \otimes J_T + \sigma_\mu^2 J_N \otimes \mathrm{I}_T$$

As for the individual error component model, we write this covariance matrix as a linear combination of idempotent and mutually orthogonal matrices. To this aim, we write:

$$B_\eta = \mathrm{I}_N \otimes J_T/T,\ B_\mu = J_T \otimes \mathrm{I}_N/N \text{ and } \bar{J} = \frac{1}{NT}J_{NT}$$

$B_\eta \times x$ computes, as before, the individual means $\bar{x}_{n.}$, $B_\mu \times x$ the time means $\bar{x}_{.t}$ and $\bar{J}x$ the overall mean $\bar{\bar{x}}$. Finally, the *within* matrix now produces deviations from the individual *and* the time means: $x_{nt} - \bar{x}_{n.} - \bar{x}_{.t} + \bar{\bar{x}}$:

$$W = \mathrm{I} - B_\eta - B_\mu + \bar{J}$$

With these notations, we get:

$$\Omega = \sigma_\nu^2 W + (T\sigma_\eta^2 + \sigma_\nu^2)B_\eta + (N\sigma_\mu^2 + \sigma_\nu^2)B_\mu - \sigma_\nu^2 \bar{J}$$

It can be easily checked that these matrices are idempotent. On the contrary, they are not all orthogonal, as $B_\eta \times B_\mu = \bar{J} \neq 0$. The product of these two matrices allows to compute the time means of the individual means, which results in the overall mean. For this reason, we use $\bar{B}_\eta = B_\eta - \bar{J}$ and $\bar{B}_\mu = B_\mu - \bar{J}$, which return respectively the individual and the time means in deviations from the overall mean. We finally obtain:

$$\Omega = \sigma_\nu^2 W + (T\sigma_\eta^2 + \sigma_\nu^2)\bar{B}_\eta + (N\sigma_\mu^2 + \sigma_\nu^2)\bar{B}_\mu + (T\sigma_\eta^2 + N\sigma_\mu^2 + \sigma_\nu^2)\bar{J}$$

2.5.2 Fixed and Random Effects Models

As for the individual effects model, the two-ways fixed effects model can be obtained in two different ways:

- by estimating by **OLS** the model that includes individual *and* time dummies,
- by estimating by **OLS** the model where all the variables have been transformed in deviations from the individual and the time means: $z_{nt} - \bar{z}_{n.} - \bar{z}_{.t} + \bar{\bar{z}}$.

For the **GLS** model the variables are pre-multiplied by $\Omega^{-0.5}$ or more simply by:

$$\sigma_\nu \Omega^{-0.5} = W + \frac{\sigma_\nu}{\sqrt{\sigma_\nu^2 + T\sigma_\eta^2}}\bar{B}_\eta + \frac{\sigma_\nu}{\sqrt{(\sigma_\nu^2 + N\sigma_\mu^2)}}\bar{B}_\mu + \frac{\sigma_\nu}{\sqrt{\sigma_\nu^2 + T\sigma_\eta^2 + N\sigma_\mu^2}}\bar{J}$$

Collecting terms, we obtain the following expression for the transformed data:

$$\tilde{z}_{nt} = z_{nt} - \theta_\eta \bar{z}_{n.} - \theta_\mu \bar{z}_{.t} + (\theta_\eta + \theta_\mu - \theta_2)\bar{\bar{z}}$$

with:

$$\begin{cases} \theta_\eta = 1 - \frac{\sigma_\nu}{\sqrt{\sigma_\nu^2 + T\sigma_\eta^2}} = 1 - \phi_\eta \\ \theta_\mu = 1 - \frac{\sigma_\nu}{\sqrt{\sigma_\nu^2 + N\sigma_\mu^2}} = 1 - \phi_\mu \\ \theta_2 = 1 - \frac{\sigma_\nu}{\sqrt{\sigma_\nu^2 + T\sigma_\eta^2 + N\sigma_\mu^2}} = 1 - \phi_2 \end{cases}$$

Example 2.8 two-ways effect model – `TobinQ` data set

We've previously stored the four random effect models in a list called `Q.models`. The two-ways effect model is obtained by setting the `effect` argument to `'twoways'`.

```
Q.models2 <- lapply(Q.models, function(x) update(x, effect = "twoways"))
sapply(Q.models2, function(x) sqrt(ercomp(x)$sigma2))
          swar   walhus  amemiya nerlove
idios 0.06970 0.06970 0.06969 0.06850
id    0.04508 0.04508 0.04573 0.04735
time  0.02093 0.02093 0.02170 0.02262
sapply(Q.models2, function(x) ercomp(x)$theta)
      swar   walhus amemiya nerlove
id    0.7472 0.7472 0.7505  0.7624
time  0.764  0.764  0.772   0.7843
total 0.6863 0.6863 0.6933  0.7085
```

The first `sapply` command extracts the standard deviations of the three components of the error. As for the individual effects model, the estimates of the variance components are very similar. The standard deviation of individual effects is more than twice the one of time effects. The second command extracts the `theta` parameters. About 75% of the individual and time means are removed from the variables.

2.6 Estimation of a Wage Equation

Example 2.9 multiple linear model – `UnionWage` data set
The estimation of a wage function is an important subject in econometrics, especially in panel data econometrics, the main covariate of interest being generally education. We use here the `UnionWage` dataset used by Vella and Verbeek (1998), who investigated the impact of union negotiations on wages and the potential endogeneity of this covariate. The data concern 545 men observed during 8 years, from 1980 to 1987.

```
data("UnionWage", package = "pglm")
pdim(UnionWage)
Balanced Panel: n = 545, T = 8, N = 4360
```

The response, `wage`, is the log of the hourly wage. The covariates are: whether wages are set during negotiations with unions `union`, the number of years of education `school`, the number of years of experience `exper` and its square, the community `com`, which identifies black `black` and Hispanic `hisp` workers, whether one lives in a rural area `rural`, the marital status `married`, having a health problem `health`, the region `region`, and the activity sector `sector`.

The *within* and OLS models are estimated, including or not occupation dummies.

```
UnionWage$exper2 <- with(UnionWage, exper ^ 2)
wages.within1 <- plm(wage ~ union + school + exper + exper2 +
                         com + rural + married + health +
                         region + sector, UnionWage)
wages.within2 <- plm(wage ~ union + school + exper + exper2 +
                         com + rural + married + health +
                         region + sector + occ, UnionWage)
wages.pooling1 <- update(wages.within1, model = "pooling")
wages.pooling2 <- update(wages.within2, model = "pooling")
```

Table 2.1 Wage Equation.

	Dependent variable:			
	log of hourly wage			
	pooling estimation		within estimation	
	(1)	(2)	(3)	(4)
union membership	0.176***	0.146***	0.080***	0.079***
	(0.017)	(0.017)	(0.020)	(0.019)
education years	0.078***	0.090***		
	(0.005)	(0.005)		
experience years	0.070***	0.076***	0.111***	0.112***
	(0.010)	(0.010)	(0.009)	(0.008)
experience years squared	−0.002***	−0.002***	−0.004***	−0.004***
	(0.001)	(0.001)	(0.001)	(0.001)
black	−0.130***	−0.155***		
	(0.023)	(0.023)		
hispanic	−0.047**	−0.059***		
	(0.022)	(0.022)		
rural residence	−0.116***	−0.131***	0.048*	0.050*
	(0.019)	(0.018)	(0.029)	(0.029)
married	0.102***	0.110***	0.038**	0.040**
	(0.015)	(0.015)	(0.018)	(0.018)
health problems	−0.035	−0.058	−0.010	−0.017
	(0.054)	(0.054)	(0.047)	(0.047)
Intercept	0.273***	−0.039		
	(0.091)	(0.076)		
region dummies	Yes	Yes	Yes	Yes
sector dummies	Yes	Yes	Yes	Yes
occupation dummies	Yes	No	Yes	No
Observations	4,360	4,360	4,360	4,360
R^2	0.278	0.264	0.192	0.190

Note: *p<0.1; **p<0.05; ***p<0.01

Estimation results are presented in Table 2.1, using the **stargazer** library (Hlavac, 2013). We use several possibilities offered by the library to improve the appearance of the table:

- the `omit` argument is used to omit two sets of coefficients corresponding to the `region` and `sector` factors; `omit.labels` indicates how the information about these covariates will be included in the table,
- the F statistic and adjusted R^2 are removed from the output using `omit.stat`,
- customized names for the response and the covariates are provided with `dep.var.labels` and `covariate.labels`
- `column.labels` and `column.separate` are used to indicate the method of estimation used for the first two and the last two models.

```
library("stargazer")
stargazer(wages.pooling2, wages.pooling1, wages.within2, wages.within1,
          omit = c("region", "sector", "occ"),
          omit.labels = c("region dummies", "sector dummies", "occupation
          dummies"),
          column.labels = c("pooling estimation", "within estimation"),
          column.separate = c(2, 2),
          dep.var.labels = "log of hourly wage",
          covariate.labels = c("union membership", "education years",
                               "experience years", "experience years squared",
                               "black", "hispanic", "rural residence",
                               "married", "health problems",
                               "Intercept"),
          omit.stat = c("adj.rsq", "f"),
          title = "Wage equation",
          label = "tab:wagesresult",
          no.space = TRUE
)
```

Table 2.1 exactly matches the results presented in Vella and Verbeek (1998) in columns (2), (1), (3), and (4).

Looking at the results, we see that the union premium is about 18% with the **OLS** model and falls to 8% for the *within* model. This indicates that the individual effects are strongly positively correlated with union membership. The return of education is about 8% more wage for one more year of education. This is a consistent result only if the education level is uncorrelated with the individual effects. If there is any correlation, the only consistent model is the *within* model; unfortunately, the within transformation eliminates all the time-invariant covariates (education, community, and rural residence).

This example illustrates the main concern about panel data econometrics, the correlation between some covariates and the two components of the error term:

- if there is no correlation, use **GLS**, which gives consistent and efficient estimators and allows estimating the coefficients for time-invariant covariates;
- if there is correlation only with the individual component of the error, use the *within* model; it provides consistent estimates, but the effect of time-invariant covariates cannot be estimated;
- if there is correlation between any covariates and both components of the error term, none of the models we have presented are consistent. Vella and Verbeek (1998) argued that the endogeneity of union membership is not limited to the time-invariant part of the error. In this case, all the models presented, including the *within* model, are inconsistent, and the authors propose a more sophisticated estimation procedure in order to obtain a consistent estimator.

3

Advanced Error Components Models

3.1 Unbalanced Panels

For unbalanced panels, the number of observations for each individual is now individual specific and denoted by T_n. We'll denote by $O = \sum_{n=1}^{N} T_n$ the total number of observations. Compared to the balanced panel case, three complications appear:

- firstly, the covariance matrix of the errors cannot be written any more as a linear combination of idempotent and mutually orthogonal matrices (for the one-way error component model, the *within* and the *between* matrices), the weights being the variances of the errors (σ_η^2 and σ_ι^2). Denoting by D_η and D_μ two matrices of individual and time dummies, matrices of the type $S = \mathrm{DD}^\top$, returning either the sum of the values for an individual or for a time series, will explicitly appear, and these matrices are not idempotent;
- secondly, for the individual effects model, the *within* transformation still consists of removing the individual mean from the variable. On the contrary, for the two-ways effects, the *within* transformation is not obtained by performing a difference with the individual and time means, as in the balanced panel case, but requires more tedious matrix algebra;
- finally, to estimate the components of the variance, we will still compute quadratic forms of the residuals of some consistent preliminary estimations, but there is no obvious choice of denominators, as there was in the balanced case.

3.1.1 Individual Effects Model

The model to be estimated can be written:

$$y = Z\gamma + \epsilon = Z\gamma + \mathrm{D}_\eta \eta + \nu$$

The fixed effects model may be estimated by regressing y on X and D_η. Like in the balanced panel case, the Frisch-Waugh theorem enables to avoid the estimation of the fixed effects. The estimation of β may be obtained by regressing in a first stage y and X on D_η, computing the residuals and then regressing in the second stage the residuals of y on those of X. As in the balanced panel case, these residuals are just the individual *within* transformation, i.e., $z_{nt} - \bar{z}_{n.}$ or $W_\eta z$ in matrix form, and the fixed effects model is simply obtained by regressing $W_\eta y$ on $W_\eta X$.

For the GLS model, the covariance matrix of the errors is:

$$\Omega = \sigma_\nu^2 \mathrm{I} + \sigma_\eta^2 \mathrm{D}_\eta \mathrm{D}_\eta^\top = \sigma_\nu^2 \left(\mathrm{I} + \frac{\sigma_\eta^2}{\sigma_\nu^2} \mathrm{D}_\eta \mathrm{D}_\eta^\top \right) = \sigma_\nu^2 \Sigma$$

Panel Data Econometrics with R, First Edition. Yves Croissant and Giovanni Millo.

Companion website: www.wiley.com/go/croissant/data-econometrics-with-R

The **GLS** estimator writes:

$$\hat{\gamma}_{\text{GLS}} = (Z^\top\Omega^{-1}Z)^{-1}Z^\top\Omega^{-1}Z$$

$D_\eta D_\eta^\top$ is a block-diagonal matrix that contains N square matrices of ones of dimension T_n. For balanced panels, $T_n = T\ \forall n$ and $S = TB$. Sz returns the sum of the values of z for each individual. Ω is also a block-diagonal matrix, with blocks Ω_n of the form:

$$\Omega_n = \sigma_\nu^2 \bar{\mathrm{I}}_{T_n} + \sigma_\eta^2 \bar{J}_{T_n} = \sigma_\nu^2 \left[\bar{\mathrm{I}}_{T_n} + \frac{\sigma_\nu^2 + T_n\sigma_\eta^2}{\sigma_\nu^2}\bar{J}_{T_n}\right]$$

with $\bar{\mathrm{I}} = \mathrm{I} - \bar{J}$.

The inverse of a block-diagonal matrix being equal to a block-diagonal matrix for which the blocks are the inverses of those of the initial matrix, it is sufficient to calculate the inverse of Ω_n. As it is a linear combination of two idempotent and orthogonal matrices, the general formula for any power of Ω_n is:

$$\Omega_n^v = \sigma_\nu^{2v}\left[\bar{\mathrm{I}}_{T_n} + \left(\frac{\sigma_\nu^2 + T_n\sigma_\eta^2}{\sigma_\nu^2}\right)^v \bar{J}_{T_n}\right]$$

In particular, the inverse is:

$$\Omega_n^{-1} = \frac{1}{\sigma_\nu^2}\left[\bar{\mathrm{I}}_{T_n} + \frac{\sigma_\nu^2}{\sigma_\nu^2 + T_n\sigma_\eta^2}\bar{J}_{T_n}\right]$$

which can also be written as $\Omega_n^{-1} = \Omega_n^{-0.5}\Omega_n^{-0.5}$, with:

$$\Omega_n^{-0.5} = \frac{1}{\sigma_\nu}\left[\bar{\mathrm{I}}_{T_n} + \frac{\sigma_\nu}{\sqrt{\sigma_\nu^2 + T_n\sigma_\eta^2}}\bar{J}_{T_n}\right]$$

The **GLS** estimator may then be obtained by applying **OLS** on variables that have been transformed by pre-multiplying them by $\Omega_n^{-0.5}$ or, equivalently, by $\sigma_\nu\Omega_n^{-0.5}$ (which will simplify notation):

$$\sigma_\nu\Omega_n^{-0.5}z_n = \left[\bar{\mathrm{I}}_{T_n} + \frac{\sigma_\nu}{\sqrt{\sigma_\nu^2 + T_n\sigma_\eta^2}}\bar{J}_{T_n}\right]z_n = \left[\mathrm{I}_{T_n} - \left(1 - \frac{\sigma_\nu}{\sqrt{\sigma_\nu^2 + T_n\sigma_\eta^2}}\bar{J}_{T_n}\right)\right]z_n$$

As in the balanced case, the transformed data can be expressed as quasi-differences, $\tilde{z}_{nt} = z_{nt} - \theta_n z_{n.}$, with:

$$\theta_n = 1 - \frac{\sigma_\nu}{\sqrt{\sigma_\nu^2 + T_n\sigma_\eta^2}} = 1 - \phi_n$$

the only difference being that now, the proportion of the individual mean that is removed is not a constant, as it depends on the number of observations for each individual.

3.1.2 Two-ways Error Component Model

For the two-ways error component model, we have:

$$y_{nt} = \alpha + \beta^\top x_{nt} + \eta_n + \mu_t + \nu_{nt}$$

or, in matrix form:

$$y = \alpha j + X\beta + D_\eta \eta + D_\mu \mu + v$$

where D_η and D_μ are matrices of respectively individual and time dummies. Pre-multiplying a vector by $S_\eta = D_\eta D_\eta^\top$ and $S_\mu = D_\mu D_\mu^\top$ returns, respectively, the individual and time sum of the variable.

$D_\eta^\top D_\eta$ and $D_\mu^\top D_\mu$ are two diagonal matrices that contain the number of observations for each individual and time-series. Pre-multiplying a vector by $B_\eta = D_\eta^\top (D_\eta^\top D_\eta)^{-1} D_\eta$ or by $B_\mu = D_\mu^\top (D_\mu^\top D_\mu)^{-1} D_\mu$ returns, respectively, the individual and the time series means. Finally, $D_\mu^\top D_\eta$ is a $T \times N$ matrix of ones and zeros, which indicates whether an observation for a specific individual and time period is present or not.

To help visualizing these matrices, we consider a panel with 3 individuals and 4 periods; the panel is unbalanced, as the first individual is not observed in the third and fourth periods, and the third one is not observed in the first period.

$$D_\eta = \begin{pmatrix} 1 & 0 & 0 \\ 1 & 0 & 0 \\ 0 & 1 & 0 \\ 0 & 1 & 0 \\ 0 & 1 & 0 \\ 0 & 1 & 0 \\ 0 & 0 & 1 \\ 0 & 0 & 1 \\ 0 & 0 & 1 \end{pmatrix} \quad D_\mu = \begin{pmatrix} 1 & 0 & 0 & 0 \\ 0 & 1 & 0 & 0 \\ 1 & 0 & 0 & 0 \\ 0 & 1 & 0 & 0 \\ 0 & 0 & 1 & 0 \\ 0 & 0 & 0 & 1 \\ 0 & 1 & 0 & 0 \\ 0 & 0 & 1 & 0 \\ 0 & 0 & 0 & 1 \end{pmatrix}$$

$$D_\eta^\top D_\eta = \begin{pmatrix} 2 & 0 & 0 \\ 0 & 4 & 0 \\ 0 & 0 & 3 \end{pmatrix} \quad D_\mu^\top D_\mu = \begin{pmatrix} 2 & 0 & 0 & 0 \\ 0 & 3 & 0 & 0 \\ 0 & 0 & 2 & 0 \\ 0 & 0 & 0 & 2 \end{pmatrix} \quad D_\mu^\top D_\eta = \begin{pmatrix} 1 & 1 & 0 \\ 1 & 1 & 1 \\ 0 & 1 & 1 \\ 0 & 1 & 1 \end{pmatrix}$$

$$D_\eta D_\eta^\top = \begin{pmatrix} 1 & 1 & 0 & 0 & 0 & 0 & 0 & 0 & 0 \\ 1 & 1 & 0 & 0 & 0 & 0 & 0 & 0 & 0 \\ 0 & 0 & 1 & 1 & 1 & 1 & 0 & 0 & 0 \\ 0 & 0 & 1 & 1 & 1 & 1 & 0 & 0 & 0 \\ 0 & 0 & 1 & 1 & 1 & 1 & 0 & 0 & 0 \\ 0 & 0 & 1 & 1 & 1 & 1 & 0 & 0 & 0 \\ 0 & 0 & 0 & 0 & 0 & 0 & 1 & 1 & 1 \\ 0 & 0 & 0 & 0 & 0 & 0 & 1 & 1 & 1 \\ 0 & 0 & 0 & 0 & 0 & 0 & 1 & 1 & 1 \end{pmatrix} \quad D_\mu D_\mu^\top = \begin{pmatrix} 1 & 0 & 1 & 0 & 0 & 0 & 0 & 0 & 0 \\ 0 & 1 & 0 & 1 & 0 & 0 & 1 & 0 & 0 \\ 1 & 0 & 1 & 0 & 0 & 0 & 0 & 0 & 0 \\ 0 & 1 & 0 & 1 & 0 & 0 & 1 & 0 & 0 \\ 0 & 0 & 0 & 0 & 1 & 0 & 0 & 1 & 0 \\ 0 & 0 & 0 & 0 & 0 & 1 & 0 & 0 & 1 \\ 0 & 1 & 0 & 1 & 0 & 0 & 1 & 0 & 0 \\ 0 & 0 & 0 & 0 & 1 & 0 & 0 & 1 & 0 \\ 0 & 0 & 0 & 0 & 0 & 1 & 0 & 0 & 1 \end{pmatrix}$$

3.1.2.1 Fixed Effects Model

The fixed effects model can be estimated regressing y on X and the two matrices associated with the effects vectors D_η and D_μ.

The application of the Frisch-Waugh theorem implies that the estimation can be performed by regressing in a first stage y, X, and D_μ on D_η and then, in a second stage, by regressing the residuals of y on those of X and D_μ, which means regressing $W_\eta y$ on $W_\eta X$ and $W_\eta D_\mu$.

Applying the same theorem again, one can regress in $W_\eta y$ and $W_\eta X$ on $W_\eta \mathrm{D}_\mu$ in the first stage, and the residuals of $W_\eta y$ on those of $W_\eta X$ in the second stage.

Residuals of a regression on $W_\eta \mathrm{D}_\mu$ are obtained by pre-multiplying the variables by the matrix:

$$W = \mathrm{I} - W_\eta \mathrm{D}_\mu (\mathrm{D}_\mu^\top W_\eta \mathrm{D}_\mu)^- W_\eta \mathrm{D}_\mu$$

where, for any matrix A, A^- is the generalized inverse of A. Finally, the two-ways error component fixed effects model may be obtained by applying to y and every column of X the following transformation:

$$(\mathrm{I} - W_\eta \mathrm{D}_\mu (\mathrm{D}_\mu^\top W_\eta \mathrm{D}_\mu)^- W_\eta \mathrm{D}_\mu) W_\eta z$$

The double-*within* transformation consists then, for unbalanced panels, in multiplying any data vector by the following matrix:

$$W = W_\eta - W_\eta \mathrm{D}_\mu (\mathrm{D}_\mu^\top W_\eta \mathrm{D}_\mu)^- W_\eta \mathrm{D}_\mu W_\eta$$

Therefore, the two-ways fixed effects model is still easy to compute even if the panel is unbalanced: all that is required is **OLS** estimation and the computation of deviations from the individual means. One proceeds as follows:

- first, the individual *within* transformation is applied to X, y and D_μ,
- next, $W_\eta X$ and $W_\eta y$ are regressed on $W\mathrm{D}_\mu$,
- finally, from these regressions, the residuals of $W_\eta X$ and $W_\eta y$ are obtained; then the residuals of the latter are regressed on those of the former.

The *within* transformation is performed on $K + T + 1$ variables, and then $K + 1$ preliminary linear estimations are performed on T covariates before the final estimation for which there are K covariates.

Note that no specific matrix computation is required and that, in particular, the matrix of individual dummies, which is often very large ($O \times N$), need not to be stored during the estimation.

3.1.2.2 Random Effects Model

The variance matrix of the errors is:

$$\Omega = \sigma_\nu^2 \mathrm{I} + \sigma_\eta^2 \mathrm{D}_\eta \mathrm{D}_\eta^\top + \sigma_\mu^2 \mathrm{D}_\mu \mathrm{D}_\mu^\top = \sigma_\nu^2 \Sigma$$

with:

$$\Sigma = \mathrm{I} + \frac{\sigma_\eta^2}{\sigma_\nu^2} \mathrm{D}_\eta \mathrm{D}_\eta^\top + \frac{\sigma_\mu^2}{\sigma_\nu^2} \mathrm{D}_\mu \mathrm{D}_\mu^\top$$

Denote $\sigma_\nu^2 \Sigma_\eta = \sigma_\nu^2 \left(\mathrm{I} + \frac{\sigma_\eta^2}{\sigma_\nu^2} \mathrm{D}_\eta \mathrm{D}_\eta^\top\right)$ the covariance matrix of the errors of the individual one-way error components model. We then have:

$$\Omega = \sigma_\nu^2 \Sigma = \sigma_\nu^2 \left(\Sigma_\eta + \frac{\sigma_\mu^2}{\sigma_\nu^2} \mathrm{D}_\mu \mathrm{D}_\mu^\top \right)$$

Σ_η is block-diagonal, with blocks: $\mathrm{I}_{T_n} + \frac{\sigma_\eta^2}{\sigma_\nu^2} J_{T_n} = \bar{\mathrm{I}}_{T_n} + \frac{\sigma_\nu^2}{\sigma_\nu^2 + T_n \sigma_\eta^2} \bar{J}_{T_n}$. $\bar{\mathrm{I}}_{T_n}$ and $\bar{J}_{T_n}$ being idempotent and orthogonal, the matrix $\Sigma_\eta^{0.5}$ (defined so that $\Sigma_\eta^{0.5} \Sigma_\eta^{0.5} = \Sigma_\eta$) is also a block-diagonal matrix with blocks: $\bar{\mathrm{I}}_{T_n} + \frac{\sqrt{\sigma_\nu^2 + T_n \sigma_\eta^2}}{\sigma_\nu} \bar{J}_{T_n}$. We then have:

$$\Sigma = \Sigma_\eta^{0.5} \Sigma_\eta^{0.5} + \frac{\sigma_\mu^2}{\sigma_\nu^2} \mathrm{D}_\mu \mathrm{D}_\mu^\top$$

$$\Sigma = \Sigma_\eta^{0.5}\left(I + \frac{\sigma_\mu^2}{\sigma_\nu^2}\Sigma_\eta^{-0.5}D_\mu D_\mu^\top \Sigma_\eta^{-0.5}\right)\Sigma_\eta^{0.5}$$

for which the inverse is:

$$\Sigma^{-1} = \Sigma_\eta^{-0.5}\left(I + \frac{\sigma_\mu^2}{\sigma_\nu^2}\Sigma_\eta^{-0.5}D_\mu D_\mu^\top \Sigma_\eta^{-0.5}\right)^{-1}\Sigma_\eta^{-0.5}$$

We then apply the following result: $(I + XX^\top)^{-1} = I - X(I + X^\top X)^{-1}X^\top$ to the matrix in brackets:

$$\left(I + \frac{\sigma_\mu^2}{\sigma_\nu^2}\Sigma_\eta^{-0.5}D_\mu D_\mu^\top \Sigma_\eta^{-0.5}\right)^{-1} = I - \frac{\sigma_\mu^2}{\sigma_\nu^2}\Sigma_\eta^{-0.5}D_\mu\left(I + \frac{\sigma_\mu^2}{\sigma_\nu^2}D_\mu^\top \Sigma_\eta^{-1}D_\mu\right)^{-1}D_\mu^\top \Sigma_\eta^{-0.5}$$

Finally, we have:

$$\Sigma^{-1} = \Sigma_\eta^{-1} - \frac{\sigma_\mu^2}{\sigma_\nu^2}\Sigma_\eta^{-1}D_\mu\left[I + \frac{\sigma_\mu^2}{\sigma_\nu^2}D_\mu^\top \Sigma_\eta^{-1}D_\mu\right]^{-1}D_\mu^\top \Sigma_\eta^{-1}$$

and the GLS estimator is:

$$\hat{\gamma} = (Z^\top \Sigma^{-1} Z)^{-1} Z^\top \Sigma^{-1} y$$

Let $\tilde{Z} = \Sigma_\eta^{-0.5} Z$ and $\tilde{D}_\mu = \Sigma_\eta^{-0.5} D_\mu$ the matrix of the covariates and of the time dummies measured in quasi-difference from the individual means. We then have:

$$Z^\top \Sigma^{-1} Z = \tilde{Z}^\top \tilde{Z} - \frac{\sigma_\mu^2}{\sigma_\nu^2}\tilde{Z}^\top \tilde{D}_\mu\left(I + \frac{\sigma_\mu^2}{\sigma_\nu^2}\tilde{D}_\mu^\top \tilde{D}_\mu\right)^{-1}\tilde{D}_\mu^\top \tilde{Z}$$

and a similar expression for $Z^\top \Sigma^{-1} y$. With the two matrices $\tilde{D}_\mu$ and $\tilde{Z}$ in hand, the computation of the estimator requires:

- computing the cross products of the two matrices $\tilde{Z}^\top \tilde{Z}$, $\tilde{Z}^\top \tilde{D}_\mu$ and $\tilde{D}_\mu^\top \tilde{D}_\mu$,
- computing the inverse of a matrix of dimension T.

These are reasonable computational tasks: note especially that the matrix of individual effects needn't be stored and that the dimension of the matrix that has to be inverted is T and not N or O and that, at least for micro-panels, T is relatively small. Note also that computation of the GLS estimator requires explicit matrix operations and it can no longer be obtained as a series of linear regressions on transformed data.

3.1.3 Estimation of the Components of the Error Variance

Remember that, in the balanced panel case, we used the result that natural estimators of $\hat{\sigma}_\iota^2$ and $\hat{\sigma}_\nu^2$ were:

$$\begin{cases} \hat{\sigma}_\iota^2 = \frac{\epsilon^\top B \epsilon}{N} \\ \hat{\sigma}_\nu^2 = \frac{\epsilon^\top W \epsilon}{N(T-1)} \end{cases}$$

Feasible estimates were obtained by replacing ϵ by the residuals $\hat{\epsilon}$ from a consistent estimation. For the balanced case, N and $N(T-1)$ were natural denominators. This is no longer the case when the panel is unbalanced, as T_n is not the same for all individuals (and N_t is not the same for all time periods).

The strategy used here consists in computing the expected values of the quadratic forms in order to obtain unbiased estimators of the variance components:

- first, for a given estimator, define the matrix M^V that transforms the errors into the residuals of the V estimator: $\hat{\epsilon}^V = M^V \epsilon$,
- compute the two quadratic forms of the *within* and *between* transformation of the residuals: $\hat{q}^V_W = \hat{\epsilon}^{V\top} W \hat{\epsilon}^V$ and $\hat{q}^V_B = \hat{\epsilon}^{V\top} B \hat{\epsilon}^V$ ($\hat{q}^V_{B_\eta} = \hat{\epsilon}^{V\top} B_\eta \hat{\epsilon}^V$ and $\hat{q}^V_{B_\mu} = \hat{\epsilon}^{V\top} B_\mu \hat{\epsilon}^V$ for the two-ways error component model).
- compute the expected values of these quadratic forms, which are functions of σ^2_ν, σ^2_η (and σ^2_μ for the two-ways model),
- equate the quadratic forms to their expected values and solve the system of two (or three equations for the two-ways error component model) for σ^2_ν, σ^2_η (and σ^2_μ in the latter case).

Different estimators are obtained using different preliminary models to obtain the residuals. Among the numerous possible choices, as previously seen on chapter 2:

- Wallace and Hussain (1969) use the residuals of the pooling estimator for the two quadratic forms,
- Amemiya (1971) use the residuals of the **OLS** estimator for the two quadratic forms,
- Swamy and Arora (1972) use the residuals of the **OLS** estimator for the first quadratic form and those of the **OLS** estimator for the second one.

The model and its estimation are:

$$\left\{ \begin{array}{l} y = Z\gamma + \epsilon = \alpha j + X\beta + \epsilon \\ y = Z\hat{\gamma} + \hat{\epsilon} = \hat{\alpha} j + X\hat{\beta} + \hat{\epsilon} \end{array} \right. \tag{3.1}$$

The intercept can be removed by pre-multiplying every element of the model by: $\bar{I} = I - \bar{J}$, which subtracts from every variable its overall mean and therefore removes the intercept, as $\bar{I}j = 0$.

$$\left\{ \begin{array}{l} \bar{I}y = \bar{I}X\beta + \bar{I}\epsilon \\ \bar{I}y = \bar{I}X\hat{\beta} + \hat{\epsilon} \end{array} \right. \tag{3.2}$$

Subtracting the expression of the model and of its estimation, we get:

$$\left\{ \begin{array}{ll} Z(\hat{\gamma} - \gamma) + \hat{\epsilon} - \epsilon & = 0 \\ \bar{I}X(\hat{\beta} - \beta) + \hat{\epsilon} - \bar{I}\epsilon & = 0 \end{array} \right. \tag{3.3}$$

The three estimators we use (**OLS**, *within*, and *between*) can be seen as **GLS** estimators of this model, with V being equal, respectively, to I, W, and B:

$$\left\{ \begin{array}{lll} \hat{\gamma} = (Z^\top VZ)^{-1}Z^\top Vy & = & \gamma + (Z^\top VZ)^{-1}Z^\top V\epsilon \\ \hat{\beta} = (X^\top\bar{I}V\bar{I}X)^{-1}X^\top\bar{I}V\bar{I}y & = & \beta + (X^\top\bar{I}V\bar{I}X)^{-1}X^\top\bar{I}V\bar{I}\epsilon \end{array} \right. \tag{3.4}$$

Using the two previous expressions, we get $\hat{\epsilon} = M^V \epsilon$ with:

$$M^V = I - Z(Z^\top VZ)^{-1}Z^\top V = \bar{I} - \bar{I}X(X^\top\bar{I}V\bar{I}X)^{-1}X^\top\bar{I}V\bar{I} \tag{3.5}$$

M^V is the matrix that transforms the error vector into the residuals vector. Note that it is not a symmetric matrix, at least unless $V = I$ (which corresponds to the pooling model). The quadratic form of the residuals with a matrix A is:

$$\hat{q}_A^V = \hat{\epsilon}^{V\top} A \hat{\epsilon}^V = \epsilon^\top \mathrm{M}^{V\top} A \mathrm{M}^V \epsilon$$

$\hat{q}_A^V$ being a scalar, it is also equal to its trace:

$$\hat{q}_A^V = \mathrm{tr}(\epsilon^\top \mathrm{M}^{V\top} A \mathrm{M}^V \epsilon)$$

Using the cyclic property of the trace operator, we get:

$$\hat{q}_A^V = \mathrm{tr}(\mathrm{M}^{V\top} A \mathrm{M}^V \epsilon\epsilon^\top)$$

from which, taking expectations, we obtain:

$$\mathrm{E}(\hat{q}_A^V) = \mathrm{tr}(\mathrm{M}^{V\top} A \mathrm{M}^V \mathrm{E}(\epsilon\epsilon^\top)) = \mathrm{tr}(\mathrm{M}^{V\top} A \mathrm{M}^V \Omega)$$

with $\Omega = \sigma_\nu^2 \mathrm{I} + \sigma_\eta^2 S_\eta + \sigma_\mu^2 S_\mu$.

Finally, we get:

$$\mathrm{E}(\hat{q}_A^V) = \sigma_\nu^2 \mathrm{tr}(\mathrm{M}^{V\top} A \mathrm{M}^V) + \sigma_\eta^2 \mathrm{tr}(\mathrm{M}^{V\top} A \mathrm{M}^V S_\eta) + \sigma_\mu^2 \mathrm{tr}(\mathrm{M}^{V\top} A \mathrm{M}^V S_\mu)$$

Replacing M^V by its expression and denoting $\Theta_A = X^\top A X$, we get:

$$\mathrm{E}(\hat{q}_A^V) = \begin{pmatrix} \mathrm{tr}(\bar{\mathrm{I}} A \bar{\mathrm{I}}) - 2\mathrm{tr}(\Theta_V^{-1} \Theta_{\bar{\mathrm{I}}A\bar{\mathrm{I}}V}) + \mathrm{tr}(\Theta_V^{-1} \Theta_{\bar{\mathrm{I}}A\bar{\mathrm{I}}}) \\ \mathrm{tr}(\bar{\mathrm{I}} A \bar{\mathrm{I}} S_\eta) - 2\mathrm{tr}(\Theta_V^{-1} \Theta_{\bar{\mathrm{I}}A\bar{\mathrm{I}}S_\eta V}) + \mathrm{tr}(\Theta_V^{-1} \Theta_{\bar{\mathrm{I}}A\bar{\mathrm{I}}} \Theta_V^{-1} \Theta_{VS_\eta V}) \\ \mathrm{tr}(\bar{\mathrm{I}} A \bar{\mathrm{I}} S_\mu) - 2\mathrm{tr}(\Theta_V^{-1} \Theta_{\bar{\mathrm{I}}A\bar{\mathrm{I}}S_\mu V}) + \mathrm{tr}(\Theta_V^{-1} \Theta_{\bar{\mathrm{I}}A\bar{\mathrm{I}}} \Theta_V^{-1} \Theta_{VS_\mu V}) \end{pmatrix}^\top \begin{pmatrix} \sigma_\nu^2 \\ \sigma_\eta^2 \\ \sigma_\mu^2 \end{pmatrix}$$

or, denoting $\Upsilon_A = Z^\top A Z$:

$$\mathrm{E}(\hat{q}_A^V) = \begin{pmatrix} \mathrm{tr}(A) - 2\mathrm{tr}(\Upsilon_V^{-1} \Upsilon_{VA}) + \mathrm{tr}(\Upsilon_V^{-1} \Upsilon_A) \\ \mathrm{tr}(A S_\eta) - 2\mathrm{tr}(\Upsilon_V^{-1} \Upsilon_{VS_\eta A}) + \mathrm{tr}(\Upsilon_V^{-1} \Upsilon_A \Upsilon_V^{-1} \Upsilon_{VS_\eta V}) \\ \mathrm{tr}(A S_\mu) - 2\mathrm{tr}(\Upsilon_V^{-1} \Upsilon_{VS_\mu A}) + \mathrm{tr}(\Upsilon_V^{-1} \Upsilon_A \Upsilon_V^{-1} \Upsilon_{VS_\mu V}) \end{pmatrix}^\top \begin{pmatrix} \sigma_\nu^2 \\ \sigma_\eta^2 \\ \sigma_\mu^2 \end{pmatrix}$$

The most common estimators are obtained by considering the quadratic forms with the *within*, *between*-individual, and *between*-time matrices. We then get the following system of equations:

$$\begin{pmatrix} \hat{q}_{\mathrm{W}}^V \\ \hat{q}_{\mathrm{B}_\eta}^V \\ \hat{q}_{\mathrm{B}_\mu}^V \end{pmatrix} = H \begin{pmatrix} \sigma_\nu^2 \\ \sigma_\eta^2 \\ \sigma_\mu^2 \end{pmatrix} \tag{3.6}$$

with

$$H = \begin{pmatrix} \mathrm{tr}(\mathrm{M}^{V\top} W \mathrm{M}^V) & \mathrm{tr}(\mathrm{M}^{V\top} W \mathrm{M}^V S_\eta) & \mathrm{tr}(\mathrm{M}^{V\top} W \mathrm{M}^V S_\mu) \\ \mathrm{tr}(\mathrm{M}^{V\top} B_\eta \mathrm{M}^V) & \mathrm{tr}(\mathrm{M}^{V\top} B_\eta \mathrm{M}^V S_\eta) & \mathrm{tr}(\mathrm{M}^{V\top} B_\eta \mathrm{M}^V S_\mu) \\ \mathrm{tr}(\mathrm{M}^{V\top} B_\mu \mathrm{M}^V) & \mathrm{tr}(\mathrm{M}^{V\top} B_\mu \mathrm{M}^V S_\eta) & \mathrm{tr}(\mathrm{M}^{V\top} B_\mu \mathrm{M}^V S_\mu) \end{pmatrix} \tag{3.7}$$

Using the following results: $\bar{\mathrm{I}}W = W$, $\mathrm{tr}(W) = O - N - T + 1$, $WS_\eta = 0$, $WS_\mu = 0$, $B_\eta S_\eta = S_\eta$, $B_\mu S_\mu = S_\mu$, $\mathrm{tr}(S_\mu) = \mathrm{tr}(S_\eta) = O$, $\mathrm{tr}(B_\eta) = N$, $\mathrm{tr}(B_\mu) = T$, $\bar{\mathrm{I}}B_\eta\bar{\mathrm{I}}S_\eta = \bar{\mathrm{I}}S_\eta$ and $\mathrm{tr}(\bar{\mathrm{I}}S_\eta) = O - \sum_n T_n/O$, $\bar{\mathrm{I}}B_\mu\bar{\mathrm{I}}S_\mu = \bar{\mathrm{I}}S_\mu$ and $\mathrm{tr}(\bar{\mathrm{I}}S_\mu) = O - \sum_t N_t/O$, $\mathrm{tr}(\bar{\mathrm{I}}B_\eta) = N - 1$, $\mathrm{tr}(\bar{\mathrm{I}}B_\mu) = T - 1$,

$\mathrm{tr}(\bar{I}B_\mu\bar{I}S_\eta) = T - \sum_n T_n^2/O$, $\mathrm{tr}(\bar{I}B_\eta\bar{I}S_\mu) = N - \sum_t N_t^2/O$, $\mathrm{tr}(B_\mu S_\eta) = T$, $\mathrm{tr}(B_\eta S_\mu) = N$, we get:

$$\left\{\begin{array}{rcl}
\mathrm{E}(\hat{q}_W^V) &=& \sigma_\nu^2[O - N - T + 1 - 2\mathrm{tr}(\Theta_V^{-1}\Theta_{\bar{I}VW}) + \mathrm{tr}(\Theta_V^{-1}\Theta_W)] \\
&& +\sigma_\eta^2[\mathrm{tr}(\Theta_V^{-1}\Theta_W\Theta_V^{-1}\Theta_{VS_\eta V})] \\
&& +\sigma_\mu^2[\mathrm{tr}(\Theta_V^{-1}\Theta_W\Theta_V^{-1}\Theta_{VS_\mu V})] \\
\mathrm{E}(\hat{q}_{B_\eta}^V) &=& \sigma_\nu^2[N - 1 - 2\mathrm{tr}(\Theta_V^{-1}\Theta_{\bar{I}VB_\eta}) + \mathrm{tr}(\Theta_V^{-1}\Theta_{\bar{B}_\eta})] \\
&& +\sigma_\eta^2\left[O - \sum_n T_n^2/O + 2\mathrm{tr}(\Theta_V^{-1}\Theta_{\bar{I}S_\eta V}) + \mathrm{tr}(\Theta_V^{-1}\Theta_{\bar{B}_\eta}\Theta_V^{-1}\Theta_{VS_\eta V})\right] \\
&& +\sigma_\mu^2\left[N - \sum_t N_t^2/O - 2\mathrm{tr}(\Theta_V^{-1}\Theta_{\bar{I}S_\mu V}) + \mathrm{tr}(\Theta_V^{-1}\Theta_{\bar{B}_\eta}\Theta_V^{-1}\Theta_{VS_\mu V})\right] \\
\mathrm{E}(\hat{q}_{B_\mu}^V) &=& \sigma_\nu^2[T - 1 - 2\mathrm{tr}(\Theta_V^{-1}\Theta_{\bar{I}VB_\mu}) + \mathrm{tr}(\Theta_V^{-1}\Theta_{\bar{B}_\mu})] \\
&& +\sigma_\eta^2\left[T - \sum_n T_n^2/O - 2\mathrm{tr}(\Theta_V^{-1}\Theta_{\bar{I}S_\eta V}) + \mathrm{tr}(\Theta_V^{-1}\Theta_{\bar{B}_\mu}\Theta_V^{-1}\Theta_{VS_\eta V})\right] \\
&& +\sigma_\mu^2\left[O - \sum_t N_t^2/O - 2\mathrm{tr}(\Theta_V^{-1}\Theta_{\bar{I}S_\mu V}) + \mathrm{tr}(\Theta_V^{-1}\Theta_{\bar{B}_\mu}\Theta_V^{-1}\Theta_{VS_\mu V})\right]
\end{array}\right.$$

or:

$$\left\{\begin{array}{rcl}
\mathrm{E}(\hat{q}_W^V) &=& \sigma_\nu^2[O - N - T + 1 - \mathrm{tr}(\Upsilon_V^{-1}\Upsilon_W)] \\
&& +\sigma_\eta^2[\mathrm{tr}(\Upsilon_V^{-1}\Upsilon_W\Upsilon_V^{-1}\Upsilon_{VS_\eta V})] \\
&& +\sigma_\mu^2[\mathrm{tr}(\Upsilon_V^{-1}\Upsilon_W\Upsilon_V^{-1}\Upsilon_{VS_\mu V})] \\
\mathrm{E}(\hat{q}_{B_\eta}^V) &=& \sigma_\nu^2[N - 2\mathrm{tr}(\Upsilon_V^{-1}\Upsilon_{B_\eta V}) + \mathrm{tr}(\Upsilon_V^{-1}\Upsilon_{B_\eta})] \\
&& +\sigma_\eta^2[O - 2\mathrm{tr}(\Upsilon_V^{-1}\Upsilon_{S_\eta V}) + \mathrm{tr}(\Upsilon_V^{-1}\Upsilon_{B_\eta}\Upsilon_V^{-1}\Upsilon_{VS_\eta V})] \\
&& +\sigma_\mu^2[N - 2\mathrm{tr}(\Upsilon_V^{-1}\Upsilon_{B_\eta S_\mu V}) + \mathrm{tr}(\Upsilon_V^{-1}\Upsilon_{B_\eta}\Upsilon_V^{-1}\Upsilon_{VS_\mu V})] \\
\mathrm{E}(\hat{q}_{B_\mu}^V) &=& \sigma_\nu^2[T - 2\mathrm{tr}(\Upsilon_V^{-1}\Upsilon_{B_\mu V}) + \mathrm{tr}(\Upsilon_V^{-1}\Upsilon_{B_\mu})] \\
&& +\sigma_\eta^2[T - 2\mathrm{tr}(\Upsilon_V^{-1}\Upsilon_{B_\mu S_\eta V}) + \mathrm{tr}(\Upsilon_V^{-1}\Upsilon_{B_\mu}\Upsilon_V^{-1}\Upsilon_{VS_\eta V})] \\
&& +\sigma_\mu^2[O - 2\mathrm{tr}(\Upsilon_V^{-1}\Upsilon_{S_\mu V}) + \mathrm{tr}(\Upsilon_V^{-1}\Upsilon_{B_\mu}\Upsilon_V^{-1}\Upsilon_{VS_\mu V})]
\end{array}\right.$$

The estimator is obtained by equating the quadratic form and its expected value:

$$\begin{pmatrix} \hat{q}_{\mathrm{W}}^{V1} \\ \hat{q}_{\mathrm{B}_\eta}^{V2} \\ \hat{q}_{\mathrm{B}_\mu}^{V3} \end{pmatrix} = H \begin{pmatrix} \sigma_\nu^2 \\ \sigma_\eta^2 \\ \sigma_\mu^2 \end{pmatrix}$$

The H matrices corresponding to the three most common estimators are presented in Figure 3.1.

Example 3.1 unbalanced panel – `Tileries` data set

To illustrate the estimation of unbalanced panels, we employ the `Tileries` data, concerning the weekly production of cement floor tiles for 25 Egyptian small-scale tileries in 1982–1983. The data are observed over 66 weeks in total, and aggregated on periods of three weeks. The number of observations for each firm ranges from 12 to 22, which can be checked using the `pdim` function.

$$\begin{pmatrix} \hat{q}^{V1}_{\mathrm{W}} \\ \hat{q}^{V2}_{\mathrm{B}_\eta} \\ \hat{q}^{V3}_{\mathrm{B}_\mu} \end{pmatrix} = H \begin{pmatrix} \sigma^2_\nu \\ \sigma^2_\eta \end{pmatrix}$$

Amemiya (1971):

$$H = \begin{pmatrix} O-N-T+1-K & 0 & 0 \\ N-1+\mathrm{tr}(\Theta_W^{-1}\Theta_{B-\bar{J}}) & O-\sum_n T_n^2/O & N-\sum_t N_t^2/O \\ T-1+\mathrm{tr}(\Theta_W^{-1}\Theta_{B_\mu-\bar{J}}) & T-\sum_n T_n^2/O & O-\sum_t N_t^2/O \end{pmatrix}$$

Wallace and Hussain (1969):

$$H = \begin{pmatrix} O-N-T+1-\mathrm{tr}(\Upsilon_I^{-1}\Upsilon_W) & \mathrm{tr}(\Upsilon_I^{-1}\Upsilon_W\Upsilon_I^{-1}\Upsilon_{S_\eta}) & \mathrm{tr}(\Upsilon_I^{-1}\Upsilon_W\Upsilon_I^{-1}\Upsilon_{S_\mu}) \\ N-\mathrm{tr}(\Upsilon^{-1}\Upsilon_{B_\eta}) & O-2\mathrm{tr}(\Upsilon_I^{-1}\Upsilon_{S_\eta})+\mathrm{tr}(\Upsilon_I^{-1}\Upsilon_{B_\eta}\Upsilon_I^{-1}\Upsilon_{S_\eta}) & N-2\mathrm{tr}(\Upsilon_I^{-1}\Upsilon_{B_\eta S_\mu})+\mathrm{tr}(\Upsilon_I^{-1}\Upsilon_{B_\eta}\Upsilon_I^{-1}\Upsilon_{S_\mu}) \\ T-\mathrm{tr}(\Upsilon_I^{-1}\Upsilon_{B_\mu}) & T-2\mathrm{tr}(\Upsilon_I^{-1}\Upsilon_{B_\mu S_\eta})+\mathrm{tr}(\Upsilon_I^{-1}\Upsilon_{B_\mu}\Upsilon_I^{-1}\Upsilon_{S_\eta}) & O-2\mathrm{tr}(\Upsilon_I^{-1}\Upsilon_{S_\mu})+\mathrm{tr}(\Upsilon_I^{-1}\Upsilon_{B_\mu}\Upsilon_I^{-1}\Upsilon_{S_\mu}) \end{pmatrix}$$

Swamy and Arora (1972):

$$H = \begin{pmatrix} O-N-T+1-K & 0 & 0 \\ N-K-1 & O-\mathrm{tr}(\Upsilon_{B_\eta}^{-1}\Upsilon_{S_\eta}) & N-\mathrm{tr}(\Upsilon_{B_\eta}^{-1}\Upsilon_{B_\eta S_\mu B_\eta}) \\ T-K-1 & T-\mathrm{tr}(\Upsilon_{B_\mu}^{-1}\Upsilon_{B_\mu S_\eta B_\mu}) & O-\mathrm{tr}(\Upsilon_{B_\mu}^{-1}\Upsilon_{S_\mu}) \end{pmatrix}$$

Figure 3.1 Estimators of the variance components for unbalanced panels.

```
data("Tileries", package = "pder")
head(Tileries, 3)
  id week   area output labor machine
1  2    1 fayoum  5.650 4.533   4.663
2  2    2 fayoum  6.522 5.347   4.234
3  2    3 fayoum  6.303 4.970   4.234
pdim(Tileries)
Unbalanced Panel: n = 25, T = 12-22, N = 483
```

We estimate a Cobb-Douglas production function where the production (`output`) depends on the quantity of two inputs, labor (`labor`) and machines (`machine`). We first check that the same fixed effects model can still be estimated either by applying **OLS** on the *within* transformed variables or using individual dummy variables:

```
Tileries <- pdata.frame(Tileries)
plm.within <- plm(log(output) ~ log(labor) + log(machine), Tileries)
y <- log(Tileries$output)
x1 <- log(Tileries$labor)
x2 <- log(Tileries$machine)
lm.within <- lm(I(y - Between(y)) ~ I(x1 - Between(x1)) + I(x2 - Between(x2)) - 1)
lm.lsdv <- lm(log(output) ~ log(labor) + log(machine) + factor(id), Tileries)
coef(lm.lsdv)[2:3]
  log(labor) log(machine)
     0.87062      0.02438
coef(lm.within)
I(x1 - Between(x1)) I(x2 - Between(x2))
            0.87062             0.02438
coef(plm.within)
  log(labor) log(machine)
     0.87062      0.02438
```

The one-way random effects model is then estimated:

```
tile.r <- plm(log(output) ~ log(labor) + log(machine), Tileries, model = "random")
summary(tile.r)
Oneway (individual) effect Random Effect Model
   (Swamy-Arora's transformation)

Call:
plm(formula = log(output) ~ log(labor) + log(machine), data = Tileries,
    model = "random")

Unbalanced Panel: n = 25, T = 12-22, N = 483

Effects:
                   var  std.dev share
idiosyncratic 0.002640 0.051377  0.81
individual    0.000623 0.024964  0.19
theta:
   Min. 1st Qu.  Median    Mean 3rd Qu.    Max.
  0.489   0.573   0.582   0.578   0.590   0.598
```

```
Residuals:
   Min. 1st Qu.  Median    Mean 3rd Qu.    Max.
-0.1866 -0.0272  0.0031  0.0000  0.0334  0.2268

Coefficients:
             Estimate Std. Error t-value Pr(>|t|)
(Intercept)    0.2779     0.0608    4.57  6.1e-06 ***
log(labor)     0.9088     0.0300   30.25  < 2e-16 ***
log(machine)   0.0240     0.0270    0.89     0.38
---
Signif. codes:
0 '***' 0.001 '**' 0.01 '*' 0.05 '.' 0.1 ' ' 1

Total Sum of Squares:    4.84
Residual Sum of Squares: 1.3
R-Squared:        0.732
Adj. R-Squared: 0.731
F-statistic: 656.318 on 2 and 480 DF, p-value: <2e-16
```

The transformation parameter is now individual specific; more precisely, it depends on the number of available observations for every individual. The θ parameter here varies from 0.49 to 0.60.

The two-ways random effect model is obtained by setting the `effect` argument to `'twoways'`.

We check that the **OLS** model cannot be obtained any more by applying **OLS** to variables where the individual and time means have been removed.

```
plm.within <- plm(log(output) ~ log(labor) + log(machine),
                  Tileries, effect = "twoways")
lm.lsdv <- lm(log(output) ~ log(labor) + log(machine) +
                  factor(id) + factor(week), Tileries)
y <- log(Tileries$output)
x1 <- log(Tileries$labor)
x2 <- log(Tileries$machine)
y <- y - Between(y, "individual") - Between(y, "time") + mean(y)
x1 <- x1 - Between(x1, "individual") - Between(x1, "time") + mean(x1)
x2 <- x2 - Between(x2, "individual") - Between(x2, "time") + mean(x2)
lm.within <- lm(y ~ x1 + x2 - 1)
coef(plm.within)
  log(labor) log(machine)
     0.86951      0.03539
coef(lm.within)
     x1      x2
0.88085 0.03554
coef(lm.lsdv)[2:3]
  log(labor) log(machine)
     0.86951      0.03539
```

Finally we estimate the time and individual random effects model, using the three methods of estimation we have described:

```
wh <- plm(log(output) ~ log(labor) + log(machine), Tileries,
          model = "random", random.method = "walhus",
          effect = "twoways")
am <- update(wh, random.method = "amemiya")
sa <- update(wh, random.method = "swar")
ercomp(sa)
                    var   std.dev share
idiosyncratic 0.002589 0.050884  0.77
individual    0.000625 0.025001  0.19
time          0.000158 0.012551  0.05
theta:
        Min. 1st Qu. Median   Mean 3rd Qu.   Max.
id    0.4934  0.5769 0.5858 0.5813  0.5941 0.6019
time  0.1962  0.3461 0.3544 0.3487  0.3625 0.3702
total 0.1665  0.3023 0.3097 0.3058  0.3186 0.3295
```

The shares of the individual and the time effects in the total error variance are now about 19 and 5% for the Swamy-Arora estimator.

```
re.models <- list(walhus = wh, amemiya = am, swar = sa)
sapply(re.models, function(x) sqrt(ercomp(x)$sigma2))
       walhus amemiya    swar
idios 0.05167 0.05088 0.05088
id    0.02778 0.03192 0.02500
time  0.01177 0.01267 0.01255
sapply(re.models, coef)
              walhus amemiya    swar
(Intercept)  0.27420 0.28560 0.26528
log(labor)   0.90778 0.90062 0.91279
log(machine) 0.02696 0.02774 0.02692
```

3.2 Seemingly Unrelated Regression

3.2.1 Introduction

Very often in economics, the phenomenon under investigation is not well described by a single equation but by a system of equations. It is particularly the case in the field of micro-econometrics of consumption or production. For example, the behavior of a producer is described by a minimum cost equation along with equations of factor demand. In this case, there are two advantages in considering the whole system of equations:

- firstly, the errors of the different equations for an observation may be correlated. In this case, even if the estimation of a single equation is consistent, it is inefficient because it does not take into account the correlation between the errors,
- secondly, economic theory may impose restrictions on different coefficients of the system, for example, the equality of two coefficients in two different equations of the system. In this case, these restrictions can be taken into account using the method of constrained least squares.

3.2.2 Constrained Least Squares

Linear restrictions on the vector of coefficients to be estimated can be represented using a restriction matrix R and a numeric vector q:

$$R\beta = q$$

For example, if the sum of the first two coefficients must equal 1 and the first and third ones should be equal, the joint restrictions can be written as:

$$\begin{pmatrix} 1 & 1 & 0 \\ 1 & 0 & -1 \end{pmatrix} \begin{pmatrix} \beta_1 \\ \beta_2 \\ \beta_3 \end{pmatrix} = \begin{pmatrix} 1 \\ 0 \end{pmatrix}$$

To estimate the constrained OLS estimator, we write the Lagrangian:

$$L = \epsilon^\top \epsilon + 2\lambda^\top (R\beta - q)$$

with $\epsilon = y - Z\gamma$ and λ the vector of Lagrange multipliers associated to the different constraints.[1] The Lagrangian can also be written as:

$$L = y^\top y - 2\beta^\top X^\top y + \beta^\top X^\top X\beta + 2\lambda(R\beta - q)$$

The first-order conditions become:

$$\begin{cases} \frac{\partial L}{\partial \beta} = -2X^\top y + 2X^\top X\beta + 2R^\top \lambda = 0 \\ \frac{\partial L}{\partial \lambda} = 2(R\beta - q) = 0 \end{cases}$$

which can also be written in matrix form:

$$\begin{pmatrix} X^\top X & R^\top \\ R & 0 \end{pmatrix} \begin{pmatrix} \beta \\ \lambda \end{pmatrix} = \begin{pmatrix} X^\top y \\ q \end{pmatrix}$$

The constrained OLS estimator can be obtained using the formula for the inverse of a partitioned matrix (see equation 2.18):

$$\begin{pmatrix} A_{11} & A_{12} \\ A_{21} & A_{22} \end{pmatrix}^{-1} = \begin{pmatrix} B_{11} & B_{12} \\ B_{21} & B_{22} \end{pmatrix} = \begin{pmatrix} A_{11}^{-1}(I + A_{12}F_2A_{21}A_{11}^{-1}) & -A_{11}^{-1}A_{12}F_2 \\ -F_2A_{21}A_{11}^{-1} & F_2 \end{pmatrix}$$

with $F_2 = (A_{22} - A_{21}A_{11}^{-1}A_{12})^{-1}$ and $F_1 = (A_{11} - A_{12}A_{22}^{-1}A_{21})^{-1}$.

We have here $F_2 = -(R(X^\top X)^{-1}R^\top)^{-1}$. The constrained estimator is then: $\hat{\beta}_c = B_{11}X^\top y + B_{12}q$, with $B_{11} = (X^\top X)^{-1}(I - R^\top (R(X^\top X)^{-1}R^\top)^{-1}R(X^\top X)^{-1})$ and $B_{12} = (X^\top X)^{-1}R^\top (R(X^\top X)^{-1}R^\top)^{-1}$

The unconstrained estimator being $\hat{\beta}_{nc} = (X^\top X)^{-1}X^\top y$, we finally get:

$$\hat{\beta}_c = \hat{\beta}_{nc} - (X^\top X)^{-1}R^\top (R(X^\top X)^{-1}R^\top)^{-1}(R\hat{\beta}_{nc} - q)$$

The difference between the constrained and the unconstrained estimators is then a linear combination of the excess of the linear constraints of the model evaluated for the unconstrained model.

1 These multipliers are multiplied by two in order to simplify the first- order conditions.

3.2.3 Inter-equations Correlation

We consider a system of L equations denoted $y_l = X_l\beta_l + \epsilon_l$, with $l = 1 \dots L$. In matrix form, the system can be written as follows:

$$\begin{pmatrix} y_1 \\ y_2 \\ \vdots \\ y_L \end{pmatrix} = \begin{pmatrix} X_1 & 0 & \dots & 0 \\ 0 & X_2 & \dots & 0 \\ \vdots & \vdots & \ddots & \vdots \\ 0 & 0 & \dots & X_L \end{pmatrix} \begin{pmatrix} \beta_1 \\ \beta_2 \\ \vdots \\ \beta_L \end{pmatrix} + \begin{pmatrix} \epsilon_1 \\ \epsilon_2 \\ \vdots \\ \epsilon_L \end{pmatrix}$$

The covariance matrix of the errors of the system is:

$$\Omega = \mathrm{E}(\epsilon\epsilon^\top) = \mathrm{E} \begin{pmatrix} \epsilon_1\epsilon_1^\top & \epsilon_1\epsilon_2^\top & \dots & \epsilon_1\epsilon_L^\top \\ \epsilon_2\epsilon_1^\top & \epsilon_2\epsilon_2^\top & \dots & \epsilon_2\epsilon_L^\top \\ \vdots & \vdots & \ddots & \vdots \\ \epsilon_L\epsilon_1^\top & \epsilon_L\epsilon_2^\top & \dots & \epsilon_L\epsilon_L^\top \end{pmatrix}$$

We suppose that the errors of two equations l and m for the same observations are correlated and that the covariance, denoted by σ_{lm}, is constant. With this hypothesis, the covariance matrix is:

$$\Omega = \begin{pmatrix} \sigma_{11}\mathrm{I} & \sigma_{12}\mathrm{I} & \dots & \sigma_{1L}\mathrm{I} \\ \sigma_{12}\mathrm{I} & \sigma_{22}\mathrm{I} & \dots & \sigma_{2L}\mathrm{I} \\ \vdots & \vdots & \ddots & \vdots \\ \sigma_{1L}\mathrm{I} & \sigma_{2L}\mathrm{I} & \dots & \sigma_{LL}\mathrm{I} \end{pmatrix}$$

Denoting by Σ the matrix of inter-equations covariances, we have:

$$\Sigma = \begin{pmatrix} \sigma_{11} & \sigma_{12} & \cdots & \sigma_{1L} \\ \sigma_{12} & \sigma_{22} & \cdots & \sigma_{2L} \\ \vdots & \vdots & \ddots & \vdots \\ \sigma_{1L} & \sigma_{2L} & \cdots & \sigma_{LL} \end{pmatrix}$$

$$\Omega = \Sigma \otimes \mathrm{I}$$

Because of the inter-equations correlations, the efficient estimator is the GLS estimator: $\hat{\beta} = (X^\top\Omega^{-1}X)^{-1}X^\top\Omega^{-1}y$. This estimator, first proposed by Zellner (1962), is known by the acronym SUR for *seemingly unrelated regression*. It can be obtained by applying OLS on transformed data, each variable being pre-multiplied by $\Omega^{-0.5}$. This matrix is simply $\Omega^{-0.5} = \Sigma^{-0.5} \otimes \mathrm{I}$. Denoting by r_{lm} the elements of $\Sigma^{-0.5}$, the transformed response and covariates are:

$$\tilde{y} = \begin{pmatrix} r_{11}y_1 + r_{12}y_2 + \dots + r_{1L}y_L \\ r_{21}y_1 + r_{22}y_2 + \dots + r_{2L}y_L \\ \vdots \\ r_{L1}y_1 + r_{L2}y_2 + \dots + r_{LL}y_L \end{pmatrix} \text{and } \tilde{X} = \begin{pmatrix} r_{11}X_1 & r_{12}X_2 & \dots & r_{1L}X_L \\ r_{21}X_1 & r_{22}X_2 & \dots & r_{2L}X_L \\ \vdots & \vdots & \ddots & \vdots \\ r_{L1}X_1 & r_{L2}X_2 & \dots & r_{LL}X_L \end{pmatrix}$$

Σ is a matrix that contains unknown parameters, which can be estimated using residuals of a consistent but inefficient preliminary estimator, like OLS. The efficient estimator is then obtained the following way:

- first, each equation is estimated separately by OLS and we note $\hat{\Xi} = (\hat{\epsilon}_1, \hat{\epsilon}_2, \dots, \hat{\epsilon}_L)$ the $N \times L$ matrix for which every column is the residual vector of one of the equations in the system,
- then, estimate the covariance matrix of the errors: $\hat{\Sigma} = \hat{\Xi}^\top \hat{\Xi}/N$,
- compute the matrix $\hat{\Sigma}^{-0.5}$ and use it to transform the response and the covariates of the model,
- finally, estimate the model by applying OLS on transformed data.

$\Sigma^{-0.5}$ can conveniently be computed using the Cholesky decomposition, i.e., computing the lower-triangular matrix C such that $CC^\top = \Sigma^{-1}$.

3.2.4 SUR With Panel Data

Applying the SUR estimator on panel data is straightforward when only the *between* or the *within* variability of the data is taken into account. In this case, one just has to apply the above formula using the variables in individual means (*between*-SUR) or in deviations from individual means (*within*-SUR). Taking into account both sources of variability requires more attention and leads to the SUR error component model proposed by Avery (1977) and Baltagi (1980). The errors of the model then present two sources of correlation:

- the correlation of the SUR model, i.e., inter-equations correlation,
- the correlation taken into account in the error component model, i.e., the intra-individual correlations.

Every observation is now characterized by three indexes: z_{lnt} is the observation of z for equation l, individual n and period t. The observations are first ordered by equation, then by individual. Denoting $\epsilon_{ln}^\top = (\epsilon_{ln1}^\top, \epsilon_{ln2}^\top, \dots, \epsilon_{lnT}^\top)$ the error vector for equation l and individual n, one gets:

$$\mathrm{E}(\epsilon_{ln}\epsilon_{mn}^\top) = \sigma_{\nu_{lm}}^2 \mathrm{I}_T + \sigma_{\eta_{lm}}^2 J_T$$

The errors concerning different individuals being uncorrelated, the correlation matrix for two equations and all individuals is:

$$\begin{aligned}
\mathrm{E}(\epsilon_l\epsilon_m^\top) &= \mathrm{I}_N \otimes (\sigma_{\nu_{lm}} \mathrm{I}_T + \sigma_{\eta_{lm}} J_T) \\
&= \sigma_{\nu_{lm}} \mathrm{I}_{NT} + \sigma_{\eta_{lm}} \mathrm{I}_N \otimes J_T \\
&= \sigma_{\nu_{lm}} (W + B) + T\sigma_{\eta_{lm}} B \\
&= \sigma_{\nu_{lm}} W + (\sigma_{\nu_{lm}} + T\sigma_{\eta_{lm}}) B \\
&= \sigma_{\nu_{lm}} W + \sigma_{\iota_{lm}} B
\end{aligned}$$

Finally, for the whole system of equations, denoting Σ_ν and Σ_ι the two matrices of dimension $L \times L$ containing the parameters $\sigma_{\nu_{lm}}$ and $\sigma_{1_{lm}}$, the covariance matrix of the errors is:

$$\Omega = \Sigma_\nu \otimes W + \Sigma_\iota \otimes B$$

The SUR error component model may be obtained by applying OLS on transformed data, every variable being pre-multiplied by $\Omega^{-0.5}$.

$$\Omega^{-0.5} = \Sigma_\nu^{-0.5} \otimes W + \Sigma_\iota^{-0.5} \otimes B \tag{3.8}$$

and may be estimated using the Cholesky decomposition of Σ_ν^{-1} and Σ_ι^{-1} (see Kinal and Lahiri, 1990).

The two error covariance matrices being unknown, the error-component SUR estimator is obtained with the following steps:

- first, each equation is estimated separately using a consistent method of estimation (for example OLS): we denote by $W\hat{\Xi}$ and $B\hat{\Xi}$ the matrices of residuals in deviation from the individual means and in individual means, respectively,

- next, we estimate the error covariance matrices: $\hat{\Sigma}_\nu = (W\hat{\Xi})^\top(W\hat{\Xi})/(N(T-1))$ and $\hat{\Sigma}_\iota = (B\hat{\Xi})^\top(B\hat{\Xi})/(N-1)$,
- we then compute the matrices $\hat{\Sigma}_\nu^{-0.5}$ and $\hat{\Sigma}_\iota^{-0.5}$ and hence, through 3.8, we obtain the transformed variables $\tilde{y}$ and $\tilde{X}$,
- finally, we apply OLS on $\tilde{y}$ and $\tilde{X}$.

Different choices of preliminary estimates lead to different SUR-error component estimators. For example, Baltagi (1980) used the method of Amemiya (1971) while Avery (1977) chose the one of Swamy and Arora (1972).

Example 3.2 SUR estimation – `TexasElectr` data set
A common application of the SUR model is the analysis of production cost. The cost function returns the minimum cost of production C for a given vector of prices of the F production factors $p^\top = (p_1, p_2, \dots, p_F)$ and the level of output q. The minimum cost function is $C(p, q)$. It has several properties:

- it is homogeneous of degree 1 with respect to the factor prices: $C(\lambda p, q) = \lambda C(p, q)$,
- the demand functions for production factors are the derivatives of the minimum cost function with respect to factor prices,[2] i.e., the gradient of the cost function: $\frac{\partial C}{\partial p}(p, q) = x(p, q)$
- the Hessian matrix of the cost function is symmetric: $\frac{\partial^2 C}{\partial p_i \partial p_j} = \frac{\partial^2 C}{\partial p_j \partial p_i}$.

The most common functional form assumed for the cost function is the translog, defined by:

$$\begin{aligned}\ln C(p,q) &= \beta_0 + \beta_q \ln q + \sum_{i=1}^{F} \beta_i \ln p_i \\ &+ 0.5\beta_{qq}\ln^2 q + 0.5\sum_{i=1}^{F}\sum_{j=1}^{F} \beta_{ij} \ln p_i \ln p_j\end{aligned}$$

Dividing total cost and factor prices by one of these prices (the first, for example), homogeneity of degree 1 with respect to prices is imposed:

$$\begin{aligned}\ln \frac{C}{p_1}(p,q) &= \beta_0 + \beta_q \ln q + \sum_{i=2}^{F} \beta_i \ln \frac{p_i}{p_1} \\ &+ 0.5\beta_{qq}\ln^2 q + 0.5\sum_{i=2}^{F}\sum_{j=2}^{F} \beta_{ij} \ln \frac{p_i}{p_1} \ln \frac{p_j}{p_1}\end{aligned}$$

Shephard's lemma implies that: $\frac{\partial \ln C}{\partial \ln p_i} = \frac{\partial C}{\partial p_i}\frac{p_i}{C} = \frac{p_i x_i}{C} = s_i$, that is, the logarithmic derivative of the cost with respect to the price of a factor equals the share of that factor in total cost. The share of factor j is then:

$$s_j = \frac{\partial \ln C}{\partial \ln p_j} = \beta_j + \sum_{i=2}^{F} \beta_{ij} \ln \frac{p_i}{p_1}$$

It is customary to divide each price and the production by their means. In this case, $\ln q$ and $\ln p_i$ are zero at the sample mean, which gives an intuitive meaning to the first-order coefficients. β_q is then the cost elasticity with respect to the production level at the sample mean, and β_i the share of factor i in the cost at the sample mean.

The data we use concern the production cost of ten electricity producers in Texas over 18 years (from 1966 to 1983). They have been analyzed by Kumbhakar (1996), Horrace and

2 This result is known as Shephard's lemma.

Schmidt (1996) and Horrace and Schmidt (2000). Three production factors are used: fuel, labor, and capital. For each factor, we observe unit factor prices (`pfuel`, `plab`, and `pcap`) and factor expenses (`expfuel`, `explab`, and `expcap`).

We first compute the prices in logarithms, we divide them by their sample mean, and we also divide them by one of the prices, here fuel price. We perform this task using the `mutate` function of the **dplyr** package.

```
data("TexasElectr", package = "pder")
library("dplyr")
TexasElectr <- mutate(TexasElectr,
                      pf = log(pfuel / mean(pfuel)),
                      pl = log(plab / mean(plab)) - pf,
                      pk = log(pcap / mean(pcap)) - pf)
```

The production is also measured in logarithms and divided by its sample mean.

```
TexasElectr <- mutate(TexasElectr, q = log(output / mean(output)))
```

We then compute total production cost by summing the expenses for the three factors and factor shares. Finally, we measure the cost in logarithms and divide it by its sample mean and by the reference price.

```
TexasElectr <- mutate(TexasElectr,
                      C = expfuel + explab + expcap,
                      sl = explab / C,
                      sk = expcap / C,
                      C = log(C / mean(C)) - pf)
```

Finally, we compute the squares and the interaction terms for the variables.

```
TexasElectr <- mutate(TexasElectr,
                      pll = 1/2 * pl ^ 2,
                      plk = pl * pk,
                      pkk = 1/2 * pk ^ 2,
                      qq = 1/2 * q ^ 2)
```

We define the three equations of the system, one for total cost and the other two for factor shares.[3]

```
cost <- C ~ pl + pk + q + pll + plk + pkk + qq
shlab <- sl ~ pl + pk
shcap <- sk ~ pl + pk
```

Factor shares being the derivatives of the cost function, the following restrictions must be imposed:

- the coefficient of `pl` in the cost equation must equal the intercept in the labor share equation,
- the coefficient of `pk` in the cost equation must equal the intercept in the capital share equation,

3 The fuel share is omitted to avoid perfect collinearity, given that the three shares sum to one.

- the coefficient of `pll` in the cost equation should equal the coefficient associated to `pl` in the labor share equation,
- the coefficient of `pkk` in the cost equation should be equal to the coefficient associated to `pk` in the capital share equation,
- the coefficient of `plk` in the cost equation should equal the coefficient of `pk` in the labor share equation and the coefficient of `pl` in the capital share equation.

We construct for this purpose a 6 (number of restrictions) by 14 (number of coefficients) matrix.

```
R <- matrix(0, nrow = 6, ncol = 14)
R[1, 2] <- R[2, 3] <- R[3, 5] <- R[4, 6] <- R[5, 6] <- R[6, 7] <- 1
R[1, 9] <- R[2, 12] <- R[3, 10] <- R[4, 11] <- R[5, 13] <- R[6, 14] <- -1
```

The first line of the matrix indicates that the second coefficient (the one associated to `pl` in the cost equation) must be equal to the ninth (the constant term in the labor share equation).

The SUR model is estimated providing a list of formulae, defining the system of equations to be estimated, as the first argument to `plm`. The different formulae in the list can be named, which makes the output more readable. The `model` argument is set to `'random'` in order to estimate the SUR error components model. Lastly, the arguments `restrict.matrix` and `restrict.rhs` allow to specify the matrix R and the vector q defining the linear constraints of the model. If, as happens here, all elements of q are zero, the `restrict.rhs` argument can be omitted.

```
z <- plm(list(cost = C ~ pl + pk + q + pll + plk + pkk + qq,
              shlab = sl ~ pl + pk,
              shcap = sk ~ pl + pk),
         TexasElectr, model = "random",
         restrict.matrix = R)
summary(z)
Oneway (individual) effect Random Effect Model
   (Swamy-Arora's transformation)
Call:
plm.list(formula = list(cost = C ~ pl + pk + q + pll + plk +
    pkk + qq, shlab = sl ~ pl + pk, shcap = sk ~ pl + pk), data = TexasElectr,
    model = "random", restrict.matrix = R)

Balanced Panel: n = 10, T = 18, N = 180

Effects:

  Estimated standard deviations of the error
        cost  shlab  shcap
id     0.1429 0.0248 0.0270
idios 0.0377 0.0195 0.0175

  Estimated correlation matrix of the individual effects
         cost shlab shcap
cost    1.0000     .     .
shlab -0.6926  1.00     .
shcap -0.0964  0.21     1
```

```
  Estimated correlation matrix of the idiosyncratic effects
         cost shlab shcap
cost   1.0000     .     .
shlab  0.2813 1.000     .
shcap -0.0766 0.204     1

 - cost
            Estimate Std. Error t-value Pr(>|t|)
(Intercept) -0.22924    0.04175   -5.49  6.2e-08 ***
pl           0.12484    0.00614   20.32  < 2e-16 ***
pk           0.31573    0.00612   51.59  < 2e-16 ***
q            0.85452    0.01200   71.20  < 2e-16 ***
pll          0.13698    0.00931   14.71  < 2e-16 ***
plk         -0.04025    0.00867   -4.64  4.3e-06 ***
pkk          0.19884    0.00832   23.90  < 2e-16 ***
qq           0.19821    0.01150   17.23  < 2e-16 ***
---
Signif. codes:
0 '***' 0.001 '**' 0.01 '*' 0.05 '.' 0.1 ' ' 1

 - shlab
            Estimate Std. Error t-value Pr(>|t|)
(Intercept)  0.12484    0.00614   20.32  < 2e-16 ***
pl           0.13698    0.00931   14.71  < 2e-16 ***
pk          -0.04025    0.00867   -4.64  4.3e-06 ***
---
Signif. codes:
0 '***' 0.001 '**' 0.01 '*' 0.05 '.' 0.1 ' ' 1

 - shcap
            Estimate Std. Error t-value Pr(>|t|)
(Intercept)  0.31573    0.00612   51.59  < 2e-16 ***
pl          -0.04025    0.00867   -4.64  4.3e-06 ***
pk           0.19884    0.00832   23.90  < 2e-16 ***
---
Signif. codes:
0 '***' 0.001 '**' 0.01 '*' 0.05 '.' 0.1 ' ' 1
```

The results indicate the presence of increasing returns to scale, as q is significantly lower than 1. Factor shares at the sample mean for labor and capital are respectively 12 and 31%.

3.3 The Maximum Likelihood Estimator

An alternative to the OLS estimator presented in the previous chapter is the maximum likelihood estimator. Contrary to the GLS estimator, the parameters are not estimated sequentially (first ϕ and then β) but simultaneously.

3.3.1 Derivation of the Likelihood Function

In order to write the likelihood of the model, the distribution of the errors must be perfectly characterized; compared with the GLS model, we then must add an hypothesis concerning the distribution of the two components of the error term, the individual η and the idiosyncratic ν effects: we'll suppose that they are both normally distributed. The likelihood is the joint density for the whole sample, which is the product of the individual densities in the case of a random sample. This is not the case here, as the T_n observations of individual n are correlated because of the common individual effect. The model to be estimated is then:

$$y_{nt} = \beta^\top x_n + \eta_n + \nu_{nt}$$

with $\eta_n \sim N(0, \sigma_\eta)$ and $\nu_{nt} \sim N(0, \sigma_\nu)$. For a given value of the individual effect, η_n, the density for y_{nt} is:

$$f(y_{nt} \mid \eta_n) = \frac{1}{\sqrt{2\pi}\sigma_\nu} e^{-\frac{1}{2}\left(\frac{y_{nt}-\beta^\top x_{nt}-\eta_n}{\sigma_\nu}\right)^2}$$

For a given value of η, the distribution of $y_n = y_{n1}, \dots, y_{nt}$ is the one of a vector of independent random deviates, and the joint distribution is therefore the product of individual densities:

$$f(y_n \mid \eta_n) = \left(\frac{1}{2\pi\sigma_\nu^2}\right)^{\frac{T_n}{2}} e^{-\frac{1}{2\sigma_\nu^2}\sum_{t=1}^{T_n}(y_{nt}-\beta^\top x_{nt}-\eta_n)^2}$$

The unconditional distribution is obtained by integrating out the individual effects η, which means that the mean value of the density is computed for all possible values of η:

$$f(y_n) = \frac{1}{\sqrt{2\pi\sigma_\eta^2}} \int_{-\infty}^{+\infty} f(y_n \mid \eta_n) e^{-\frac{1}{2}\left(\frac{\eta}{\sigma_\eta}\right)^2} d\eta = \frac{1}{\sqrt{2\pi\sigma_\eta^2}} \left(\frac{1}{2\pi\sigma_\nu^2}\right)^{\frac{T_n}{2}} \int_{-\infty}^{+\infty} e^{-\frac{1}{2}A} d\eta$$

with, denoting $\epsilon_{nt} = y_{nt} - \beta^\top x_{nt}$, $\bar{\epsilon}_n = \bar{y}_n - \beta^\top x_n$ and $\sigma_{\iota n}^2 = \sigma_\nu^2 + T_n\sigma_\eta^2$:

$$\begin{aligned} A &= \sum_{t=1}^{T} \frac{(\epsilon_{nt} - \eta)^2}{\sigma_\nu^2} + \frac{\eta^2}{\sigma_\eta^2} \\ &= \frac{1}{\sigma_\nu^2}\left(\frac{\sigma_{\iota n}^2}{\sigma_\eta^2}\eta^2 - 2T_n\bar{\epsilon}_{n.}\eta + \sum_t \epsilon_{nt}^2\right) \\ &= \frac{1}{\sigma_\nu^2}\left(\frac{\sigma_{\iota n}}{\sigma_\eta}\eta - T_n\bar{\epsilon}_{n.}\frac{\sigma_\eta}{\sigma_{\iota n}}\right)^2 + \frac{1}{\sigma_\nu^2}\left(\sum_t \epsilon_{nt}^2 - T_n^2\bar{\epsilon}_{n.}^2\frac{\sigma_\eta^2}{\sigma_{\iota n}^2}\right) \end{aligned}$$

Denoting by z^2 the first term, we have $dz = \frac{\sigma_{\iota n}}{\sigma_\nu\sigma_\eta}d\eta$ and the joint density is then (denoting $\phi_n = \frac{\sigma_\nu}{\sigma_{\iota n}}$):

$$f(y_n) = \left(\frac{1}{2\pi\sigma_\nu^2}\right)^{\frac{T_n}{2}} \phi_n e^{-\frac{1}{2\sigma_\nu^2}\left(\sum_t \epsilon_{nt}^2 - T_n^2\bar{\epsilon}_{n.}^2\frac{\sigma_\eta^2}{\sigma_{\iota n}^2}\right)}$$

For the second term, we have:

$$\sum_t \epsilon_{nt}^2 - T_n^2\bar{\epsilon}_{n.}^2\frac{\sigma_\eta^2}{\sigma_{\iota n}^2} = \sum_t \epsilon_{nt}^2 - T_n(1-\phi_n^2)\bar{\epsilon}_{n.}^2 = \sum_t (\epsilon_{nt} - (1-\phi_n)\bar{\epsilon}_{n.})^2$$

so that the joint density for an individual is finally:

$$f(y_n) = \left(\frac{1}{2\pi\sigma_\nu^2}\right)^{\frac{T_n}{2}} \phi_n e^{-\frac{1}{2\sigma_\nu^2}\sum_t(\epsilon_{nt}-(1-\phi_n)\bar{\epsilon}_{n.})^2}$$

The contribution of the n-th individual to the log likelihood function is simply the logarithm of the joint density:

$$\ln L_n = -\frac{T_n}{2}\ln 2\pi - \frac{T_n}{2}\ln\sigma_\nu^2 + \frac{1}{2}\ln\phi_n^2 - \frac{1}{2\sigma_\nu^2}\sum_t (\epsilon_{nt} - (1-\phi_n)\bar{\epsilon}_{n.})^2$$

The log likelihood function is then obtained by summing over all the individuals of the panel:

$$\ln L = -\frac{\sum_n T_n}{2}\ln 2\pi - \frac{\sum_n T_n}{2}\ln\sigma_\nu^2 + \frac{1}{2}\sum_n \ln\phi_n^2 - \frac{1}{2\sigma_\nu^2}\sum_n\sum_t (\epsilon_{nt} - (1-\phi_n)\bar{\epsilon}_{n.})^2$$

or, more simply in the special case of a balanced panel:

$$\ln L = -\frac{NT}{2}\ln 2\pi - \frac{NT}{2}\ln\sigma_\nu^2 + \frac{N}{2}\ln\phi^2 - \frac{1}{2\sigma_\nu^2}\sum_n\sum_t (\epsilon_{nt} - (1-\phi)\bar{\epsilon}_{n.})^2$$

Note also that:

$$\sum_n\sum_t (\epsilon_{nt} - (1-\phi)\bar{\epsilon}_{n.})^2 = \sum_n\sum_t (\epsilon_{nt} - \bar{\epsilon}_{n.})^2 + \phi^2\sum_n T_n\bar{\epsilon}_{n.}^2 = \epsilon^\top W\epsilon + \phi^2\epsilon^\top B\epsilon$$

3.3.2 Computation of the Estimator

The first derivatives of the log likelihood are, denoting $\tilde{z}_{nt} = z_{nt} - (1-\phi)\bar{z}_{n.}$:

$$\frac{\partial \ln L}{\partial \beta} = -\frac{2}{\sigma_\nu^2}(\tilde{X}^\top\tilde{y} - (\tilde{X}^\top\tilde{X})\beta) \tag{3.9}$$

$$\frac{\partial \ln L}{\partial \sigma_\nu^2} = -\frac{NT}{2\sigma_\nu^2} + \frac{1}{2\sigma_\nu^4}(\epsilon^\top W\epsilon + \phi^2\epsilon^\top B\epsilon) \tag{3.10}$$

$$\frac{\partial \ln L}{\partial \phi^2} = \frac{N}{2\phi^2}\frac{\epsilon^\top B\epsilon}{2\sigma_\nu^2} \tag{3.11}$$

Solving (3.9), we obtain:

$$\hat{\beta} = (\tilde{X}^\top\tilde{X})^{-1}\tilde{X}^\top\tilde{y} \tag{3.12}$$

The estimator of σ_ν^2 is simply obtained by using (3.10) as the residual variance of the model estimated on the transformed data:

$$\hat{\sigma}_\nu^2 = \frac{\hat{\epsilon}^\top W\hat{\epsilon} + \hat{\phi}^2\hat{\epsilon}^\top B\hat{\epsilon}}{NT} \tag{3.13}$$

Finally, using (3.11) and (3.13), the transformation parameter is:

$$\hat{\phi}^2 = \frac{\hat{\epsilon}^\top W\hat{\epsilon}}{(T-1)\hat{\epsilon}^\top B\hat{\epsilon}} \tag{3.14}$$

The estimation can be performed iteratively. Starting from an estimator of β (for example the *within* estimator), we calculate $\hat{\phi}^2$ using the formula given by 3.14. We then transform the response and the covariates using this estimator of ϕ^2 and we compute a second estimation of β using (3.12). These computations are repeated until the convergence of $\hat{\beta}$ and $\hat{\phi}^2$. σ_ν^2 is then estimated using (3.13).

Example 3.3 maximum likelihood estimator – `RiceFarms` data set
The maximum likelihood estimator is available in the **pglm** package. The `pglm` function enables maximum likelihood estimation of generalized linear models for panel data.[4] We have to specify

4 See chapter 9.

the distribution of the errors of the model, here normal, by setting the argument `family` to `'gaussian'`.

```
data("RiceFarms", package = "splm")
Rice <- pdata.frame(RiceFarms, index = "id")
library("pglm")
rice.ml <- pglm(log(goutput) ~ log(seed) + log(totlabor) + log(size),
                data = Rice, family = gaussian)
```

```
summary(rice.ml)
--------------------
Maximum Likelihood estimation
Newton-Raphson maximisation, 5 iterations
Return code 2: successive function values within tolerance limit
Log-Likelihood: -460.5
6  free parameters
Estimates:
              Estimate Std. error t value Pr(> t)
(Intercept)     5.3125     0.2038   26.07 < 2e-16 ***
log(seed)       0.2200     0.0283    7.76 8.2e-15 ***
log(totlabor)   0.2855     0.0311    9.20 < 2e-16 ***
log(size)       0.5280     0.0326   16.17 < 2e-16 ***
sd.id           0.1190     0.0171    6.95 3.7e-12 ***
sd.idios        0.3637     0.0086   42.28 < 2e-16 ***
---
Signif. codes:
0 '***' 0.001 '**' 0.01 '*' 0.05 '.' 0.1 ' ' 1
--------------------
```

The coefficients are very similar to those obtained with the GLS estimator. The two parameters called `sd.idios` and `sd.id` are the estimated standard deviations of the idiosyncratic and of the individual parts of the error. These values are also almost equal to those obtained using the GLS estimator.

3.4 The Nested Error Components Model

3.4.1 Presentation of the Model

The nested random effect model is relevant when the individuals can be put together in different groups. For example, with a panel of firms, groups may be constituted by regions or production sectors.

In this chapter, we'll restrict ourselves to panels with two characteristics:

- panels without time effects,
- balanced panels inside each group, which means that, for every group, the number of observations for each individual is the same.

The number of individuals and the length of time series for two groups may be different. This is why this model, presented in Baltagi et al. (2001) is called the unbalanced nested error component model, even if its unbalancedness must be understood in the very restrictive sense we've just described.

Three effects will now be considered: the usual individual η and idiosyncratic ν effects, but also a new one that represents group effects λ. Denoting by D_λ the matrix of group dummies:

$$y = \alpha + X\beta + D_\eta \eta + D_\lambda \lambda + \nu$$

Ω is block-diagonal with G (the number of groups) blocks of the following shape:

$$\Omega_g = \sigma_\eta^2[I_{N_g} \otimes J_{T_g}] + \sigma_\lambda^2[J_{N_g} \otimes J_{T_g}] + \sigma_\nu^2[I_{N_g} \otimes I_{T_g}]$$

Replacing J_R by $R\bar{J}_R$ and I_R by $\bar{I}_R + \bar{J}_R$, this can be rewritten as a linear combination of three symmetric, idempotent, and orthogonal matrices which sum to I:

$$\Omega_g = [\sigma_\nu^2 I_{N_g} \otimes \bar{I}_{T_g}] + (\sigma_\nu^2 + T\sigma_\eta^2)[\bar{I}_{N_g} \otimes \bar{J}_{T_g}] + (\sigma_\nu^2 + T\sigma_\eta^2 + NT\sigma_\lambda^2)[\bar{J}_{N_g} \otimes \bar{J}_{T_g}]$$

where:

- $I_{N_g} \otimes \bar{I}_{T_g}$ is the *within*-individual transformation,
- $\bar{I}_{N_g} \otimes \bar{J}_{T_g}$ is the *between*-individual transformation *measured as a difference with the group mean*,
- $\bar{J}_{N_g} \otimes \bar{J}_{T_g}$ is the *between*-group transformation.

This expression enables to easily find the expression for $\sigma_\nu \Omega_g^{-0.5}$, denoting $\phi_\eta = \frac{\sigma_\nu}{\sqrt{\sigma_\nu^2 + T\sigma_\eta^2}}$ and $\phi_\lambda = \frac{\sigma_\nu}{\sqrt{\sigma_\nu^2 + T\sigma_\eta^2 + NT\sigma_\lambda^2}}$:

$$\sigma_\nu \Omega_g^{-0.5} = [I_N \otimes \bar{I}_T] + \phi_\eta[\bar{I}_N \otimes \bar{J}_T] + \phi_\lambda[\bar{J}_{N_g} \otimes \bar{J}_T]$$

which finally writes:

$$\sigma_\nu \Omega_g^{-0.5} = [I_{N_g} \otimes I_{T_g}] - \theta_\eta[I_{N_g} \otimes \bar{J}_{T_g}] - \theta_\lambda[\bar{J}_{N_g} \otimes \bar{J}_{T_g}]$$

with $\theta_\eta = 1 - \phi_\eta$ and $\theta_\lambda = \phi_\eta - \phi_\lambda$.

The model can therefore be estimated by OLS on transformed variables for which part of the individual and the group mean (respectively θ_η and θ_λ have been subtracted).

3.4.2 Estimation of the Variance of the Error Components

We proceed along the lines of section 3.1.3. Using residuals from a preliminary estimation V denoted $\hat{\epsilon}^V = M^V \epsilon$, we compute a quadratic form of $\hat{\epsilon}^V$ with a matrix A $\hat{q}_A^V$.

$$E(\hat{q}_A^V) = \sigma_\nu^2 \text{tr}(M^{V\top} A M^V) + \sigma_\eta^2 \text{tr}(M^{V\top} A M^V S_\eta) + \sigma_\lambda^2 \text{tr}(M^{V\top} A M^V S_\lambda)$$

Replacing M^V by its expression and denoting $\Theta_A = X^\top A X$, we obtain:

$$E(\hat{q}_A^V) = \begin{pmatrix} \text{tr}(\bar{I}A\bar{I}) - 2\text{tr}(\Theta_V^{-1}\Theta_{\bar{I}A\bar{I}V}) + \text{tr}(\Theta_V^{-1}\Theta_{\bar{I}A\bar{I}}) \\ \text{tr}(\bar{I}A\bar{I}S_\eta) - 2\text{tr}(\Theta_V^{-1}\Theta_{\bar{I}A\bar{I}S_\eta V}) + \text{tr}(\Theta_V^{-1}\Theta_{\bar{I}A\bar{I}}\Theta_V^{-1}\Theta_{VS_\eta V}) \\ \text{tr}(\bar{I}A\bar{I}S_\lambda) - 2\text{tr}(\Theta_V^{-1}\Theta_{\bar{I}A\bar{I}S_\lambda V}) + \text{tr}(\Theta_V^{-1}\Theta_{\bar{I}A\bar{I}}\Theta_V^{-1}\Theta_{VS_\lambda V}) \end{pmatrix}^\top \begin{pmatrix} \sigma_\nu^2 \\ \sigma_\eta^2 \\ \sigma_\lambda^2 \end{pmatrix}$$

or, denoting $\Upsilon_A = Z^\top A Z$:

$$E(\hat{q}_A^V) = \begin{pmatrix} \mathrm{tr}(A) - 2\mathrm{tr}(\Upsilon_V^{-1}\Upsilon_{VA}) + \mathrm{tr}(\Upsilon_V^{-1}\Upsilon_A) \\ \mathrm{tr}(AS_\eta) - 2\mathrm{tr}(\Upsilon_V^{-1}\Upsilon_{VS_\eta A}) + \mathrm{tr}(\Upsilon_V^{-1}\Upsilon_A\Upsilon_V^{-1}\Upsilon_{VS_\eta V}) \\ \mathrm{tr}(AS_\lambda) - 2\mathrm{tr}(\Upsilon_V^{-1}\Upsilon_{VS_\lambda A}) + \mathrm{tr}(\Upsilon_V^{-1}\Upsilon_A\Upsilon_V^{-1}\Upsilon_{VS_\lambda V}) \end{pmatrix}^\top \begin{pmatrix} \sigma_\nu^2 \\ \sigma_\eta^2 \\ \sigma_\lambda^2 \end{pmatrix}$$

The most popular estimators are obtained by computing the three quadratic forms with the *within*-individual, *between*-individual and *between*-group matrices. We then get the following system of equations:

$$\begin{pmatrix} \hat{q}_{\mathrm{W}}^V \\ \hat{q}_{\mathrm{B}_\eta}^V \\ \hat{q}_{\mathrm{B}_\lambda}^V \end{pmatrix} = H \begin{pmatrix} \sigma_\nu^2 \\ \sigma_\eta^2 \\ \sigma_\lambda^2 \end{pmatrix} \tag{3.15}$$

with:

$$\begin{pmatrix} \mathrm{tr}(M^{V\top}WM^V) & \mathrm{tr}(M^{V\top}WM^VS_\eta) & \mathrm{tr}(M^{V\top}WM^VS_\lambda) \\ \mathrm{tr}(M^{V\top}B_\eta M^V) & \mathrm{tr}(M^{V\top}B_\eta M^VS_\eta) & \mathrm{tr}(M^{V\top}B_\eta M^VS_\lambda) \\ \mathrm{tr}(M^{V\top}B_\lambda M^V) & \mathrm{tr}(M^{V\top}B_\lambda M^VS_\eta) & \mathrm{tr}(M^{V\top}B_\lambda M^VS_\lambda) \end{pmatrix} \tag{3.16}$$

Using the following results: $\bar{\mathrm{I}}W = W$, $\mathrm{tr}(W) = O - N - T + 1$, $WS_\eta = 0$, $WS_\mu = 0$, $B_\eta S_\eta = S_\eta$, $B_\mu S_\mu = S_\mu$, $\mathrm{tr}(S_\mu) = \mathrm{tr}(S_\eta) = O$, $\mathrm{tr}(B_\eta) = N$, $\mathrm{tr}(B_\mu) = T$, $\bar{\mathrm{I}}B_\eta\bar{\mathrm{I}}S_\eta = \bar{\mathrm{I}}S_\eta$, $\mathrm{tr}(\bar{\mathrm{I}}S_\eta) = O - \sum_n T_n/O$, $\bar{\mathrm{I}}B_\mu\bar{\mathrm{I}}S_\mu = \bar{\mathrm{I}}S_\mu$, $\mathrm{tr}(\bar{\mathrm{I}}S_\mu) = O - \sum_t N_t/O$, $\mathrm{tr}(\bar{\mathrm{I}}B_\eta) = N - 1$, $\mathrm{tr}(\bar{\mathrm{I}}B_\mu) = T - 1$, $\mathrm{tr}(\bar{\mathrm{I}}B_\mu\bar{\mathrm{I}}S_\eta) = T - \sum_n T_n^2/O$, $\mathrm{tr}(\bar{\mathrm{I}}B_\eta\bar{\mathrm{I}}S_\mu) = N - \sum_t N_t^2/O$, $\mathrm{tr}(B_\mu S_\eta) = T$, $\mathrm{tr}(B_\eta S_\mu) = N$.

We finally obtain:

$$\left\{\begin{aligned}
E(\hat{q}_{\mathrm{W}}^V) &= \sigma_\nu^2[O - N - T + 1 - 2\mathrm{tr}(\Theta_V^{-1}\Theta_{\bar{\mathrm{I}}VW}) + \mathrm{tr}(\Theta_V^{-1}\Theta_W)] \\
&\quad + \sigma_\eta^2[\mathrm{tr}(\Theta_V^{-1}\Theta_W\Theta_V^{-1}\Theta_{VS_\eta V})] \\
&\quad + \sigma_\mu^2[\mathrm{tr}(\Theta_V^{-1}\Theta_W\Theta_V^{-1}\Theta_{VS_\mu V})] \\
E(\hat{q}_{B_\eta}^V) &= \sigma_\nu^2[N - 1 - 2\mathrm{tr}(\Theta_V^{-1}\Theta_{\bar{\mathrm{I}}VB_\eta}) + \mathrm{tr}(\Theta_V^{-1}\Theta_{\bar{B}_\eta})] \\
&\quad + \sigma_\eta^2\left[O - \textstyle\sum_n T_n^2/O + 2\mathrm{tr}(\Theta_V^{-1}\Theta_{\bar{\mathrm{I}}S_\eta V}) + \mathrm{tr}(\Theta_V^{-1}\Theta_{\bar{B}_\eta}\Theta_V^{-1}\Theta_{VS_\eta V})\right] \\
&\quad + \sigma_\mu^2\left[N - \textstyle\sum_t N_t^2/O - 2\mathrm{tr}(\Theta_V^{-1}\Theta_{\bar{\mathrm{I}}S_\mu V}) + \mathrm{tr}(\Theta_V^{-1}\Theta_{\bar{B}_\eta}\Theta_V^{-1}\Theta_{VS_\mu V})\right] \\
E(\hat{q}_{B_\mu}^V) &= \sigma_\nu^2[T - 1 - 2\mathrm{tr}(\Theta_V^{-1}\Theta_{\bar{\mathrm{I}}VB_\mu}) + \mathrm{tr}(\Theta_V^{-1}\Theta_{\bar{B}_\mu})] \\
&\quad + \sigma_\eta^2\left[T - \textstyle\sum_n T_n^2/O - 2\mathrm{tr}(\Theta_V^{-1}\Theta_{\bar{\mathrm{I}}S_\eta V}) + \mathrm{tr}(\Theta_V^{-1}\Theta_{\bar{B}_\mu}\Theta_V^{-1}\Theta_{VS_\eta V})\right] \\
&\quad + \sigma_\mu^2\left[O - \textstyle\sum_t N_t^2/O - 2\mathrm{tr}(\Theta_V^{-1}\Theta_{\bar{\mathrm{I}}S_\mu V}) + \mathrm{tr}(\Theta_V^{-1}\Theta_{\bar{B}_\mu}\Theta_V^{-1}\Theta_{VS_\mu V})\right]
\end{aligned}\right.$$

or:

$$\begin{cases} \mathrm{E}(\hat{q}_{W}^{V}) = \sigma_\nu^2[O - N - T + 1 - \mathrm{tr}(\Upsilon_V^{-1}\Upsilon_W)] \\ \quad +\sigma_\eta^2[\mathrm{tr}(\Upsilon_V^{-1}\Upsilon_W\Upsilon_V^{-1}\Upsilon_{VS_\eta V})] \\ \quad +\sigma_\mu^2[\mathrm{tr}(\Upsilon_V^{-1}\Upsilon_W\Upsilon_V^{-1}\Upsilon_{VS_\mu V})] \\ \mathrm{E}(\hat{q}_{B_\eta}^{V}) = \sigma_\nu^2[N - 2\mathrm{tr}(\Upsilon_V^{-1}\Upsilon_{B_\eta V}) + \mathrm{tr}(\Upsilon_V^{-1}\Upsilon_{B_\eta})] \\ \quad +\sigma_\eta^2[O - 2\mathrm{tr}(\Upsilon_V^{-1}\Upsilon_{S_\eta V}) + \mathrm{tr}(\Upsilon_V^{-1}\Upsilon_{B_\eta}\Upsilon_V^{-1}\Upsilon_{VS_\eta V})] \\ \quad +\sigma_\mu^2[N - 2\mathrm{tr}(\Upsilon_V^{-1}\Upsilon_{B_\eta S_\mu V}) + \mathrm{tr}(\Upsilon_V^{-1}\Upsilon_{B_\eta}\Upsilon_V^{-1}\Upsilon_{VS_\mu V})] \\ \mathrm{E}(\hat{q}_{B_\mu}^{V}) = \sigma_\nu^2[T - 2\mathrm{tr}(\Upsilon_V^{-1}\Upsilon_{B_\mu V}) + \mathrm{tr}(\Upsilon_V^{-1}\Upsilon_{B_\mu})] \\ \quad +\sigma_\eta^2[T - 2\mathrm{tr}(\Upsilon_V^{-1}\Upsilon_{B_\mu S_\eta V}) + \mathrm{tr}(\Upsilon_V^{-1}\Upsilon_{B_\mu}\Upsilon_V^{-1}\Upsilon_{VS_\eta V})] \\ \quad +\sigma_\mu^2[O - 2\mathrm{tr}(\Upsilon_V^{-1}\Upsilon_{S_\mu V}) + \mathrm{tr}(\Upsilon_V^{-1}\Upsilon_{B_\mu}\Upsilon_V^{-1}\Upsilon_{VS_\mu V})] \end{cases}$$

Baltagi et al. (2001) have proposed a variant of the Amemiya (1971) estimator (where the *within* estimator is used for the three quadratic forms), the Wallace and Hussain (1969) estimator (the **OLS** estimator is used for the three quadratic forms) and of the Swamy and Arora (1972) estimator (the *within*, *between*-individual and *between*-group are used respectively for the *within*, *between*-individual, and *between*-group quadratic forms). The detailed formulas are presented in Figure 3.2.

Example 3.4 nested error component model – `Produc` data set
Baltagi et al. (2001) have estimated the nested error component model extending the work of Baltagi and Pinnoi (1995). This article, inspired by Munnell (1990), aims at analyzing the effect of public capital on production. The dataset consists on 48 American states for the period 1970–1986. The observations are nested, as the states can be grouped in 9 regions, which contain between 3 and 8 states. The panel is therefore unbalanced, as the number of individuals differs from one group to another, but the number of time series is the same for all the individuals inside a group (and in fact here for every individual), which is a necessity in order to be able to estimate the model.

A Cobb-Douglas production function is estimated; the state output `gsp` is explained by the private capital stock `pc` and non-agricultural labor `emp`, but also by three measures of the public capital stock:

- roads and highways `hwy`,
- water infrastructure `water`,
- other public buildings and infrastructure `util`.

The state unemployment rate is also used as a covariate in order to take into account the business cycle of every state. All the covariates, except for the unemployment rate, are in logarithms.

In order to estimate the model, one has to indicate which variable is the group index. This can be done using several variants:

- if the first three columns of the data frame are the individual, time, *and* group indexes, the structure of the panel is directly understood by `plm`,
- the `index` argument of `plm` or of `pdata.frame` can also be used to indicate which variable is the group index, naming this variable if the other indexes are not indicated.

$$\begin{pmatrix} \hat{q}_{\mathrm{W}}^{V1} \\ \hat{q}_{(\mathrm{B}_\eta-\mathrm{B}_\lambda)}^{V2} \\ \hat{q}_{\mathrm{B}_\lambda}^{V3} \end{pmatrix} = H \begin{pmatrix} \sigma_\nu^2 \\ \sigma_\eta^2 \\ \sigma_\lambda^2 \end{pmatrix}$$

Amemiya (1971):

$$A = \begin{pmatrix} O-N-K & 0 & 0 \\ N-G+\mathrm{tr}(\Theta_W^{-1}\Theta_{B_\eta-B_\lambda}) & O-\sum_g T_g & 0 \\ G-1+\mathrm{tr}(\Theta_W\Theta_{B_\lambda-\bar{J}}) & \sum_g T_g - \sum_g N_g T_g^2/O & O-\sum_g N_g^2 T_g^2/O \end{pmatrix}$$

Wallace and Hussain (1969):

$$A = \begin{pmatrix} O-N-\mathrm{tr}(\Upsilon_I^{-1}\Upsilon_W) & \mathrm{tr}(\Upsilon_I^{-1}\Upsilon_W\Upsilon_I^{-1}\Upsilon_{S_\eta}) & \mathrm{tr}(\Upsilon_I^{-1}\Upsilon_W\Upsilon_I^{-1}\Upsilon_{S_\lambda}) \\ N-G-\mathrm{tr}(\Upsilon^{-1}\Upsilon_{(B_\eta-B_\lambda)}) & O-\sum_g T_g - 2\mathrm{tr}(\Upsilon_I^{-1}\Upsilon_{(B_\eta-B_\lambda)S_\eta}) + \mathrm{tr}(\Upsilon_I^{-1}\Upsilon_{(B_\eta-B_\lambda)}\Upsilon_I^{-1}\Upsilon_{S_\eta}) & \mathrm{tr}(\Upsilon_I^{-1}\Upsilon_{(B_\eta-B_\lambda)}\Upsilon_I^{-1}\Upsilon_{S_\lambda}) \\ G-\mathrm{tr}(\Upsilon_I^{-1}\Upsilon_{B_\lambda}) & \sum_g T_g - 2\mathrm{tr}(\Upsilon_I^{-1}\Upsilon_{B_\lambda S_\eta}) + \mathrm{tr}(\Upsilon_I^{-1}\Upsilon_{B_\lambda}\Upsilon_I^{-1}\Upsilon_{S_\eta}) & O-2\mathrm{tr}(\Upsilon_I^{-1}\Upsilon_{S_\lambda}) + \mathrm{tr}(\Upsilon_I^{-1}\Upsilon_{B_\lambda}\Upsilon_I^{-1}\Upsilon_{S_\lambda}) \end{pmatrix}$$

Swamy and Arora (1972):

$$A = \begin{pmatrix} O-N-K & 0 & 0 \\ N-G-K & O-\sum_g T_g - \mathrm{tr}(\Theta_{(B_\eta-B_\lambda)}^{-1}\Theta_{S_\eta(B_\eta-B_\lambda)}) & 0 \\ G-K & \sum_g T_g - \mathrm{tr}(\Upsilon_{B_\lambda}^{-1}\Upsilon_{B_\lambda S_\eta}) & O-\mathrm{tr}(\Upsilon_{B_\lambda}^{-1}\Upsilon_{S_\lambda}) \end{pmatrix}$$

Figure 3.2 Error components estimators for the nested error component model.

To illustrate these different possibilities, we use the `RiceFarms` data set from **plm**.

```
data("RiceFarms", package = "plm")
head(RiceFarms, 2)
      id size status varieties bimas seed urea phosphate
1 101001    3  owner     mixed mixed   90  900        80
2 101001    2  owner      trad mixed   40  600         0
  pesticide pseed purea pphosph hiredlabor famlabor
1      6000    80    75      75       2875       40
2      3000    70    75      75       2110       45
  totlabor  wage goutput noutput price        region
1     2915 68.49    7980    6800    60 wargabinangun
2     2155 60.09    4083    3500    60 wargabinangun
```

The individual index is `id` (the fist column), we use `region`, the last column of the data frame as the group index. The three lines below give the same results:

```
R1 <- pdata.frame(RiceFarms, index = c(id = "id", time = NULL, group = "region"))
R2 <- pdata.frame(RiceFarms, index = c(id = "id", group = "region"))
R3 <- pdata.frame(RiceFarms, index = c("id", group = "region"))
head(index(R1))
      id time        region
1 101001    1 wargabinangun
2 101001    2 wargabinangun
3 101001    3 wargabinangun
4 101001    4 wargabinangun
5 101001    5 wargabinangun
6 101001    6 wargabinangun
```

For the `Produc` data frame, it is easier to describe the structure of the sample as the first three columns are the individual, time, and group indexes.

To estimate the nested error component model, the `model` must be set to `'nested'`. We first estimate the Swamy and Arora (1972) model:

```
data("Produc", package = "plm")
nswar <- plm(log(gsp) ~ log(pc) + log(emp) + log(hwy) + log(water) +
                 log(util) + unemp, data = Produc,
             model = "random", effect = "nested",
             random.method = "swar", index = c(group = "region"))
summary(nswar)
Nested effects Random Effect Model
   (Swamy-Arora's transformation)

Call:
plm(formula = log(gsp) ~ log(pc) + log(emp) + log(hwy) + log(water) +
    log(util) + unemp, data = Produc, effect = "nested", model = "random",
    random.method = "swar", index = c(group = "region"))

Balanced Panel: n = 48, T = 17, N = 816
```

```
Effects:
                   var std.dev share
idiosyncratic 0.00135 0.03676  0.19
individual    0.00428 0.06541  0.60
group         0.00146 0.03815  0.21
theta:
         Min. 1st Qu.  Median    Mean 3rd Qu.    Max.
id    0.86493 0.86493 0.86493 0.86493 0.86493 0.86493
group 0.03961 0.04669 0.05714 0.05578 0.06458 0.06458

Residuals:
   Min. 1st Qu.  Median    Mean 3rd Qu.    Max.
-0.1062 -0.0248 -0.0018 -0.0001  0.0198  0.1828

Coefficients:
             Estimate Std. Error t-value Pr(>|t|)
(Intercept)  2.089211   0.145702   14.34  < 2e-16 ***
log(pc)      0.274124   0.020544   13.34  < 2e-16 ***
log(emp)     0.739838   0.025750   28.73  < 2e-16 ***
log(hwy)     0.072736   0.022025    3.30    0.001 **
log(water)   0.076453   0.013858    5.52  4.6e-08 ***
log(util)   -0.094374   0.016773   -5.63  2.5e-08 ***
unemp       -0.006163   0.000903   -6.82  1.8e-11 ***
---
Signif. codes:
0 '***' 0.001 '**' 0.01 '*' 0.05 '.' 0.1 ' ' 1

Total Sum of Squares:    43
Residual Sum of Squares: 1.12
R-Squared:      0.974
Adj. R-Squared: 0.974
F-statistic: 5025.33 on 6 and 809 DF, p-value: <2e-16
```

We then update the model in order to use the two other estimators of the variances of the components of the error. The results are summarized using the `screenreg` function of the **texreg** package.

```
library("texreg")
namem <- update(nswar, random.method = "amemiya")
nwalhus <- update(nswar, random.method = "walhus")
iswar <- update(nswar, effect = "individual")
iwith <- update(nswar, model = "within", effect = "individual")
screenreg(list("fe-id" = iwith, "re-id" = iswar,
               "Swamy_Arora" = nswar, "Wallas-Hussein" = nwalhus,
               "Amemiya" = namem), digits = 3)
```

	fe-id	re-id	Swamy_Arora	Wallas-Hussein	Amemiya
log(pc)	0.235 ***	0.273 ***	0.274 ***	0.273 ***	0.264 ***
	(0.026)	(0.020)	(0.021)	(0.021)	(0.022)
log(emp)	0.801 ***	0.749 ***	0.740 ***	0.742 ***	0.758 ***
	(0.030)	(0.025)	(0.026)	(0.026)	(0.027)
log(hwy)	0.077 *	0.062 **	0.073 **	0.075 ***	0.072 **
	(0.031)	(0.022)	(0.022)	(0.022)	(0.024)
log(water)	0.079 ***	0.076 ***	0.076 ***	0.076 ***	0.076 ***
	(0.015)	(0.014)	(0.014)	(0.014)	(0.014)
log(util)	-0.115 ***	-0.098 ***	-0.094 ***	-0.095 ***	-0.102 ***
	(0.018)	(0.017)	(0.017)	(0.017)	(0.017)
unemp	-0.005 ***	-0.006 ***	-0.006 ***	-0.006 ***	-0.006 ***
	(0.001)	(0.001)	(0.001)	(0.001)	(0.001)
(Intercept)		2.168 ***	2.089 ***	2.082 ***	2.131 ***
		(0.143)	(0.146)	(0.150)	(0.160)
R^2	0.946	0.961	0.974	0.972	0.968
Adj. R^2	0.942	0.961	0.974	0.972	0.968
Num. obs.	816	816	816	816	816
s_idios		0.037	0.037	0.038	0.037
s_id		0.082	0.065	0.067	0.083
s_gp			0.038	0.052	0.047

*** p < 0.001, ** p < 0.01, * p < 0.05

4

Tests on Error Component Models

The double dimensionality of panel data allows for much richer specifications than simple cross sections or time series. This is both a blessing and a curse, given how much more complicated the specification may become. In fact, all possible features from either cross sections or time series, like distance-decaying correlation in – respectively – space or time, can coexist with individual (time), time-(individual-) invariant heterogeneity. Moreover, diagnostic tests will usually have a hard time distinguishing between different forms of persistence along the same dimension unless explicitly designed to take the "other" effect into account.

The specification problem of panel models is typically associated with the presence or absence of individual effects, i.e., with the need to account for unobserved heterogeneity. Given that in the vast majority of cases it will be inappropriate to rule out individual heterogeneity altogether, the related issue emerges of whether it is safe to assume that the latter is uncorrelated with the explanatory variables (and therefore to proceed in a random effects framework) or rather to proceed estimating out (transforming out) the individual effects in a fixed effects fashion. Hence, tests for individual effects under either of the two approaches and Hausman-type tests for determining which one is appropriate are among the most popular diagnostic procedures in this field.

Next to the fundamental specification issues with individual effects , the remainder errors can in turn be correlated: either in time, in which case it will be crucial to distinguish time-decaying persistence of idiosyncratic shocks from the time-invariant persistence deriving from the presence of an individual effect; or in space, and then the issue becomes whether correlation simply descends from participating in the same cross section or, provided the data are referenced in some space (e.g., in geography), whether nearby observations are more correlated than distant ones.

For these reasons, a rich toolbox of diagnostic and specification testing procedures has been developed, which will be the subject of this chapter, presented roughly in the order given above up to the issue of cross-sectional correlation. On the converse, spatial correlation proper will be the subject of a separate chapter.

4.1 Tests on Individual and/or Time Effects

In order to test whether either individual or time effects are present, two approaches are possible:

- the first is to start from estimating said effects out (*within* model) and then perform a zero restriction test,
- the second is to start from the OLS model and to infer about the presence of the effects drawing on the OLS residuals.

Panel Data Econometrics with R, First Edition. Yves Croissant and Giovanni Millo.

Companion website: www.wiley.com/go/croissant/data-econometrics-with-R

4.1.1 F Tests

The sum of squared residuals and the degrees of freedom for the *within* model are: $\hat{\epsilon}_{\text{W}}^{\top}\hat{\epsilon}_{\text{W}}$ and $N(T-1)-K$. Let the null hypothesis be the absence of individual effects so that the restricted model is pooled OLS where the sum of squared residuals and the degrees of freedom are, respectively, $\hat{\epsilon}_{\text{OLS}}^{\top}\hat{\epsilon}_{\text{OLS}}$ and $NT-K-1$. Under H_0, the test statistic:

$$\frac{\hat{\epsilon}_{\text{OLS}}^{\top} W \hat{\epsilon}_{\text{OLS}} - \hat{\epsilon}_{\text{W}}^{\top}\hat{\epsilon}_{\text{W}}}{\hat{\epsilon}_{\text{W}}^{\top} W \hat{\epsilon}_{\text{W}}} \frac{NT-K-N+1}{N-1}$$

follows a Fisher-Snedecor F with $N-1$ and $NT-K-N+1$ degrees of freedom.

The test of the null hypothesis of no individual and time effects is obtained by using the two-ways *within* model and the pooling model:

$$\frac{\hat{\epsilon}_{\text{OLS}}^{\top}\hat{\epsilon}_{\text{OLS}} - \hat{\epsilon}_{2\text{W}}^{\top} W^2 \hat{\epsilon}_{2\text{W}}}{\hat{\epsilon}_{2\text{W}}^{\top} W^2 \hat{\epsilon}_{2\text{W}}} \frac{NT-K-N-T+1}{N+T-1}$$

Finally, the test of the null hypothesis of, say, no time effects, but in the presence of individual effects is:

$$\frac{\hat{\epsilon}_{\text{W}}^{\top} W \hat{\epsilon}_{\text{W}} - \hat{\epsilon}_{2\text{W}}^{\top} W^2 \hat{\epsilon}_{2\text{W}}}{\hat{\epsilon}_{2\text{W}}^{\top} W^2 \hat{\epsilon}_{2\text{W}}} \frac{NT-K-N-T+1}{T-1}$$

4.1.2 Breusch-Pagan Tests

The Breusch and Pagan (1980) test is a Lagrange multipliers test based on the OLS residuals. It is based on the score vector $g(\theta) = \frac{\partial \ln L}{\partial \theta}$, i.e., the vector of partial derivatives of the log-likelihood function from the restricted model. The variance of the score vector is the information matrix:

$$I(\theta) = \text{E}\left(-\frac{\partial \ln L}{\partial \theta \partial \theta^{\top}}\right)(\theta)$$

We estimate a restricted model characterized by a parameter vector $\hat{\theta}$; under H_0, we have:

$$g(\hat{\theta}) \sim N(0, \text{V}(\hat{\theta}))$$

or, denoting by $\hat{g}$ and $\hat{\text{V}}$ the score and its estimated variance in the restricted model:

$$\hat{g}^{\top}\hat{\text{V}}^{-1}\hat{g}$$

which is distributed as a χ^2 where the degrees of freedom are equal to the number of restrictions. We'll first derive the test for the one-way individual error component model, for which the log-likelihood function is:

$$\begin{aligned}\ln L = & -\frac{NT}{2}\ln 2\pi - \frac{N(T-1)}{2}\ln \sigma_\nu^2 - \frac{N}{2}\ln(\sigma_\nu^2 + T\sigma_\eta^2) \\ & -\frac{\epsilon^{\top} W_\eta \epsilon}{2\sigma_\nu^2} - \frac{\epsilon^{\top} B_\eta \epsilon}{2(\sigma_\nu^2 + T\sigma_\eta^2)}\end{aligned}$$

The gradient is then:

$$g(\theta) = \begin{pmatrix} \frac{\partial \ln L}{\partial \sigma_\nu^2} \\ \frac{\partial \ln L}{\partial \sigma_\eta^2} \end{pmatrix} = \begin{pmatrix} -\frac{N(T-1)}{2\sigma_\nu^2} - \frac{N}{2(\sigma_\nu^2+T\sigma_\eta^2)} + \frac{\epsilon^{\top} W_\eta \epsilon}{2\sigma_\nu^4} + \frac{\epsilon^{\top} B_\eta \epsilon}{2(\sigma_\nu^2+T\sigma_\eta^2)^2} \\ -\frac{NT}{2(\sigma_\nu^2+T\sigma_\eta^2)} + \frac{T\epsilon^{\top} B_\eta \epsilon}{2(\sigma_\nu^2+T\sigma_\eta^2)^2} \end{pmatrix}$$

To derive the variance, we start by calculating the matrix of second derivatives $H(\theta) = \frac{\partial \ln L}{\partial\theta\partial\theta^\top}$:

$$H(\theta) = \begin{pmatrix} -\frac{N(T-1)}{2\sigma_\nu^4} + \frac{N}{2(\sigma_\nu^2+T\sigma_\eta^2)^2} - \frac{\epsilon^\top W_\eta \epsilon}{\sigma_\nu^6} - \frac{\epsilon^\top B_\eta \epsilon}{(\sigma_\nu^2+T\sigma_\eta^2)^3} & \frac{NT}{2(\sigma_\nu^2+T\sigma_\eta^2)^2} - \frac{T\epsilon^\top B_\eta \epsilon}{(\sigma_\nu^2+T\sigma_\eta^2)^3} \\ \frac{NT}{2(\sigma_\nu^2+T\sigma_\eta^2)^2} - \frac{T\epsilon^\top B_\eta \epsilon}{(\sigma_\nu^2+T\sigma_\eta^2)^3} & \frac{NT^2}{2(\sigma_\nu^2+T\sigma_\eta^2)^2} - \frac{T^2\epsilon^\top B_\eta \epsilon}{(\sigma_\nu^2+T\sigma_\eta^2)^3} \end{pmatrix}$$

To compute the expectation of this matrix, we note that $\mathrm{E}(\epsilon^\top W_\eta \epsilon) = N(T-1)\sigma_\nu^2$ and $\mathrm{E}(\epsilon^\top B_\eta \epsilon) = N(\sigma_\nu^2 + T\sigma_\eta^2)$:

$$\mathrm{E}(H(\theta)) = \begin{pmatrix} -\frac{N(T-1)}{2\sigma_\nu^4} - \frac{N}{2(\sigma_\nu^2+T\sigma_\eta^2)^2} & -\frac{NT}{2(\sigma_\nu^2+T\sigma_\eta^2)^2} \\ -\frac{NT}{2(\sigma_\nu^2+T\sigma_\eta^2)^2} & -\frac{NT^2}{2(\sigma_\nu^2+T\sigma_\eta^2)^2} \end{pmatrix}$$

To compute the test statistic, we impose the null hypothesis: $\mathrm{H}_0 : \sigma_\eta^2 = 0$ (absence of individual effects). In this case, the estimator for the parameters is OLS and that of $\hat{\sigma}_\nu^2$ is $\hat{\epsilon}^\top\hat{\epsilon}/NT$. The score and its estimated variance are then:

$$g(\hat{\theta}) = \begin{pmatrix} 0 \\ -\frac{NT}{2\hat{\sigma}_\nu^2}\left(\frac{\hat{\epsilon}^\top B_\eta \hat{\epsilon}}{N\sigma_\nu^2} - 1\right) \end{pmatrix}$$

$$\mathrm{E}(-H(\hat{\theta})) = \frac{NT}{2\hat{\sigma}_\nu^4}\begin{pmatrix} 1 & 1 \\ 1 & T \end{pmatrix}$$

whose inverse is:

$$I(\theta) = \frac{2\hat{\sigma}_\nu^4}{NT(T-1)}\begin{pmatrix} T & -1 \\ -1 & 1 \end{pmatrix}$$

Finally, the test statistic is:

$$\mathrm{LM}_\eta = \left(-\frac{NT}{2\hat{\sigma}_\nu^2}\left(\frac{\hat{\epsilon}^\top B_\eta \hat{\epsilon}}{1 - N\hat{\sigma}_\nu^2}\right)\right)^2 \times \frac{2\hat{\sigma}_\nu^4}{NT(T-1)} = \frac{NT}{2(T-1)}\left(1 - \frac{\hat{\epsilon}^\top B_\eta \hat{\epsilon}}{N\hat{\sigma}_\nu^2}\right)^2$$

Or, replacing $\hat{\sigma}_\nu^2$ by $\hat{\epsilon}^\top\hat{\epsilon}/NT$:

$$\mathrm{LM}_\eta = \frac{NT}{2(T-1)}\left(T\frac{\hat{\epsilon}^\top B_\eta \hat{\epsilon}}{\hat{\epsilon}^\top\hat{\epsilon}} - 1\right)^2$$

which is asymptotically distributed as a χ^2 with 1 degree of freedom.

The test of the time effect is likewise computed:

$$\mathrm{LM}_\mu = \frac{NT}{2(N-1)}\left(N\frac{\hat{\epsilon}^\top B_\mu \hat{\epsilon}}{\hat{\epsilon}^\top\hat{\epsilon}} - 1\right)^2$$

The Breusch-Pagan test extends easily to the two-ways error component model, as the statistic can be written as the sum of the two previous statistics:

$$\mathrm{LM}_{\eta\mu} = \frac{NT}{2(T-1)}\left(T\frac{\hat{\epsilon}^\top B_\eta \hat{\epsilon}}{\hat{\epsilon}^\top\hat{\epsilon}} - 1\right)^2 + \frac{NT}{2(N-1)}\left(N\frac{\hat{\epsilon}^\top B_\mu \hat{\epsilon}}{\hat{\epsilon}^\top\hat{\epsilon}} - 1\right)^2$$

and follows a χ^2 with two degrees of freedom under the null hypothesis of no individual and time effects.

For unbalanced panels, the relevant statistics are:[1]

$$\begin{cases} \mathrm{LM}_\eta &= \frac{O^2}{2(\sum_n T_n^2 - O)}\left(\frac{\hat{\epsilon}^\top S_\eta \hat{\epsilon}}{\hat{\epsilon}^\top \hat{\epsilon}} - 1\right)^2 \\ \mathrm{LM}_\mu &= \frac{O^2}{2(\sum_t N_t^2 - O)}\left(\frac{\hat{\epsilon}^\top S_\mu \hat{\epsilon}}{\hat{\epsilon}^\top \hat{\epsilon}} - 1\right)^2 \\ \mathrm{LM}_{\eta\mu} &= \mathrm{LM}_\eta + \mathrm{LM}_\mu \end{cases}$$

These statistics present two problems. The first one is that the alternative hypothesis is that the effects' variance is non-zero, i.e., strictly positive or negative; when a variance must be non-negative. For the one-way error component model, Honda (1985) and King and Wu (1997) proposed a one-sided test based on the square root of the above statistic, which is then normally distributed. The Honda statistic is then $\sqrt{\mathrm{LM}_\eta}$ and its 5% critical value is 1.64 (and likewise for the test of no time effects). For the two-ways error components model, Honda (1985) proposed to use $\frac{\sqrt{\mathrm{LM}_\eta}+\sqrt{\mathrm{LM}_\mu}}{\sqrt{2}}$ as Baltagi et al. (1992) and King and Wu (1997) use:

$$\frac{\sqrt{T-1}}{\sqrt{N+T-2}}\sqrt{\mathrm{LM}_\eta} + \frac{\sqrt{N-1}}{\sqrt{N+T-2}}\sqrt{\mathrm{LM}_\mu}$$

The second problem is due to the fact that $T\frac{\hat{\epsilon}^\top B_\eta \hat{\epsilon}}{\hat{\epsilon}^\top \hat{\epsilon}}$ or $N\frac{\hat{\epsilon}^\top B_\mu \hat{\epsilon}}{\hat{\epsilon}^\top \hat{\epsilon}}$ may be lower than 1. In this case, Baltagi et al. (1992), following Gourieroux et al. (1982), proposed to replace the statistic by 0. The modified statistic is then defined by:

$$\frac{NT}{2(N-1)}\left(\max\left(0, T\frac{\hat{\epsilon}^\top B_\eta \hat{\epsilon}}{\hat{\epsilon}^\top \hat{\epsilon}} - 1\right), \max\left(0, N\frac{\hat{\epsilon}^\top B_\mu \hat{\epsilon}}{\hat{\epsilon}^\top \hat{\epsilon}} - 1\right)\right)^2$$

which follows a mixed χ^2 distribution: $\chi^2_m \sim \left(\frac{1}{4}\right)\chi^2(0) + \left(\frac{1}{2}\right)\chi^2(1) + \left(\frac{1}{4}\right)\chi^2(2)$

Example 4.1 F and LM tests – `RiceFarms` data set
A F test for the presence of individual effects is implemented in the function `pFtest`, which compares the nested models OLS and *within*. Under the null of no individual effects, the statistic is distributed as an F with degrees of freedom equal to the number of individuals minus 1 on the numerator and to the degrees of freedom of the *within* model on the denominator. We first estimate the relevant models:

```
data("RiceFarms", package = "splm")
Rice <- pdata.frame(RiceFarms, index = "id")
rice.w <- plm(log(goutput) ~ log(seed) + log(totlabor) + log(size), Rice)
rice.p <- update(rice.w, model = "pooling")
rice.wd <- plm(log(goutput) ~ log(seed) + log(totlabor) + log(size), Rice,
               effect = "twoways")
```

and we then supply the testing function with two fitted models:

```
pFtest(rice.w, rice.p)

  F test for individual effects
```

1 See Baltagi and Li (1990).

```
data:  log(goutput) ~ log(seed) + log(totlabor) + log(size)
F = 1.7, df1 = 170, df2 = 850, p-value = 3e-06
alternative hypothesis: significant effects
```

The `formula-data` syntax may also be used:

```
pFtest(log(goutput) ~ log(seed) + log(totlabor) + log(size), Rice)
```

Unsurprisingly, the absence of individual effects is strongly rejected.

To test the absence of individual *and* time effects, one would use:

```
pFtest(rice.wd, rice.p)

  F test for twoways effects

data:  log(goutput) ~ log(seed) + log(totlabor) + log(size)
F = 4.3, df1 = 180, df2 = 850, p-value <2e-16
alternative hypothesis: significant effects
```

or

```
pFtest(log(goutput) ~ log(seed) + log(totlabor) + log(size), Rice,
       effect = "twoways")
```

To test the absence of time effects allowing for the presence of individual effects, we compare the individual and the two-ways effect *within* models:

```
pFtest(rice.wd, rice.w)

  F test for twoways effects

data:  log(goutput) ~ log(seed) + log(totlabor) + log(size)
F = 70, df1 = 5, df2 = 850, p-value <2e-16
alternative hypothesis: significant effects
```

Once more, the null hypothesis is very strongly rejected.

The Breusch and Pagan (1980) test can be computed using the function `plmtest`. The argument is either an OLS model or a `formula-data` pair. By default, the Honda (1985) version is computed. The direction of the effects, as usual, is determined by the `effect` argument.

```
plmtest(rice.p)

  Lagrange Multiplier Test - (Honda) for balanced
  panels

data:  log(goutput) ~ log(seed) + log(totlabor) + log(size)
normal = 4.8, p-value = 7e-07
alternative hypothesis: significant effects
plmtest(log(goutput)~log(seed)+log(totlabor)+log(size), Rice)
```

```
  Lagrange Multiplier Test - (Honda) for balanced
  panels

data:  log(goutput) ~ log(seed) + log(totlabor) + log(size)
normal = 4.8, p-value = 7e-07
alternative hypothesis: significant effects
plmtest(rice.p, effect = "time")

  Lagrange Multiplier Test - time effects (Honda) for
  balanced panels

data:  log(goutput) ~ log(seed) + log(totlabor) + log(size)
normal = 59, p-value <2e-16
alternative hypothesis: significant effects
plmtest(rice.p, effect = "twoways")

  Lagrange Multiplier Test - two-ways effects (Honda)
  for balanced panels

data:  log(goutput) ~ log(seed) + log(totlabor) + log(size)
normal = 45, p-value <2e-16
alternative hypothesis: significant effects
```

The two useful extensions proposed by Baltagi et al. (1992) can be applied to test the existence of individual and time effects, setting the argument `type` to `'kw'` or `'ghm'` to use respectively the techniques proposed by King and Wu (1997) and Gourieroux et al. (1982).

```
plmtest(rice.p, effect = "twoways", type = "kw")

  Lagrange Multiplier Test - two-ways effects (King
  and Wu) for balanced panels

data:  log(goutput) ~ log(seed) + log(totlabor) + log(size)
normal = 59, p-value <2e-16
alternative hypothesis: significant effects
plmtest(rice.p, effect = "twoways", type = "ghm")

  Lagrange Multiplier Test - two-ways effects
  (Gourieroux, Holly and Monfort) for balanced panels

data:  log(goutput) ~ log(seed) + log(totlabor) + log(size)
chibarsq = 3500, df0 = 0.00, df1 = 1.00, df2 = 2.00,
w0 = 0.25, w1 = 0.50, w2 = 0.25, p-value <2e-16
alternative hypothesis: significant effects
```

4.2 Tests for Correlated Effects

We have seen that if the model errors are not correlated with the explanatory variables, then both estimators, fixed as well as random effects, are consistent. To compare them, we keep assuming that the idiosyncratic component of the error term ($\mathrm{E}(X^\top \nu) = 0$) is uncorrelated to the regressors. Two situations are then possible:

- $\mathrm{E}(X^\top \eta) = 0$: the individual effects are not correlated with the explanatory variables; in this case, both estimators are consistent, but the random effects estimator is more efficient than the fixed effects.
- $\mathrm{E}(X^\top \eta) \neq 0$: the individual effects are correlated with the explanatory variables; in this case, the fixed effects estimator, which estimates out the individual effects, is consistent. On the contrary, the random effects estimator is inconsistent because one component of the composite error, the individual effect, is correlated with the explanatory variables.

4.2.1 The Mundlak Approach

In order to clarify the relationship between the two estimators, Mundlak (1978) considered the following model:

$$y_{nt} = x_{nt}^\top \beta + \psi_n + \nu_{nt}$$

with

$$\psi_n = \bar{x}_{n.}^\top \pi + \eta_n$$

The individual effects are therefore correlated with the explanatory variables, being they equal to the sum of a linear combination of the individual means of said variables and of an error term ψ_n. The model to be estimated is then written, in matrix form, as:

$$y = X\beta + BX\pi + (\mathrm{I}_N \otimes J_T)\eta + \nu$$

The error term $\epsilon = (\mathrm{I}_N \otimes J_T)\eta + \nu$ has the usual properties of the error components model, i.e., zero mean and a variance equal to:

$$\Omega = \sigma_\nu^2 \mathrm{I}_{NT} + \sigma_\eta^2 (\mathrm{I}_N \otimes J_T) = \sigma_\nu^2 W + \sigma_\iota^2 B$$

The GLS model is estimated by applying OLS on the data transformed pre-multiplying each variable by $\Sigma^{-0.5} = W + \phi B$, with $\phi = \frac{\sigma_\nu}{\sigma_\iota}$.

We then have $\tilde{y} = Wy + \phi By$, $\tilde{X} = WX + \phi BX$ and $\tilde{BX} = \phi BX$. The GLS estimator is then written as:

$$\begin{pmatrix} \hat{\beta} \\ \hat{\pi} \end{pmatrix} = \begin{bmatrix} X^\top WX + \phi^2 X^\top BX & \phi^2 X^\top BX \\ \phi^2 X^\top BX & \phi^2 X^\top BX \end{bmatrix}^{-1} \begin{pmatrix} X^\top Wy + \phi^2 X^\top By \\ \phi^2 X^\top By \end{pmatrix}$$

Using the formula of the inverse of a partitioned matrix (see equation 2.18), we get:

$$\begin{pmatrix} \hat{\beta} \\ \hat{\pi} \end{pmatrix} = \begin{bmatrix} (X^\top WX)^{-1} & -(X^\top WX)^{-1} \\ -(X^\top WX)^{-1} & (X^\top WX)^{-1} + \frac{1}{\phi^2}(X^\top BX)^{-1} \end{bmatrix} \begin{pmatrix} X^\top Wy + \phi^2 X^\top By \\ \phi^2 X^\top By \end{pmatrix}$$

$$\begin{pmatrix} \hat{\beta} \\ \hat{\pi} \end{pmatrix} = \begin{pmatrix} (X^\top WX)^{-1} X^\top Wy \\ (X^\top BX)^{-1} X^\top By - (X^\top WX)^{-1} X^\top Wy \end{pmatrix} = \begin{pmatrix} \hat{\beta}_{\mathrm{W}} \\ \hat{\beta}_{\mathrm{B}} - \hat{\beta}_{\mathrm{W}} \end{pmatrix}$$

and:

$$\mathrm{V}\begin{pmatrix} \hat{\beta} \\ \hat{\pi} \end{pmatrix} = \sigma_\nu^2 \begin{pmatrix} (X^\top WX)^{-1} & -(X^\top WX)^{-1} \\ -(X^\top WX)^{-1} & (X^\top WX)^{-1} + \frac{1}{\phi^2}(X^\top BX)^{-1} \end{pmatrix}$$

The fundamental result of Mundlak (1978) is therefore that, if one correctly accounts for the correlation between the error terms and the explanatory variables, the **GLS** estimator is the *within* estimator.

4.2.2 Hausman Test

This results also suggests a way to test for the presence of correlation; in fact, testing for no correlation corresponds to testing for: $\text{H}_0 : \pi = 0$. Under H_0, we have:

$$\hat{\pi}^\top \hat{\text{V}}(\hat{\pi})^{-1}\hat{\pi}$$

which is distributed as a χ^2 with K degrees of freedom. Well, we have $\hat{\pi} = \hat{\beta}_\text{B} - \hat{\beta}_\text{W}$ and $\text{V}(\hat{\pi}) = \text{V}(\hat{\beta}_\text{W}) + \text{V}(\hat{\beta}_\text{B})$.

This test statistic is one version of the test proposed by Hausman (1978). The general principle consists in comparing two models A and B where:

- under H_0: A and B are both consistent, but B is more efficient than A,
- under H_1: only A is consistent.

The idea of the test is that if H_0 is true, then the estimated coefficients from the two models shall not diverge; under the alternative, they will. The test is therefore based on $\hat{\beta}_\text{A} - \hat{\beta}_\text{B}$ and Hausman showed that, under H_0, the variance of this difference is simply: $\text{V}(\hat{\beta}_\text{A} - \hat{\beta}_\text{B}) = \text{V}(\hat{\beta}_\text{A}) - \text{V}(\hat{\beta}_\text{B})$.

The most common version of this test is based on comparing the *within* and the **GLS** estimators. The difference between the two is: $\hat{q} = \hat{\beta}_\text{W} - \hat{\beta}_\textbf{GLS}$. Under the hypothesis of no correlation between errors and explanatory variables, we have plim $\hat{q} = 0$. The variance of $\hat{q}$ is:

$$\text{V}(\hat{q}) = \text{V}(\hat{\beta}_\text{W}) + \text{V}(\hat{\beta}_\textbf{GLS}) - 2\text{cov}(\hat{\beta}_\text{W}, \hat{\beta}_\textbf{GLS})$$

To determine these variances and covariances, we write the two estimators as functions of the errors: $\hat{\beta}_\textbf{GLS} = (X^\top\Omega^{-1}X)^{-1}X^\top\Omega^{-1}\epsilon$ and $\hat{\beta}_\text{W} = (X^\top WX)^{-1}X^\top W\epsilon$. We then have: $\text{V}(\hat{\beta}_\textbf{GLS}) = (X^\top\Omega^{-1}X)^{-1}$, $\text{V}(\hat{\beta}_\text{W}) = \sigma_\nu^2(X^\top WX)^{-1}$ and $\text{cov}(\hat{\beta}_\text{W}, \hat{\beta}_\textbf{GLS}) = (X^\top\Omega^{-1}X)^{-1}$. The variance of $\hat{q}$ is then simply:

$$\text{V}(\hat{q}) = \sigma_\nu^2(X^\top WX)^{-1} - (X^\top\Omega^{-1}X)^{-1}$$

and the test statistic becomes:

$$\hat{q}^\top\text{V}(\hat{q})^{-1}\hat{q}$$

which, under H_0, is distributed as a χ^2 with K degrees of freedom.

4.2.3 Chamberlain's Approach

Chamberlain (1982) proposed a more general model than that of Mundlak (1978). In his model, the individual effects are not assumed to be a linear function of the means of the explanatory variables anymore, but of their values over the whole time period.

Denote $y_n^\top = (y_{n1}, \dots, y_{nT})$ the vector of length T containing the values of the explanatory variables for the n-th individual, and X_n the $T \times K$ matrix containing the values of K explanatory variables for the T observation periods for the n-th individual. $x_n = \text{vec}(X_n)$ is a vector of length

$T \times K$ obtained by stacking the columns of X_n. The model is then written as:

$$y_n = \mathrm{I}_T \otimes \beta^\top x_n + \psi_n + \nu_{nt} \tag{4.1}$$

with:

$$\psi_n = \gamma^\top x_n + \eta_n \tag{4.2}$$

Substituting (4.2) in (4.1), we get:

$$y_n = (\mathrm{I}_n \otimes \beta^\top + j\gamma^\top)x_n + \eta_n + \nu_{nt} = \Pi x_n + \mu_n + \nu_{nt} \tag{4.3}$$

The parameter matrix Π, of dimension $T \times (T \times K)$, contains two types of parameters:

- the vector of parameters β, which measure the marginal effect of the explanatory variables on the response,
- the vector of parameters γ, measuring the marginal effect of the explanatory variables in each period on the individual effect.

The γ vector is only marginally interesting *per se*, but its estimation allows to consistently estimate β. If $n = 3$ and $K = 2$, the Π matrix takes the form:

$$\begin{pmatrix} \beta_1 & \beta_2 & 0 & 0 & 0 & 0 \\ 0 & 0 & \beta_1 & \beta_2 & 0 & 0 \\ 0 & 0 & 0 & 0 & \beta_1 & \beta_2 \end{pmatrix} + \begin{pmatrix} \gamma_{11} & \gamma_{12} & \gamma_{21} & \gamma_{22} & \gamma_{31} & \gamma_{32} \\ \gamma_{11} & \gamma_{12} & \gamma_{21} & \gamma_{22} & \gamma_{31} & \gamma_{32} \\ \gamma_{11} & \gamma_{12} & \gamma_{21} & \gamma_{22} & \gamma_{31} & \gamma_{32} \end{pmatrix}$$

We then have a system of T equations containing the same explanatory variables x_n. In this case, the generalized least squares estimator is the SUR estimator. The explanatory variables being the same across equations, this can be obtained simply by estimating each individual equation separately by OLS. If the assumptions of the fixed effects model hold, then the estimation of each column of the Π matrix will yield just about equal coefficients, with the exception of those situated on the diagonal.

4.2.3.1 Unconstrained Estimator

The coefficients of the tth row of Π are then:

$$\pi_t = \left(\sum_n x_n x_n^\top\right)^{-1} \left(\sum_n x_n y_{nt}\right)$$

More generally, we can write the estimator of Π in two different ways. The first consists in defining:

$$X = \begin{pmatrix} x_1^\top \\ x_2^\top \\ \vdots \\ x_N^\top \end{pmatrix} \text{ and } y = \begin{pmatrix} y_1^\top \\ y_2^\top \\ \vdots \\ y_N^\top \end{pmatrix}$$

We have then:

$$\hat{\Pi}^\top = (X^\top X)^{-1} X^\top y$$

In order to analyze the properties of this estimator, it is easier to consider the vector of coefficients $\pi = \mathrm{Vec}\,\Pi^\top$ obtained by stacking the rows of Π. Defining:

$$X = \begin{pmatrix} \mathrm{I}_T \otimes x_1^\top \\ \mathrm{I}_T \otimes x_2^\top \\ \vdots \\ \mathrm{I}_T \otimes x_N^\top \end{pmatrix} \text{ et } y = \begin{pmatrix} y_1 \\ y_2 \\ \vdots \\ y_N \end{pmatrix}$$

$$\hat{\pi} = \left(\mathrm{I}_T \otimes \left(\sum_n x_n x_n^\top \right)^{-1} \right) \sum_n (\mathrm{I}_T \otimes x_n) y_n$$

Denoting $s_{xx} = \sum_{n=1}^N x_n x_n^\top / N$ and substituting, in the last expression, y by its expression in the "true" model:

$$\hat{\pi} - \pi = \frac{1}{N}(\mathrm{I}_T \otimes s_{xx}^{-1}) \sum_n (\mathrm{I}_T \otimes x_n)\epsilon_n = \frac{1}{N} \sum_n (\mathrm{I}_T \otimes s_{xx}^{-1} x_n)\epsilon_n$$

$$\sqrt{N}(\hat{\pi} - \pi) = \frac{1}{\sqrt{N}} \sum_n (\mathrm{I}_T \otimes s_{xx}^{-1} x_n)\epsilon_n$$

The limiting distribution of $\sqrt{N}(\hat{\pi} - \pi)$ is the same as that of:

$$\frac{1}{\sqrt{N}} \sum n (\mathrm{I}_T \otimes \sigma_{xx}^{-1} x_n)\epsilon_n$$

with $\sigma_{xx} = \mathrm{E}(x_n x_n^\top)$ because $\sum n x_n x_n^\top / N \rightarrow \mathrm{E}(x_n x_n^\top)$.

As $\mathrm{E}([\mathrm{I}_T \otimes x_n]\epsilon_n) = 0$, the central limit theorem implies that:

$$\sqrt{N}(\hat{\pi} - \pi) \sim N(0, \Omega)$$

$$\begin{aligned} \Omega &= \mathrm{V}\left(\frac{1}{\sqrt{N}} \sum n (\mathrm{I}_T \otimes \sigma_{xx}^{-1} x_n)\epsilon_n \right) \\ &= \frac{1}{N}\mathrm{E}\left(\sum n (\mathrm{I}_T \otimes \sigma_{xx} x_n)\epsilon_n \sum n \epsilon_n^\top (\mathrm{I}_T \otimes x_n^\top \sigma_{xx}) \right) \\ &= \frac{1}{N}\mathrm{E}\left(\sum n (\mathrm{I}_T \otimes \sigma_{xx} x_n)\epsilon_n \epsilon_n^\top (\mathrm{I}_T \otimes x_n^\top \sigma_{xx}) \right) \\ &= \frac{1}{N}\mathrm{E}\left(\sum n \epsilon_n \epsilon_n^\top \otimes (\sigma_{xx} x_n x_n^\top \sigma_{xx}) \right) \\ &= \mathrm{E}(\epsilon_n \epsilon_n^\top \otimes (\sigma_{xx} x_n x_n^\top \sigma_{xx})) \end{aligned}$$

If the error variance of n is not correlated with $x_n x_n^\top$, this matrix simplifies to:

$$\Omega_1 = \mathrm{E}(\epsilon_n \epsilon_n^\top \otimes \sigma_{xx})$$

Finally, if the errors are homoscedastic, we get an even simpler expression:

$$\Omega_2 = \sigma \otimes \sigma_{xx}$$

with $\sigma = \mathrm{E}(\epsilon \epsilon^\top)$.

An estimator of Ω can be obtained considering the sample equivalent. Denoting by $\hat{\epsilon}_n$ the estimation residuals, we get:

$$\hat{\Omega} = \frac{1}{N} \sum n (\hat{\epsilon}_n \hat{\epsilon}_n^\top \otimes s_x^{-1} x_n x_n^\top s_x^{-1})$$

$$\hat{\Omega}_1 = \frac{1}{N} \sum n(\hat{\epsilon}_n \hat{\epsilon}_n^{\top} \otimes s_x^{-1})$$

$$\hat{\Omega}_2 = \frac{1}{N} \sum_{n=1}^{N} \hat{\epsilon}_n \hat{\epsilon}_n^{\top} \otimes s_x^{-1}$$

4.2.3.2 Constrained Estimator

In a second time, Chamberlain (1982) utilizes the asymptotic least squares estimator in order to obtain an estimator of the structural coefficients of the model, denoted by θ. These are made of the K coefficients associated to the explanatory variables of equation (4.1) and of the $K \times T$ coefficients from equation (4.2) concerning the individual effects. There are therefore $K \times (T+1)$ structural coefficients, while the number of coefficients in the matrix Π is $K \times T^2$. The relation between the two coefficient vectors can be expressed as $\pi = F\theta$, F being a matrix of dimensions $(K \times T^2) \times (K \times (T+1))$.

The restricted model is obtained employing the method of asymptotic least squares, consisting in minimizing a quadratic form in the deviations between $\hat{\pi}$ and $F\hat{\theta}$:

$$(\hat{\pi} - F\theta)^{\top} A(\hat{\pi} - F\theta)$$

The first-order conditions for a minimum can be written as:

$$-2F^{\top} A(\hat{\pi} - F\theta) = 0$$

which yields the following estimator:

$$\hat{\theta} = (F^{\top} AF)^{-1} F^{\top} A\hat{\pi}$$

This estimator is consistent regardless of the weighting matrix employed. Just as with the generalized method of moments, the estimator is efficient if A^{-1} is the covariance matrix of the residuals' vector. If the hypotheses are verified ($\pi = F\theta$), the latter can be written as: $A^{-1} = \mathrm{V}(\hat{\pi}) = \Omega/N$.

4.2.3.3 Fixed Effects Models

If the hypotheses $\pi = F\theta$ hold, the minimum of the objective function from the asymptotic least squares method is distributed as a χ^2 with degrees of freedom equal to the difference in length between the two vectors π and θ, i.e., $T^2 \times K - (T+1) \times K = (T^2 - T - 1) \times K$. This enables the computation of a test of the restrictions on the coefficients implied by the fixed effects model.

Angrist and Newey (1991) have shown that these restrictions can more simply be tested using the results of T artificial regressions of the fixed effect model's residuals of a specific period on the covariates for every period. Denoting R_t^2 the coefficient of determination of the artifactual regression of residuals of period t, we have:

$$(NT - KT) \sum_{t=1}^{T} R_t^2$$

which follows a χ^2 with $(T^2 - T - 1) \times K$ if the underlying hypotheses of the *within* model are relevant.

Example 4.2 Hausman test – `RiceFarms` data set

The Hausman test is performed with the `phtest` function, which can either take as arguments two estimated models (here: the *within* and the GLS) or use the `formula` – `data` interface:

```
data("RiceFarms", package = "splm")
Rice <- pdata.frame(RiceFarms, index = "id")
rice.w <- plm(log(goutput) ~ log(seed) + log(totlabor) + log(size), Rice)
rice.r <- update(rice.w, model = "random")
phtest(rice.w, rice.r)

  Hausman Test

data:  log(goutput) ~ log(seed) + log(totlabor) + log(size)
chisq = 3.8, df = 3, p-value = 0.3
alternative hypothesis: one model is inconsistent
```

Under the hypothesis of no correlation between the regressors and the individual effects, the statistic is distributed as a χ^2 with three degrees of freedom. This hypothesis, with a p-value of 29%, is not rejected at the 5% confidence level. One could get the same result following the Mundlak (1978) approach, drawing on the difference between the *within* and *between* estimators:

```
rice.b <- update(rice.w, model = "between")
cp <- intersect(names(coef(rice.b)), names(coef(rice.w)))
dcoef <- coef(rice.w)[cp] - coef(rice.b)[cp]
V <- vcov(rice.w)[cp, cp] + vcov(rice.b)[cp, cp]
as.numeric(t(dcoef) %*% solve(V) %*% dcoef)
[1] 3.773
```

This result is confirmed by the correlation coefficient between the individual effects (estimated by the fixed effects of the *within* model) and the individual means of the explanatory variable, which is obtained by applying the `between` function to the series.[2]

```
cor(fixef(rice.w), between(log(Rice$goutput)))
[1] 0.448
```

The correlation is positive but moderate.

The Chamberlain test is available in the function `piest`. It is computed using the usual `formula-data` interface.

```
data("RiceFarms", package = "splm")
pdim(RiceFarms, index = "id")
Balanced Panel: n = 171, T = 6, N = 1026
piest(log(goutput) ~ log(seed) + log(totlabor) + log(size),
      RiceFarms, index = "id")

  Chamberlain's pi test

data:  log(goutput) ~ log(seed) + log(totlabor) + log(size)
chisq = 110, df = 87, p-value = 0.03
```

2 Note that we use `between` and not `Between`, the latter returning a vector of the same length as the series with the individual means repeated T_n times.

The variant of the Chamberlain test proposed by Angrist and Newey (1991) is available with the `aneweytest` function which uses the same interface.

```
aneweytest(log(goutput) ~ log(seed) + log(totlabor) + log(size),
           RiceFarms, index = "id")

  Angrist and Newey's test of within model

data:  log(goutput) ~ log(seed) + log(totlabor) + log(size)
chisq = 140, df = 87, p-value = 2e-04
```

The restrictions implied by the *within* model are rejected by both tests at the 5% level, although they are not rejected at the 1% level for Chamberlain's version of the test.

4.3 Tests for Serial Correlation

A model with individual effects has composite errors that are serially correlated by definition. The presence of the time-invariant error component gives rise to serial correlation that does not die out over time; thus standard tests applied on pooled data usually end up rejecting the null of spherical residuals. There may also be serial correlation of the time-decaying kind in the idiosyncratic error terms, e.g., as an AR(1) process. By "testing for serial correlation" we mean testing for this latter kind of dependence.

For these reasons, the subjects of testing for individual error components and for serially correlated idiosyncratic errors are closely related. In particular, simple (*marginal*) tests for one direction of departure from the hypothesis of spherical errors usually have power against the other one: in case it is present, they are substantially biased toward rejection. *Joint* tests are correctly sized and have power against both directions but usually do not give any information about which one actually caused rejection. *Conditional* tests for serial correlation that take into account the error components are correctly sized under presence of both departures from sphericity and have power only against the alternative of interest. While most powerful if correctly specified, the latter, based on the likelihood framework, are crucially dependent on normality and homoscedasticity of the errors.

In **plm** a number of joint, marginal, and conditional ML-based tests are provided, plus some semi-parametric alternatives that are robust versus heteroscedasticity and free from distributional assumptions.

More tests can be obtained by comparing nested models, in a likelihood ratio test framework, or by restriction on a more general model, in the Wald test framework.

4.3.1 Unobserved Effects Test

The unobserved effects test *à la Wooldridge* (see Wooldridge, 2010, 10.4.4), is a semi-parametric test for the null hypothesis that $\sigma_\eta^2 = 0$, i.e., that there are no unobserved effects in the residuals. Given that under the null, the covariance matrix of the residuals for each individual is diagonal, the test statistic is based on the average of elements in the upper (or lower) triangle of its estimate, diagonal excluded: $N^{-1/2} \sum_{n=1}^{N} \sum_{t=1}^{T-1} \sum_{s=t+1}^{T} \hat{\epsilon}_{nt}\hat{\epsilon}_{ns}$ (where $\hat{\epsilon}$ are the pooled OLS residuals), which must be "statistically close" to zero under the null, scaled by its standard deviation:

$$W = \frac{\sum_{n=1}^{N} \sum_{t=1}^{T-1} \sum_{s=t+1}^{T} \hat{\epsilon}_{nt}\hat{\epsilon}_{ns}}{\left[\sum_{n=1}^{N} \left(\sum_{t=1}^{T-1} \sum_{s=t+1}^{T} \hat{\epsilon}_{nt}\hat{\epsilon}_{ns}\right)^2\right]^{1/2}}$$

This test is (N-) asymptotically distributed as a standard normal regardless of the distribution of the errors. It does also not rely on homoscedasticity.

It has power both against the standard random effects specification, where the unobserved effects are constant within every group, as well as against any kind of serial correlation. As such, it "nests" both individual effects and serial correlation tests, trading some power against more specific alternatives in exchange for robustness.

While not rejecting the null favors the use of pooled OLS, rejection may follow from serial correlation of different kinds, and in particular, quoting Wooldridge (2010, 10.4.4), "should not be interpreted as implying that the random effects error structure *must* be true".

Example 4.3 unobserved effects test – `RiceFarms` data set
Below, the test is applied to the rice farms data:

```
data("RiceFarms", package="plm")
Rice <- pdata.frame(RiceFarms, index = "id")
fm <- log(goutput) ~ log(seed) + log(totlabor) + log(size)
pwtest(fm, Rice)

	Wooldridge's test for unobserved individual effects

data:  formula
z = 2.9, p-value = 0.003
alternative hypothesis: unobserved effect
```

The null hypothesis of no unobserved effects is rejected.

4.3.2 Score Test of Serial Correlation and/or Individual Effects

The Wooldridge testing procedure will detect very general forms of persistence in the errors but give few directions toward a finer specification. If one is willing to make more specific parametric hypotheses, a maximum likelihood approach will allow to detect the features of persistence in a finer way.

The random effects model can be extended to having idiosyncratic errors that follow an autoregressive process of order 1 (AR(1)):

$$y = \alpha + X\beta + \epsilon$$
$$\epsilon = (\mathrm{I}_N \otimes j_T)\eta + \nu$$
$$\nu_t = \psi \nu_{t-1} + \zeta_t$$

where the now familiar hypotheses of the random effects model apply, i.e., individual effects η are independent of both the regressors and the idiosyncratic errors ν. Moreover, normality is assumed for both error components.

The specification analysis of the above model requires telling apart the time-invariant individual effects from the time-decaying persistence due to the AR(1) component. As observed, the presence of individual effects may affect tests for serial correlation and vice versa. Several alternative strategies can be used, based on the following tools:

- a joint test, which has power against both alternatives,
- marginal tests, testing the null hypothesis of no serial correlation while maintaining the hypothesis of no individual effects under both the null and the alternative hypothesis,

- locally robust tests, i.e., marginal tests with a correction that makes them robust to local deviations from the maintained hypothesis,
- conditional test, i.e., testing the null hypothesis of no serial correlation, the hypothesis of the presence of individual effects being maintained under both the null and the alternative hypothesis.

The advantage of the robust tests is that the unconstrained model (**RE-AR(1)** for random effect model with first-order auto-regressive errors) need not be estimated.

Baltagi and Li (1991) and Baltagi and Li (1995) proposed a joint test of no serial correlation and no individual effects. The test statistic is:

$$\mathrm{LM}_{\eta\psi} = \frac{NT^2}{2(T-1)(T-2)}[A^2 - 4AB + 2TB^2]$$

with $A = \hat{\epsilon}^\top S_\eta \hat{\epsilon} / \hat{\epsilon}^\top \hat{\epsilon}$ and $B = \hat{\epsilon}^\top \hat{\epsilon}_1 / \hat{\epsilon}^\top \hat{\epsilon}$, $\hat{\epsilon}$ being the **OLS** residuals.

Baltagi and Li (1995) also proposed a marginal test of serial correlation, the maintained hypothesis being the absence of individual effects:

$$\mathrm{LM}_{\psi} = \frac{NT^2}{2(T-1)}B$$

Symmetrically, the marginal test of individual effects, with the maintained hypothesis of no serial correlation is simply the Breusch-Pagan test:

$$\mathrm{LM}_{\eta} = \frac{NT^2}{2(T-1)}A$$

They also proposed a conditional test of serial correlation, the maintained hypothesis being the presence of individual effects. This latter test (LM_4 in the original paper) is based on the residuals of the random effect model estimated by the maximum likelihood method. Under the null of serially uncorrelated errors, the test turns out to be identical for both the alternative of **AR(1)** and **MA(1)** processes.

Bera et al. (2001) derive locally robust tests both for individual random effects and for first-order serial correlation in residuals as "corrected" versions of the standard LM test LM_ψ and LM_η. They write respectively:

$$\mathrm{LM}^*_{\psi} = \frac{NT^2}{(T-1)(1-2/T)}(B + A/T)^2$$

and

$$\mathrm{LM}_{\eta} = \frac{NT}{2(T-1)(1-2/T)}(A + 2B)^2$$

While still dependent on normality and homoscedasticity, these are robust to *local* departures from the hypotheses of, respectively, no serial correlation or no individual effects. Although suboptimal, these tests may help detecting the right direction of the departure from the null, thus complementing the use of joint tests. Moreover, being based on pooled **OLS** residuals, the **BSY** tests are computationally far less demanding than the conditional test of Baltagi and Li (1995).

On the other hand, the statistical properties of these locally corrected tests are inferior to those of the non-corrected counterparts when the latter are correctly specified. If there is no serial correlation, then the optimal test for random effects is the likelihood-based **LM** test of Breusch and Pagan (with refinements by Honda, see `plmtest`), while if there are no individual effects, the optimal test for serial correlation is the Breusch-Godfrey test. If the presence of a random effect is taken for granted, then the optimal test for serial correlation is the likelihood-based conditional **LM** test of Baltagi and Li (1995) (see `pbltest`).

Example 4.4 LM tests for random effects and/or serial correlation – RiceFarms data set

From the LM test in a previous example, the Rice Farming data show evidence of individual effects; and from the Hausman test, the latter seem to comply with the hypothesis that they are uncorrelated with the covariates. We now investigate the presence of serial correlation in the context of the RE-AR(1) model outlined above.

The joint LM test for random effects *and* serial correlation under normality and homoscedasticity of the idiosyncratic errors has been derived by Baltagi and Li (1991) and Baltagi and Li (1995) and is implemented as an option in `pbsytest`, by setting the `test` to `'J'`. In the case of the rice farming model, the test strongly rejects. Rejection of the joint test, though, gives no information on the direction of the departure from the null hypothesis, i.e., is rejection due to the presence of serial correlation, of random effects, or of both?

Bera et al. (2001)'s *locally robust* tests can provide statistical evidence about the direction of misspecification in the (doubly) restricted model. In fact, the presence of the "other" effect (individual effects when testing for serial correlation and vice versa) will not influence the test statistic as long as the magnitude is moderate. How much of the "other" effect is tolerated before the test statistic becomes biased, however, is an empirical question and will be case-specific, although the simulations in the original paper can provide a rough assessment.

The locally robust BSY tests for, respectively, serial correlation or individual effects are implemented in the function `pbsytest`, by setting the `test` argument to `'AR'` (default) or `'RE'`. The test for random effects is implemented in the one-sided version, which takes into account that the variance of the random effect must be non-negative.

```
bsy.LM <- matrix(ncol=3, nrow = 2)
tests <- c("J", "RE", "AR")
dimnames(bsy.LM) <- list(c("LM test", "p-value"), tests)
for(i in tests) {
    mytest <- pbsytest(fm, data = Rice, test = i)
    bsy.LM[1:2, i] <- c(mytest$statistic, mytest$p.value)
    }
round(bsy.LM, 6)
             J     RE    AR
LM test 62.65 0.3351 39.23
p-value  0.00 0.3688  0.00
```

The robust tests allow us to discriminate between time-invariant error persistence (random effects) and time-decaying persistence (autoregressive errors), concluding in favor of the second.

Finally, the optimal conditional test of Baltagi and Li for serial correlation, allowing for random effects of any magnitude, is computed using the `pbltest` function, using the residuals of the random effects maximum likelihood estimator:

```
pbltest(fm, Rice, alternative = "onesided")

Baltagi and Li one-sided LM test

data:  fm
z = 6.1, p-value = 6e-10
alternative hypothesis: AR(1)/MA(1) errors in RE panel model
```

Serial correlation is detected, i.e., we conclude that $\psi \neq 0$ in the encompassing model.

4.3.3 Likelihood Ratio Tests for AR(1) and Individual Effects

Likelihood ratio (LR) tests for restrictions are based on the likelihoods from the general and the restricted model. The test statistic is simply twice the difference of the values of the log-likelihood function:

$$2[\ln L(\hat{\theta}) - \ln L(\tilde{\theta})] \sim \chi^2_m$$

where $\hat{\theta}$ is the full vector of parameter estimates from the unrestricted model and $\tilde{\theta}$ from the restricted one, and m is the number of restrictions.

A likelihood ratio test for serial correlation in the idiosyncratic residuals can be done, in general, as a nested models test comparing the model with spherical idiosyncratic residuals with the more general alternative featuring AR(1) residuals. If both estimated models allow for random effects, then the test will become conditional on the latter feature.

Thus,

$$\mathrm{LR}_{\psi|\eta} = 2[\ln L(\hat{\psi}, \hat{\eta}, \hat{\beta}) - \ln L(\tilde{\eta}, \tilde{\beta})] \sim \chi^2_1$$

and symmetrically for $\mathrm{LR}_{\eta|\psi}$.

Example 4.5 Likelihood ratio tests – `Grunfeld` data set
Maximum likelihood estimation of linear models with or without either random individual effects or serially correlated errors can be estimated, e.g., with functionality from the **nlme** package. In its notation, the RE specification is a model with only one random effects regressor: the intercept. Below we report coefficients of Grunfeld's model estimated by GLS and then by ML.

```
library("nlme")
data(Grunfeld, package = "plm")
reGLS <- plm(inv ~ value + capital, data = Grunfeld, model = "random")
reML <- lme(inv ~ value + capital, data = Grunfeld, random = ~1 | firm)
rbind(coef(reGLS), fixef(reML))
     (Intercept)  value capital
[1,]      -57.83 0.1098  0.3081
[2,]      -57.86 0.1098  0.3082
```

Linear models with groupwise structures of time dependence may be fitted by `gls`, specifying the correlation structure in the `correlation` option:

```
lmAR1ML <- gls(inv ~ value + capital, data = Grunfeld,
    correlation = corAR1(0, form = ~ year | firm))
```

and analogously the random effects panel with, e.g., AR(1) errors (see Baltagi, 2013, Ch. 5) may be fit by `lme` specifying an additional random intercept:

```
reAR1ML <- lme(inv ~ value + capital, data = Grunfeld,
    random = ~ 1 | firm, correlation = corAR1(0, form = ~ year | firm))
summary(reAR1ML)
Linear mixed-effects model fit by REML
 Data: Grunfeld
   AIC  BIC logLik
  2095 2115  -1041
```

```
Random effects:
 Formula: ~1 | firm
        (Intercept) Residual
StdDev:       78.04     72.8

Correlation Structure: AR(1)
 Formula: ~year | firm
 Parameter estimate(s):
   Phi
0.8238
Fixed effects: inv ~ value + capital
             Value Std.Error  DF t-value p-value
(Intercept) -40.28    30.694 188  -1.312  0.1911
value         0.09     0.008 188  11.770  0.0000
capital       0.31     0.032 188   9.737  0.0000
 Correlation:
        (Intr) value
value   -0.239
capital -0.280 -0.125

Standardized Within-Group Residuals:
     Min        Q1       Med        Q3       Max
-2.40759 -0.31847  0.04847  0.19863  3.30040

Number of Observations: 200
Number of Groups: 10
```

Let us compare either with the restricted alternative. The **GLS** model without correlation in the residuals is the same as **OLS**, and one could well use `lm` for the restricted model. Here we estimate it by `gls`.

```
lmML <- gls(inv ~ value + capital, data = Grunfeld)
anova(lmML, lmAR1ML)
        Model df  AIC  BIC logLik   Test L.Ratio p-value
lmML        1  4 2400 2413  -1196
lmAR1ML     2  5 2095 2111  -1042 1 vs 2   307.3  <.0001
```

The **AR**(1) test on the random effects model is to be done in much the same way, using the random effects model objects estimated above:

```
anova(reML, reAR1ML)
        Model df  AIC  BIC logLik   Test L.Ratio p-value
reML        1  5 2206 2222  -1098
reAR1ML     2  6 2095 2114  -1041 1 vs 2     113  <.0001
```

A likelihood ratio test for random effects compares the specifications with and without random effects and spherical idiosyncratic errors:

```
anova(lmML, reML)
     Model df  AIC  BIC logLik   Test L.Ratio p-value
lmML     1  4 2400 2413  -1196
reML     2  5 2206 2222  -1098 1 vs 2   196.4  <.0001
```

The random effects, AR(1) errors model in turn nests the AR(1) pooling model; therefore, a likelihood ratio test for random effects sub AR(1) errors may be carried out, again, by comparing the two autoregressive specifications:

```
anova(lmAR1ML, reAR1ML)
        Model df  AIC  BIC logLik   Test L.Ratio p-value
lmAR1ML     1  5 2095 2111  -1042
reAR1ML     2  6 2095 2114  -1041 1 vs 2   2.134   0.144
```

whence we see that the Grunfeld model specification doesn't seem to need any random effects once we control for serial correlation in the data.

4.3.4 Applying Traditional Serial Correlation Tests to Panel Data

A general testing procedure for serial correlation in fixed effects (FE), random effects (RE), and pooled-OLS panel models alike can be based on considerations in (Wooldridge, 2010, 10.7.2). For the random effects model, Wooldridge (2010) observes that under the null of homoscedasticity and no serial correlation in the idiosyncratic errors, the residuals from the quasi-demeaned regression must be spherical as well. Else, as the individual effects are wiped out in the demeaning, any remaining serial correlation must be due to the idiosyncratic component. Hence, a simple way of testing for serial correlation is to apply a standard serial correlation test to the quasi-demeaned model. The same applies in a pooled model, w.r.t.the original data.

The FE case is different. It is well known that if the original model's errors are uncorrelated, then FE residuals are negatively serially correlated, with $\text{cor}(\hat{\epsilon}_{nt}, \hat{\epsilon}_{ns}) = -1/(T-1)$ for each t, s (see Wooldridge, 2010, 10.5.4). This correlation clearly dies out as T increases, so this kind of AR test is applicable to *within* model objects only for T "sufficiently large". Baltagi and Li (1995) derive a basically analogous T-asymptotic test for first-order serial correlation in a FE panel model as a Breusch-Godfrey LM test on *within* residuals (see Baltagi and Li, 1995, par. 2.3 and formula 12). They also observe that the test on *within* residuals can be used for testing on the RE model, as "the *within* transformation wipes out the individual effects, whether fixed or random." Generalizing the Durbin-Watson test to FE models by applying it to fixed effects residuals is documented in Bhargava et al. (1982). On the converse, in short panels the test gets severely biased toward rejection (or, as the induced correlation is negative, toward acceptance in the case of the one-sided Durbin-Watson test with `alternative` set to `'greater'`). See below for a serial correlation test applicable to "short" FE panel models.

Example 4.6 Breusch-Godfrey and Durbin Watson tests – `RiceFarms` data set

The functions `pbgtest` and `pdwtest` re-estimate the relevant quasi-demeaned model by OLS and apply, respectively, standard Breusch-Godfrey and Durbin-Watson tests from package **lmtest** to the residuals:

```
rice.re <- plm(fm, Rice, model='random')
pbgtest(rice.re, order = 2)

Breusch-Godfrey/Wooldridge test for serial
correlation in panel models

data:  fm
chisq = 36, df = 2, p-value = 2e-08
alternative hypothesis: serial correlation in idiosyncratic errors
pdwtest(rice.re, order = 2)
```

```
Durbin-Watson test for serial correlation in panel
models

data:  fm
DW = 1.7, p-value = 5e-07
alternative hypothesis: serial correlation in idiosyncratic errors
```

The tests share the features of their OLS counterparts, in particular the `pbgtest` allows testing for higher-order serial correlation, which can be of particular interest for quarterly data. As the functions are simple wrappers toward `bgtest` and `dwtest`, all arguments from the latter two apply and may be passed on through the '...' operator.

As observed above, applying the `pbgtest` and `pdwtest` functions to an FE model is appropriate only if the time dimension is long enough. In the frequent case of "short" panels, one of the two testing procedures due to Wooldridge (2010) and described in the next section should be used instead.

4.3.5 Wald Tests for Serial Correlation using *within* and First-differenced Estimators

4.3.5.1 Wooldridge's *within*-based Test

Due to the demeaning procedure, under the null of no serial correlation in the errors, the residuals of an FE model must be negatively serially correlated, with $\text{cor}(\hat{\epsilon}_{nt}, \hat{\epsilon}_{ns}) = -1/(T-1)$ for each t, s. Wooldridge suggests basing a test for this null hypothesis on a pooled regression of FE residuals on their first lag:

$$\hat{\epsilon}_{nt} = \alpha + \psi \hat{\epsilon}_{nt-1} + \zeta_{nt}$$

Rejecting the restriction $\delta = -1/(T-1)$ makes us conclude against the original null of no serial correlation.

The function carrying out this procedure estimates the FE model, retrieves residuals, then estimates an auxiliary (pooled) AR(1) model, and tests the above-mentioned restriction on ψ. Internally, a heteroscedasticity- and autocorrelation-consistent covariance matrix (`vcovHC`, see next chapter) is used, as originally prescribed. The test is applicable to any FE panel model and in particular to "short" panels with small T and large N.

Example 4.7 serial correlation tests for fixed effects models – `EmplUK` data set

In the following example, Wooldridge's *within*-based serial correlation test is applied to the EmplUK data:

```
data("EmplUK", package = "plm")
pwartest(log(emp) ~ log(wage) + log(capital), data = EmplUK)

Wooldridge's test for serial correlation in FE
panels

data:  plm.model
F = 310, df1 = 1, df2 = 890, p-value <2e-16
alternative hypothesis: serial correlation
```

We strongly reject the null of no serial correlation. If the evidence of persistence is too strong, one should wonder whether the residuals are stationary at all, and whether a specification in differences might be preferable. In the next section we will see a similar test that can be seen as a specification device in this sense.

4.3.5.2 Wooldridge's First-difference-based Test

In the context of the first difference model, Wooldridge (2010, 10.6.3) proposes a serial correlation test that can also be seen as a specification test to choose the most efficient estimator between fixed effects (*within*) and first difference (FD).

The starting point is the observation that if the idiosyncratic errors of the original model ϵ_{nt} are uncorrelated, the errors of the (first) differenced model $\epsilon^{\text{FD}}_{nt} \equiv \epsilon_{nt} - \epsilon_{nt-1}$ will be correlated, with $\text{cor}(\epsilon_{nt}, \epsilon_{nt-1}) = -0.5$, while any time-invariant effect is wiped out in the differencing. So a serial correlation test for models with individual effects of any kind can be based on estimating the auxiliary model

$$\hat{\epsilon}^{\text{FD}}_{nt} = \psi \hat{\epsilon}^{\text{FD}}_{nt-1} + \zeta_{nt}$$

and testing the restriction $\psi = -0.5$, corresponding to the null of no serial correlation in the original model. Drukker (2003) provides Monte Carlo evidence of the good empirical properties of the test.

On the other extreme (see Wooldridge, 2010, 10.6.1), if the differenced errors are uncorrelated, then ϵ_{nt} is a random walk. In this latter case, the most efficient estimator is the first difference (FD) one; in the former case, it is the fixed effects one (*within*).

Example 4.8 Wooldridge's first difference test – `EmplUK` data set

We apply the test in the context of the `EmplUK` data, a large and short panel with strong individual heterogeneity. We want to test for serial correlation in both *within* and first-differenced errors. The function `pwfdtest` allows testing either hypothesis: the default behavior `h0='fd'` is to test for serial correlation in *first-differenced* errors:

```
pwfdtest(log(emp) ~ log(wage) + log(capital), data = EmplUK)

Wooldridge's first-difference test for serial
correlation in panels

data:  plm.model
F = 0.93, df1 = 1, df2 = 750, p-value = 0.3
alternative hypothesis: serial correlation in differenced errors
```

while specifying `h0='fe'` the null hypothesis becomes no serial correlation in *original* errors, which is similar to the `pwartest`.

```
pwfdtest(log(emp) ~ log(wage) + log(capital), data = EmplUK,
    h0 = "fe")

Wooldridge's first-difference test for serial
correlation in panels

data:  plm.model
F = 130, df1 = 1, df2 = 750, p-value <2e-16
alternative hypothesis: serial correlation in original errors
```

Not rejecting one of the two is evidence in favor of using the estimator corresponding to `h0`. In this case, the original residuals show evidence of serial correlation, which disappears after first differencing. The results point at a unit root in the errors.

Example 4.9 Wooldridge's first difference test – `RiceFarms` data set
Results will not always be so clear-cut. For the `RiceFarms` example,

```
W.fd <- matrix(ncol = 2, nrow =2)
H0 <- c("fd", "fe")
dimnames(W.fd) <- list(c("test", "p-value"), H0)
for(i in H0) {
    mytest <- pwfdtest(fm, Rice, h0 = i)
    W.fd[1, i] <- mytest$statistic
    W.fd[2, i] <- mytest$p. value
    }
round(W.fd, 6)
              fd          fe
test      176.4 19.492371
p-value     0.0  0.000012
```

The truth clearly lies in the middle (both rejected, although one more strongly than the other); in this case, whichever estimator is chosen will have serially correlated errors: therefore it will be advisable to use an autocorrelation-robust covariance matrix.

4.4 Tests for Cross-sectional Dependence

Next to the more familiar issue of serial correlation, a growing body of literature has been dealing with cross-sectional dependence in panels, which can arise, e.g., if individuals respond to common shocks (as in common factor models) or if spatial diffusion processes are present, relating individuals in a way depending on a measure of distance (as in spatial models).

If cross-sectional dependence is present, the consequence is, at a minimum, inefficiency of the usual estimators and invalid inference when using the standard covariance matrix. This is the case, for example, in unobserved effects models when cross-sectional dependence is due to an unobservable factor structure but with factors uncorrelated with the regressors. In this case the *within* or **RE** are still consistent, although inefficient (see De Hoyos and Sarafidis, 2006). If the unobserved factors are correlated with the regressors, which can seldom be ruled out, consequences are more serious: the estimators will be inconsistent.

4.4.1 Pairwise Correlation Coefficients

Correlation in the cross-section can take very diverse shapes. The most common testing procedures are based on considering the population of all possible pairwise correlations between pairs of distinct individual units, estimating each one independently, by exploiting the time dimension of the data, and then calculating some synthetic measure or test statistic. The basic tool for assessing pairwise correlation between individual units n and m for a double-indexed vector z_{nm} is the product-moment correlation coefficient, defined as

$$\hat{\rho}_{nm} = \frac{\sum_{t=1}^{T} \hat{z}_{nt}\hat{z}_{mt}}{\left(\sum_{t=1}^{T} \hat{z}_{nt}^2\right)^{1/2} \left(\sum_{t=1}^{T} \hat{z}_{mt}^2\right)^{1/2}}$$

A descriptive assessment of the degree of cross-sectional correlation in the given sample can then be based on the average of individual correlation coefficients: $\hat{\rho} = 1/(N(N-1)) \sum_{n=1}^{N} \sum_{m=1}^{n-1} \hat{\rho}_{nm}$. If individual correlations are some positive and some negative, this solution has the problem that coefficients with different signs compensate, yielding a statistic that underestimates the true level of dependence in the data. Therefore, another common procedure is to average the absolute values of individual coefficients: $\hat{\rho}_{\text{ABS}} = 1/(N(N-1)) \sum_{n=1}^{N} \sum_{m=1}^{n-1} |\hat{\rho}_{nm}|$.

4.4.2 CD-type Tests for Cross-sectional Dependence

A number of statistics for testing the null hypothesis of no cross-sectional dependence in model errors can be based on $\hat{\rho}$. The function `pcdtest` implements both the calculation of $\hat{\rho}$ and $\hat{\rho}_{\text{ABS}}$, and a family of cross-sectional dependence tests that can be applied in different settings, ranging from those where T grows large with N fixed to "short" panels with a big N dimension and a few time periods. All are based on (transformations of) the product-moment correlation coefficient of a model's residuals, defined as above. The Breusch-Pagan LM test, based on the squares of ρ_{nm}, is valid for $T \to \infty$ with N fixed:

$$\text{LM}_{\rho} = \sum_{n=1}^{N-1} \sum_{m=n+1}^{N} T_{nm} \hat{\rho}_{nm}^2$$

where in the case of an unbalanced panel only pairwise complete observations are considered, and $T_{nm} = \min(T_n, T_m)$ with T_n being the number of observations for individual n; else, if the panel is balanced, $T_{nm} = T$ for each n, m. The test is distributed as $\chi^2_{N(N-1)/2}$. It is inappropriate whenever the N dimension is "large." A scaled version, applicable also if $T \to \infty$ and *then* $N \to \infty$ (as in some pooled time series contexts), is defined as:

$$\text{SCLM} = \sqrt{\frac{1}{N(N-1)}} \left(\sum_{n=1}^{N-1} \sum_{m=n+1}^{N} \sqrt{T_{nm}} \hat{\rho}_{nm}^2 \right)$$

and distributed as a standard normal.

Pesaran's (2004) CD test

$$\text{CD} = \sqrt{\frac{2}{N(N-1)}} \left(\sum_{n=1}^{N-1} \sum_{m=n+1}^{N} \sqrt{T_{nm}} \hat{\rho}_{nm} \right)$$

based on $\hat{\rho}_{nm}$ without squaring (also distributed as a standard Normal) is appropriate both for N- and T-asymptotics. It has good properties in samples of any practically relevant size and is robust to a variety of settings. The only big drawback is that the test loses power against the alternative of cross-sectional dependence if the average correlation is zero, even if individual coefficients are non-zero. Such a situation is not uncommon and can arise for example in the presence of an unobserved factor structure with factor loadings averaging zero, that is, where some units react positively to common shocks, others negatively. Another case where the test will lose power is if the data are cross-sectionally demeaned, or when the model contains time-specific dummies (see Sarafidis and Wansbeek, 2012, p. 27). In these instances, the absolute correlation coefficient $\hat{\rho}_{\text{ABS}}$ is likely to turn out much bigger than $\hat{\rho}$.

Example 4.10 tests for cross-sectional dependence – `RDSpillovers` data set
Eberhardt et al. (2013) consider the returns of own research and development (R&D) in the production function of European firms. They account for common factors and spillover effects; they find evidence that when controlling for such features, the effect of own R&D is not

significant any more and conclude that the value of R&D derives from a complex mix of own sources and spillovers from other firms. They estimate various specifications of a standard production function (where output is a function of labor and capital) augmented with R&D expenditure.

```
data("RDSpillovers", package = "pder")
fm.rds <- lny ~ lnl + lnk + lnrd
```

Pairwise-correlations-based tests are originally meant to use the residuals of separate estimation of one time-series regression for each cross-sectional unit. In the original example of Eberhardt et al. (2013), this is done in the first of the heterogeneous static specifications of Table 7, the mean groups model. This is also the default behavior of `pcdtest`.[3] The default version of the `test` is `'cd'`, which is appropriate in a large panel setting like this.

```
pcdtest(fm.rds, RDSpillovers)

Pesaran CD test for cross-sectional dependence in
panels

data:  lny ~ lnl + lnk + lnrd
z = 29, p-value <2e-16
alternative hypothesis: cross-sectional dependence
```

The residuals from separate time series regressions show strong evidence of error cross-sectional dependence.

If a different model specification (*within*, RE, ...) is assumed consistent, one can resort to its residuals for testing[4] by specifying the relevant `model` type. The main argument of this function may be either `plm` or a `formula` and a `data.frame`; in the second case, unless `model` is set to `NULL`, all usual parameters relative to the estimation of a `plm` model may be passed on. The test is compatible with any consistent `plm` model for the data at hand, with any specification of `effect`; e.g., specifying `effect = 'time'` or `effect = 'twoways'` allows to test for residual cross-sectional dependence after the introduction of time fixed effects to account for common shocks. Let us consider the static two-way fixed effects specification in Eberhardt et al. (2013, Table 5):

```
rds.2fe <- plm(fm.rds, RDSpillovers, model = "within", effect = "twoways")
pcdtest(rds.2fe)

Pesaran CD test for cross-sectional dependence in
panels

data:  lny ~ lnl + lnk + lnrd
z = -1.5, p-value = 0.1
alternative hypothesis: cross-sectional dependence
```

3 If the time dimension is insufficient and `model=NULL`, the function defaults to estimation of a *within* model and issues a warning.

4 This is also the only solution when the time dimension's length is insufficient for estimating the heterogeneous model.

As observed, the test loses its power if time fixed effects are included in the model specification. Now the test does not reject the hypothesis of no cross-sectional correlation. One can get an idea of what is happening by comparing $\hat{\rho}$ and $\hat{\rho}_{\mathbf{ABS}}$:

```
cbind("rho"   = pcdtest(rds.2fe, test = "rho")$statistic,
      "|rho|"= pcdtest(rds.2fe, test = "absrho")$statistic)
          rho   |rho|
rho -0.004879 0.5021
```

whence it can be seen how substantial cross-sectional dependence is present, but the addition of time effects has centered the mean of correlation coefficients on zero so that positive and negative $\hat{\rho}_{nm}$s compensate.

4.4.3 Testing Cross-sectional Dependence in a pseries

Next to testing for cross-sectional correlation in model residuals, tests in the CD family can be employed in preliminary statistical assessments as well, in order to determine whether the dependent and explanatory variables show any correlation to begin with. To this end, the `pcdtest` function has a `pseries` method, meaning that it can be fed a `pseries` object as well. One can either calculate the descriptive statistics $\hat{\rho}$ and $\hat{\rho}_{\mathbf{ABS}}$ or resort to a formal test.

Example 4.11 cross-sectional dependence test for a pseries – `HousePricesUS` data set

Holly et al. (2010) analyze changes in real house prices in 49 US states between 1975 and 2003 to assess to which extent they are driven by fundamentals like disposable per capita income, net borrowing costs, and population growth (see also the full replication in Millo, 2015). The empirical analysis proceeds from the initial assessment of spatial dependence of the variables; here we reproduce the cross-sectional correlation assessment for the dependent variable, the house price index (1980=100), taken in first differences of logs.

```
data("HousePricesUS", package = "pder")
php <- pdata.frame(HousePricesUS)
```

```
cbind("rho"    = pcdtest(diff(log(php$price)), test = "rho")$statistic,
      "|rho|" = pcdtest(diff(log(php$price)), test = "absrho")$statistic)
       rho   |rho|
rho 0.3942 0.4247
```

The overall averages of $\hat{\rho}$ and $\hat{\rho}_{\mathbf{ABS}}$ are quite large in magnitude and very close to each other, indicating substantial positive correlation.

To investigate whether this behavior is geographically uniform or not, one can drill down to the regional level. *within* and *between* regions correlation tables can be constructed by means of the `pcdtest` function, setting `test = 'rho'` for the average correlation coefficient. A function `cortab` is provided that automates this procedure, constructing suitable W matrices from the provided grouping index and calculating $\hat{\rho}$ (default) or $\hat{\rho}_{\mathbf{ABS}}$ (if `test = 'absrho'`) for each region (main diagonal) and each pair of regions (off-diagonal).

```
regions.names <- c("New Engl", "Mideast", "Southeast", "Great Lks",
                   "Plains", "Southwest", "Rocky Mnt", "Far West")
corr.table.hp <- cortab(diff(log(php$price)), grouping = php$region,
                        groupnames = regions.names)
colnames(corr.table.hp) <- substr(rownames(corr.table.hp), 1, 5)
round(corr.table.hp, 2)
          New E Midea South Great Plain South Rocky Far W
New Engl   0.80    NA    NA    NA    NA    NA    NA    NA
Mideast    0.68  0.66    NA    NA    NA    NA    NA    NA
Southeast  0.40  0.35  0.81    NA    NA    NA    NA    NA
Great Lks  0.27  0.20  0.62  0.61    NA    NA    NA    NA
Plains     0.40  0.32  0.57  0.53  0.52    NA    NA    NA
Southwest  0.07 -0.05  0.28  0.39  0.35  0.52    NA    NA
Rocky Mnt -0.03 -0.11  0.52  0.53  0.40  0.57  0.70    NA
Far West   0.13  0.17  0.52  0.42  0.29  0.31  0.46  0.57
```

The preliminary spatial dependence analysis highlights the correlation between neighboring regions and also some cases of correlation with distant ones, as is the case for California and some more developed states on the East Coast. According to the authors, this is evidence of factor-related dependence: common shocks to technology stimulate growth in the most advanced states irrespective of geographic proximity.

The significance of cross-sectional correlation in a `pseries` can also be assessed through a formal test, exactly as done above for model residuals. Given that, beside stationarity (which in our example was ensured by first differencing the data), the properties of the CD test rest on the hypothesis of no serial correlation, we follow Pesaran (2004)'s suggestion to remove any by specifying a univariate AR(2) model of the variable of interest and proceed testing the residuals of the latter for cross-sectional dependence. This is made easy by the lagging functionalities of **plm**. In the following, we test cross-sectional correlation in log house prices drawing on the residuals of an AR(2) model in order to control for any persistence in the data:

```
pcdtest(diff(log(price)) ~ diff(lag(log(price))) + diff(lag(log(price), 2)),
        data = php)

Pesaran CD test for cross-sectional dependence in
panels

data:  diff(log(price)) ~ diff(lag(log(price))) + diff(lag(log(price),      2))
z = 59, p-value <2e-16
alternative hypothesis: cross-sectional dependence
```

The test strongly rejects the null hypothesis, confirming substantial cross-sectional comovement in house prices.

5

Robust Inference and Estimation for Non-spherical Errors

5.1 Robust Inference

In this chapter we focus on relaxing the hypothesis of independence and homoscedasticity of the remainder errors. Independent and identically distributed (i.i.d.) errors can seldom be taken for granted in the mostly non-experimental contexts of econometrics. In the so-called *robust approach* to model diagnostics, one relaxes the hypothesis of homoscedastic and independent errors from the beginning, and consequently uses an appropriate estimator for the parameters' covariance matrix, instead of testing for departures from sphericity after estimation, as is customary in the classical approach.

In panel data, error correlation often descends from *clustering* issues: the group (firm, individual, country) and the time dimension define natural clusters; observations sharing a common individual unit, or time period, are likely to share common characters, violating the independence assumption and potentially biasing inference. In particular, variance estimates derived under the random sampling assumption are typically biased downward, possibly leading to false significance of model parameters. Although clustering can often be an issue in cross-sectional data too, especially when employing data at different levels of aggregation (Moulton, 1986, 1990), it is such an obvious feature in panels that a number of robust covariance estimators have been devised for the most common situations: within-individual and/or -time period correlation, the former of either time-constant or time-decaying type, and cross correlation between different individuals over time.

Next to the panel-specific implementation of the well-known heteroscedasticity-consistent covariance, there are a number of other robust covariance estimators specifically devised for panel data. We will now review the general idea of sandwich estimation, its application in a panel setting, and lastly the best known covariance estimators for the most common cases of nonsphericity in the errors and their implementations in **plm**.

5.1.1 Robust Covariance Estimators

Consider a linear model $y = Z\gamma + \epsilon$ and the OLS estimator $\hat{\gamma}_{\text{OLS}} = (Z^\top Z)^{-1} Z^\top y$. If the error terms ϵ are independent and identically distributed, then the estimated covariance matrix of estimators takes the familiar textbook form: $\hat{V}(\hat{\gamma}) = \hat{\sigma}^2 (Z^\top Z)^{-1}$, where $\hat{\sigma}^2$ is an estimate of the error variance. This is the classical case, also known as *spherical errors*, and the relative formulation of $\hat{V}(\hat{\gamma}_{\text{OLS}})$ is often referred to as "OLS covariance".

Panel Data Econometrics with R, First Edition. Yves Croissant and Giovanni Millo.

Companion website: www.wiley.com/go/croissant/data-econometrics-with-R

Let us consider robust estimation in the context of the simple linear model outlined above. The problem at hand is to estimate the covariance matrix of the **OLS** estimator relaxing the assumptions of serial correlation and/or homoscedasticity without imposing any particular structure to the errors' variance or interdependence. The **OLS** parameters' covariance matrix with a general error covariance Ω is:

$$V(\hat{\gamma}) = (Z^\top Z)^{-1}(Z^\top[\sigma^2\Omega]Z)(Z^\top Z)^{-1}$$

According to the seminal work of White (1980), in order to consistently estimate $V(\hat{\gamma})$, it is not necessary to estimate all the $N(N+1)/2$ unknown elements in the Ω matrix but only the $K(K+1)/2$ ones in

$$\sum_{n=1}^{N}\sum_{m=1}^{N}\sigma_{nm}z_n z_m^\top$$

which may be called the *meat* of the sandwich, the two $(Z^\top Z)^{-1}$ being the *bread*. All that is required are *pointwise consistent* estimates of the errors, which is satisfied by consistency of the estimator for γ (see Greene, 2003). In the heteroscedasticity case, correlation between different observations is ruled out, and the *meat* reduces to

$$S_0 = \sum_{n=1}^{N}\sigma_n^2 z_n z_n^\top$$

where the N unknown σ_n^2s can be substituted by $\hat{e}_n^2$ (see White, 1980). In the serial correlation case, the natural estimation counterpart would be $\sum_{n=1}^{N}\sum_{m=1}^{N}\hat{e}_n\hat{e}_m z_n z_m^\top$ but this structure proves too general to achieve convergence. Newey and West (1987) devise a heteroscedasticity and-autocorrelation consistent estimator that works based on the assumption of correlation dying out as the distance between observations increases. The Newey-West **HAC** estimator for the *meat* takes that of White and adds a sum of covariances between the different residuals, smoothed out by a kernel function giving weights decreasing with distance:

$$S_0 + \sum_{n=1}^{N}\sum_{m=1}^{N}\omega_l\hat{e}_t\hat{e}_{t-l}(z_t z_{t-l}^\top + z_{t-l}z_t^\top)$$

with ω_l the weight from the kernel smoother. For the latter, Newey and West (1987) chose the well-known Bartlett kernel function: $\omega_l = 1 - \frac{l}{L+1}$. The lag l is usually truncated well below sample size: one popular rule of thumb is $L = N^{1/4}$ (see Greene, 2003; Driscoll and Kraay, 1998).

In the following we will consider the extensions of this framework for a panel data setting where, thanks to added dimensionality, various combinations of the two above structures will turn out to be able to accommodate very general types of dependence.

5.1.1.1 Cluster-robust Estimation in a Panel Setting

Clustering estimators extend the sandwich principle to panel data. Besides heteroscedasticity, the added dimensionality allows to obtain robustness against totally unrestricted time-wise or cross-sectional correlation, provided this is along the "smaller" dimension. In the case of "large-N" (*wide*) panels, the big cross-sectional dimension allows robustness against serial correlation (Arellano, 1987); in "large-T" (*long*) panels, on the converse, robustness to cross-sectional correlation can be attained drawing on the large number of time periods observed. As a general rule, the estimator is asymptotic in the number of clusters.

Imposing cross-sectional (serial) independence in fact restricts all covariances between observations belonging to different individuals (time periods) to zero, yielding an error covariance matrix that is block-diagonal, with blocks Σ_n of the form:

$$\Sigma_n = \begin{bmatrix} \sigma^2_{n1} & \sigma_{n1,n2} & \cdots & \cdots & \sigma_{n1,nT} \\ \sigma_{n2,n1} & \sigma^2_{n2} & & & \vdots \\ \vdots & & \ddots & & \vdots \\ \vdots & & & \sigma^2_{nT-1} & \sigma_{nT-1,nT} \\ \sigma_{nT,n1} & \cdots & \cdots & \sigma_{nT,nT-1} & \sigma^2_{nT} \end{bmatrix} \tag{5.1}$$

and the consistency relies on the cross-sectional dimension being "large enough" with respect to the number of free covariance parameters in the diagonal blocks. The other case is symmetric.

White's heteroscedasticity-consistent covariance matrix has been extended to clustered data by Liang and Zeger (1986) and to econometric panel data by Arellano (1987). Observations can be clustered by the individual index, which is the most popular use of this estimator and is appropriate in *large, short* panels because it is based on N-asymptotics, or by the time index, which is based on T-asymptotics and therefore appropriate for *long* panels. In the first case, the covariance estimator is robust against cross-sectional heteroscedasticity and also against serial correlation of arbitrary form; in the second case, symmetrically, against time-wise heteroscedasticity and cross-sectional correlation. Arellano's original estimator, an instance of the first case, has the form:

$$V_{\mathbf{CX}} = (Z^\top Z)^{-1} \sum_{n=1}^{N} Z_n^\top \epsilon_n \epsilon_n^\top Z_n (Z^\top Z)^{-1} \tag{5.2}$$

It is of course still feasible to rule out serial correlation and compute an estimator that is robust to heteroscedasticity only, based on the following error structure:

$$\Sigma_n = \begin{bmatrix} \sigma^2_{n1} & \cdots & \cdots & 0 \\ 0 & \sigma^2_{n2} & & \vdots \\ \vdots & & \ddots & 0 \\ 0 & \cdots & \cdots & \sigma^2_{nT} \end{bmatrix} \tag{5.3}$$

in which case the original White estimator applies:

$$V_{\mathbf{WH}} = (Z^\top Z)^{-1} \sum_{n=1}^{N} \sum_{t=1}^{T} \hat{\epsilon}^2_{nt} z_{nt} z_{nt}^\top (Z^\top Z)^{-1} \tag{5.4}$$

The case of clustering by time period is symmetric to that along the other dimension: data are assumed to be serially independent and allowed to have arbitrary heteroscedasticity and an unrestricted cross-sectional dependence structure.

$$V_{\mathbf{CT}} = (Z^\top Z)^{-1} \sum_{t=1}^{T} Z_t^\top \hat{\epsilon}_t \hat{\epsilon}_t^\top Z_t (Z^\top Z)^{-1} \tag{5.5}$$

Example 5.1 clustered standard errors for pooled models – `Produc` data set
Munnell (1990) analyzes the impact of public infrastructure on economic activity by drawing on a sample of 48 US states (all continental states minus the District of Columbia) over 17 years, 1970–1986. She specifies a Cobb-Douglas production function that relates the gross social product (`gsp`) of a given state to the input of public capital (`pcap`), private capital (`pc`) and labor (`emp`); she also includes the state unemployment rate (`unemp`) to capture business cycle effects:

```
library("plm")
data("Produc", package = "plm")
fm <- log(gsp) ~ log(pcap) + log(pc) + log(emp) + unemp
```

The function `coeftest` from package **lmtest** produces a compact coefficients table allowing for a flexible choice of the covariance matrix. We calculate a heteroscedasticity-robust diagnostic table for two statistically equivalent models. First, pooled OLS by `lm`:

```
lmmod <- lm(fm, Produc)
library("lmtest")
library("sandwich")
coeftest(lmmod, vcov = vcovHC)

t test of coefficients:

             Estimate Std. Error t value Pr(>|t|)
(Intercept)  1.64330    0.07161   22.95  < 2e-16 ***
log(pcap)    0.15501    0.01870    8.29  4.7e-16 ***
log(pc)      0.30919    0.01263   24.48  < 2e-16 ***
log(emp)     0.59393    0.01979   30.01  < 2e-16 ***
unemp       -0.00673    0.00135   -4.99  7.5e-07 ***
---
Signif. codes:
0 '***' 0.001 '**' 0.01 '*' 0.05 '.' 0.1 ' ' 1
```

Next, we compute pooled OLS by `plm`. The `coeftest` function complies with `plm` objects, so the same syntax as above can be employed. In turn, the `summary.plm` method is itself compliant with providing a custom covariance (a note about using a nonstandard covariance will be issued):

```
plmmod <- plm(fm, Produc, model = "pooling")
summary(plmmod, vcov = vcovHC)
Pooling Model

Note: Coefficient variance-covariance matrix supplied: vcovHC

Call:
plm(formula = fm, data = Produc, model = "pooling")

Balanced Panel: n = 48, T = 17, N = 816

Residuals:
     Min.   1st Qu.    Median   3rd Qu.      Max.
-0.231762 -0.061037 -0.000102  0.050852  0.351113

Coefficients:
             Estimate Std. Error t-value Pr(>|t|)
(Intercept)  1.64330    0.24418    6.73  3.2e-11 ***
log(pcap)    0.15501    0.06012    2.58     0.01 *
log(pc)      0.30919    0.04623    6.69  4.2e-11 ***
```

```
log(emp)        0.59393     0.06861     8.66  < 2e-16 ***
unemp          -0.00673     0.00309    -2.18     0.03 *
---
Signif. codes:
0 '***' 0.001 '**' 0.01 '*' 0.05 '.' 0.1 ' ' 1

Total Sum of Squares:    850
Residual Sum of Squares: 6.29
R-Squared:        0.993
Adj. R-Squared: 0.993
F-statistic: 2778.06 on 4 and 47 DF, p-value: <2e-16
```

Coefficients are obviously the same, but the estimated standard errors will turn out different. In particular, the standard error of the coefficient on `pcap` is much larger, and while still significant at the 5% level, it is not any more at the 1% level. This is because the classes of the model objects to be tested are different, and so are the default settings of the `vcovHC.lm` and `vcovHC.plm` methods. Only if one overrides the defaults, here, specifying the `method` as `'white1'` and the small sample correction as `'HC3'`, the `lm` results will be replicated. Therefore, thanks to object orientation, if applying the generic robust method `vcovHC` to a `panelmodel` object, one will get a result that is likely to be "sensible" for the most common applications.

Clustering in Non-Panels Clustering can occur in non-panel settings too. Whenever a grouping index of some sort is provided and there is reason to believe that errors are dependent within groups defined by that index, the clustered standard errors can be employed to account for heteroscedasticity across groups and for within group correlation of any kind, not limited to proper serial correlation in time. One example is when a regression is augmented with variables at a higher level of aggregation.

The seminal example is in Moulton (1986, 1990): if some regressors are observed at group level, as is the case, e.g., when adding local GDP to individual data drawn from different geographical units, then standard errors have to be adjusted for intra-group correlation.

Froot (1989), in the context of financial data, discusses sampling firms from different industries, assumed mutually independent. In his application, clustering is employed to account for within-industry dependence, while it would be meaningless across the "other" dimension.

Any dataset mixing different levels of detail is prone to this issue. In such cases, panel data methods can seamlessly be employed on cross-sectional datasets by specifying the relevant grouping variable as the first element of the `index`. The second one will obviously be left blank as there would be no meaningful second dimension.

Example 5.2 Clustered standard errors for non-panel data – `Hedonic` data set
Harrison and Rubinfeld (1978) consider the median values of owner-occupied homes in a cross section of 506 census tracts from 92 towns in the Boston area. Values are explained by a combination of tract- and town-level variables. Crime rate (`crim`), pollution (`nox`), average number of rooms (`rm`) and age, distance to employment centers (`dis`), and proportion of blacks in the population are observed at tract level. Other variables such as proportion of industrial dwellings (`indus`), distance to radial highways (`rad`), property tax rate, and pupil-to-teacher ratio in local schools (`ptratio`) are observed at town level, thus leading to the Moulton problem. The town identifier for each tract (`townid`) allows to account for clustering within each town, which may comprise from 1 to 30 tracts. We estimate OLS with HC SEs by lm:

```
data("Hedonic", package = "plm")
hfm <- mv ~ crim + zn + indus + chas + nox + rm + age + dis +
    rad + tax + ptratio + blacks + lstat
```

```
hlmmod <- lm(hfm, Hedonic)
coeftest(hlmmod, vcov = vcovHC)

t test of coefficients:

             Estimate Std. Error t value Pr(>|t|)
(Intercept)  9.76e+00   1.74e-01   56.21  < 2e-16 ***
crim        -1.19e-02   2.85e-03   -4.16  3.8e-05 ***
zn           8.03e-05   3.89e-04    0.21  0.83670
indus        2.41e-04   1.84e-03    0.13  0.89589
chasyes      9.14e-02   3.71e-02    2.46  0.01413 *
nox         -6.38e-03   1.26e-03   -5.08  5.4e-07 ***
rm           6.33e-03   2.11e-03    3.00  0.00282 **
age          8.98e-05   6.07e-04    0.15  0.88252
dis         -1.91e-01   4.08e-02   -4.69  3.5e-06 ***
rad          9.57e-02   2.03e-02    4.71  3.2e-06 ***
tax         -4.20e-04   1.17e-04   -3.59  0.00036 ***
ptratio     -3.11e-02   4.17e-03   -7.47  3.7e-13 ***
blacks       3.64e-01   1.54e-01    2.36  0.01884 *
lstat       -3.71e-01   3.94e-02   -9.43  < 2e-16 ***
---
Signif. codes:
0 '***' 0.001 '**' 0.01 '*' 0.05 '.' 0.1 ' ' 1
```

and pooled **OLS** by `plm`; then we compare White and clustered standard errors:.

```
hplmmod <- plm(hfm, Hedonic, model = "pooling", index = "townid")

sign.tab <- cbind(coef(hlmmod), coeftest(hlmmod, vcov = vcovHC)[,4],
                  coeftest(hplmmod, vcov = vcovHC)[, 4])
dimnames(sign.tab)[[2]] <- c("Coefficient", "p-values, HC", "p-val., cluster")
round(sign.tab, 3)
            Coefficient p-values, HC p-val., cluster
(Intercept)       9.756        0.000           0.000
crim             -0.012        0.000           0.000
zn                0.000        0.837           0.882
indus             0.000        0.896           0.933
chasyes           0.091        0.014           0.064
nox              -0.006        0.000           0.003
rm                0.006        0.003           0.090
age               0.000        0.883           0.914
dis              -0.191        0.000           0.004
rad               0.096        0.000           0.000
tax               0.000        0.000           0.005
ptratio          -0.031        0.000           0.000
blacks            0.364        0.019           0.235
lstat            -0.371        0.000           0.000
```

Proximity to the Charles River, average number of rooms, and the proportion of blacks in the population are not significant any more after clustering by town.

5.1.1.2 Double Clustering

Double clustering methods have originated in the financial literature (Petersen, 2009; Cameron et al., 2011; Thompson, 2011) and are motivated by the need to account for *persistent shocks* (another name for individual, time-invariant error components) and at the same time for cross-sectional or spatial correlation. The former feature, persistent shocks, is usually dealt with in the econometric literature by parametric estimation of random effects models; the latter through spatial panels, where again it is estimated parametrically imposing a structure to the dependence, or common factor models. As Cameron et al. (2011) observe, though, double clustering, as all robustified inference of this kind, relies on much weaker assumptions as regards the data-generating process than parametric modeling of dependence does. In fact, this estimator combining both individual and time clustering relies on a combination of the asymptotics of each: the minimum number of clusters along the two dimensions must go to infinity (which will be especially appropriate for data-rich financial applications, less so in the smaller samples that are frequently encountered in economics). Apart from this, any dependence structure is allowed within each group *or* within each time period, while cross-serial correlations between observations belonging to different groups *and* time periods are ruled out.

Cameron et al. (2011) have shown how the double-clustered estimator is simply calculated by summing up the group-clustering and the time-clustering ones, then subtracting the standard White estimator in order to avoid double counting the error variances along the diagonal:

$$V_{\mathbf{CXT}} = V_{\mathbf{CX}} + V_{\mathbf{CT}} - V_{\mathbf{WH}} \tag{5.6}$$

In order to control for the effect of common shocks, Thompson (2011) proposes to add to the sum of covariances one more term, related to the covariances between observations from any group at different points in time. Given a maximum lag L, this will be the sum over $l = 1, \dots L$ of the following generic term:

$$V_{\mathbf{CT},l} = \sum_{t=1}^{T} Z_t^{\top} \hat{\epsilon}_t \hat{\epsilon}_{t-l}^{\top} Z_{t-l} \tag{5.7}$$

representing the covariance between pairs of observations from any group distanced l periods in time. As the correlation between observations belonging to the *same* group at different points in time has already been captured by the group-clustering term, to avoid double counting one must subtract the within-groups part:

$$V_{\mathbf{WH},l} = \sum_{t=1}^{T} \sum_{n=1}^{N} [z_{nt} \hat{\epsilon}_{nt} \hat{\epsilon}_{n,t-l}^{\top} z_{n,t-l}^{\top}] \tag{5.8}$$

for each l. The resulting estimator

$$V_{\mathbf{CXT},L} = V_{\mathbf{CX}} + V_{\mathbf{CT}} - V_{\mathbf{WH}} + \sum_{l=1}^{L} [V_{\mathbf{CT},l} + V_{\mathbf{CT},l}^{\top}] - \sum_{l=1}^{L} [V_{\mathbf{WH},l} + V_{\mathbf{WH},l}^{\top}] \tag{5.9}$$

is robust to cross-sectional and time-wise correlation inside, respectively, time periods and groups *and* to the cross-serial correlation between observations belonging to different groups, up to the L-th lag.

5.1.1.3 Panel Newey-west and scc

As mentioned above, in a time series context Newey and West (1987) have proposed an estimator that is robust to serial correlation as well as to heteroscedasticity. This estimator, based on the hypothesis of the serial correlation dying out "quickly enough," takes into account the covariance between units by weighting it through a kernel-smoothing function giving less weight as they get more distant and adding it to the standard White estimator.

A panel version of the original Newey-West estimator can be obtained as:

$$\begin{aligned} \mathbf{V}_{\mathbf{NW},L} &= \mathbf{V}_{\mathbf{WH}} + \sum_{l=1}^{L} \omega_l \left[\sum_{t=1}^{T} \sum_{n=1}^{N} [z_{nt} \hat{\epsilon}_{nt} \hat{\epsilon}^{\top}_{n,t-l} z^{\top}_{n,t-l} \right. \\ &\quad \left. + \sum_{t=1}^{T} \left[\sum_{n=1}^{N} [z_{nt} \hat{\epsilon}_{nt} \hat{\epsilon}^{\top}_{n,t-l} z^{\top}_{n,t-l}]^{\top} \right] \right. \\ &= \mathbf{V}_{\mathbf{WH}} + \sum_{l=1}^{L} \omega_l [\mathbf{V}_{\mathbf{WH},l} + \mathbf{V}^{\top}_{\mathbf{W}l}] \end{aligned} \tag{5.10}$$

As can readily be seen, the Newey-West non-parametric estimator closely resembles the double clustering plus lags, the difference being that instead of adding a (possibly truncated) sum of unweighted lag terms, the latter downweighs the correlation between "distant" terms through a kernel-smoothing function.

Driscoll and Kraay (1998) have adapted the Newey-West estimator to a panel time series context where not only serial correlation between residuals from the same individual in different time periods is taken into account but also *cross-serial* correlation between different individuals in different times and, within the same period, cross-sectional correlation (see also Arellano, 2003).

The Driscoll and Kraay estimator, labeled **scc** (as in "spatial correlation consistent"), is defined as the time-clustering version of Arellano plus a sum of lagged covariance terms, weighted by a distance-decreasing kernel function ω_l:

$$\begin{aligned} \mathbf{V}_{\mathbf{SCC},L} &= \mathbf{V}_{\mathbf{CT}} + \sum_{l=1}^{L} \omega_l \left[\sum_{t=1}^{T} Z_t^{\top} \hat{\epsilon}_t \hat{\epsilon}^{\top}_{t-l} Z_{t-l} + \sum_{t=1}^{T} [Z_t^{\top} \hat{\epsilon}_t \hat{\epsilon}^{\top}_{t-l} Z_{t-l}]^{\top} \right] \\ &= \mathbf{V}_{\mathbf{CT}} + \sum_{l=1}^{L} \omega_l [\mathbf{V}_{\mathbf{CT},l} + \mathbf{V}^{\top}_{\mathbf{CT},l}] \end{aligned} \tag{5.11}$$

The "**scc**" covariance estimator requires the data to be a mixing sequence, i.e., roughly speaking, to have serial and cross-serial dependence dying out quickly enough with the T dimension, which is therefore supposed to be fairly large: Driscoll and Kraay (1998), based on Monte Carlo simulation, put the practical minimum at $T > 20 - 25$; the N dimension is irrelevant in this respect and is allowed to grow at any rate relative to T.

As is apparent from Equation 5.1.1.3, if the maximum lag order is set to 0 (no serial or cross-serial dependence is allowed) the **scc** estimator becomes the cross-section version (time-clustering) of the Arellano estimator $\mathbf{V}_{\mathbf{CT}}$. On the other hand, if the cross-serial terms are all unweighted (i.e., if $\omega_l = 1 \forall l$), then $\mathbf{V}_{\mathbf{SCC},L|w=1} = \mathbf{V}_{\mathbf{CT},L}$.

A Comprehensive Definition Let us now look systematically at the similarities between the above formulas, embedding them into an encompassing one (see Millo, 2017b). A comprehensive formulation can be written in terms of White's heteroscedasticity-consistent covariance

Table 5.1 Covariance structures as combinations of the basic building blocks.

double-clustering	$V_{CXT} = V_{CX} + V_{CT} - V_{WH}$
time-clustering + shocks	$V_{CT,L} = V_{CT} + \sum_{l=1}^{L}[V_{CT,l} + V_{CT,l}^{T}]$
panel Newey-West	$V_{NW,L} = V_{WH} + \sum_{l=1}^{L}\omega_l[V_{WH,l} + V_{WH,l}^{T}]$
Driscoll and Kraay's **scc**	$V_{SCC,L} = V_{CT} + \sum_{l=1}^{L}\omega_l[V_{CT,l} + V_{CT,l}^{T}]$
double-clustering + shocks	$V_{CXT,L} = V_{CT} + \sum_{l=1}^{L}[V_{CT,l} + V_{CT,l}^{T}] + V_{CX}$
	$- V_{WH} - \sum_{l=1}^{L}[V_{WH,l} + V_{WH,l}^{T}]$
	$= V_{CT,L} + V_{CX} - V_{NW,L\|w=1}$

matrix V_{WH}, the group-clustering and time-clustering ones V_{CX} and V_{CT}, and an appropriate kernel-weighted sum of their lags:

$$V_{CXT,L|w} = V_{CT} + \sum_{l=1}^{L}\omega_l[V_{CT,l} + V_{CT,l}^{T}] + V_{CX} - V_{WH} - \sum_{l=1}^{L}\omega_l[V_{WH,l} + V_{WH,l}^{T}] \quad (5.12)$$

The different estimators are in turn particularizations of the above and can be expressed in terms of the same basic common components, as shown in Table 5.1. A function `vcovG` making either V_W, V_{CX}, or $V_{CT}$is provided at user level, mainly for educational purposes, and is used internally to construct all other estimators.[1]

Higher-level functions are provided to produce the double-clustering and kernel-smoothing estimators by (possibly weighted) sums of the former terms. The general tool in this respect, in turn based on `vcovG`, is `vcovSCC`, which computes weighted sums of $V_{.,l}$ according to a weighting function that is by default the Bartlett kernel. The default values will yield the Driscoll and Kraay estimator, $V_{SCC,L}$. As the **scc** estimator differs from the (one-way) time-shocks-robust version of the double-clustering a la Cameron et al. (2011) only by the distance-decaying weighting of the covariances between different periods so that $V_{CT,L} = V_{SCC,L|\omega=1}$, no weighting (equivalent to passing the constant 1 as the weighting function: `wj=1`) will produce the building blocks for double clustering, according to formula 5.9.

Convenient wrappers are provided as the tool of choice for the end user: `vcovNW` computes the panel Newey-West estimator $V_{NW,L}$; `vcovDC` the double-clustering one V_{CXT}.

Example 5.3 Newey-West and double-clustering estimators – `Produc` data set
Reconsidering the Munnell (1990) example, one might want to account for both the spatial correlation between states observed in the same time period and for the serial correlation within the same state and across different ones. To this end, one may supply the `vcovSCC` function to the `vcov` argument in `coeftest`:

```
coeftest(plmmod, vcov=vcovSCC)

t test of coefficients:

            Estimate Std. Error t value Pr(>|t|)
(Intercept)  1.64330    0.15035   10.93  < 2e-16 ***
```

1 `vcovG` can be used for calculating $V_{WH,l}$, $V_{CT,l}$, or $V_{CX,l}$ or, leaving the default lag at 0, to calculate V_{WH}, V_{CT}, or V_{CX}. It takes as arguments a clustering dimension (`cluster`), a function of the errors corresponding to $E(\hat{e})$ (`inner`), and a lag order. The `inner` argument can accept either one of two strings `'cluster'` or `'white'`, specifying respectively $E(\hat{e}) = \hat{e}\hat{e}^{T}$ and $E(\hat{e}) = diag(\hat{e}^{T}\hat{e})$, or a user-supplied function. For example, specifying `vcovG(plmmod, cluster = "group", inner = "cluster", l = 0)` is equivalent to set `vcovHC(plmmod)` and will produce the Arellano estimator.

```
log(pcap)       0.15501     0.03697    4.19  3.1e-05 ***
log(pc)         0.30919     0.00764   40.45  < 2e-16 ***
log(emp)        0.59393     0.03870   15.35  < 2e-16 ***
unemp          -0.00673     0.00254   -2.65   0.0082 **
---
Signif. codes:
0 '***' 0.001 '**' 0.01 '*' 0.05 '.' 0.1 ' ' 1
```

or possibly, if allowing for double clustering,

```
coeftest(plmmod, vcov=vcovDC)
```

(results omitted, see the next example). More complicated structures allowing for two-way clustering and error persistence in the sense of Thompson (2011) can be obtained by combination, as illustrated above. Below, the case of double clustering plus four periods of persistent (unweighted) shocks a la Thompson (2011) (notice that the weighting function `wj` has been defined as the constant 1 but must still be a function of two arguments):

```
myvcovDCS <- function(x, maxlag = NULL, ...) {
    w1 <- function(j, maxlag) 1
    VsccL.1 <- vcovSCC(x, maxlag = maxlag, wj = w1, ...)
    Vcx <- vcovHC(x, cluster = "group", method = "arellano", ...)
    VnwL.1 <- vcovSCC(x, maxlag = maxlag, inner = "white", wj = w1, ...)
    return(VsccL.1 + Vcx - VnwL.1)
}
coeftest(plmmod, vcov=function(x) myvcovDCS(x, maxlag = 4))

t test of coefficients:

             Estimate Std. Error t value Pr(>|t|)
(Intercept)   1.64330    0.27694    5.93  4.4e-09 ***
log(pcap)     0.15501    0.06612    2.34    0.019 *
log(pc)       0.30919    0.03265    9.47  < 2e-16 ***
log(emp)      0.59393    0.07244    8.20  9.5e-16 ***
unemp        -0.00673    0.00375   -1.80    0.073 .
---
Signif. codes:
0 '***' 0.001 '**' 0.01 '*' 0.05 '.' 0.1 ' ' 1
```

Example 5.4 computing an array of standard errors – `Produc` data set

In the following applied example, still considering the Munnell (1990) model, we take advantage of the capabilities of the R language for compactly presenting the complete array of standard error estimates for each estimator in Table 5.1 by defining a vector of covariance functions and then looping on it.[2]

Looping on a vector of functions is a useful consequence of R treating functions as a data type. For the sake of clarity, let us predefine some functions for calculating the different covariance

2 One must nevertheless keep in mind that the sample size and the number of clusters in either cross section or time might prove inadequate for some estimators, as reported in the reference papers (see in particular Thompson, 2011).

estimators with the appropriate parameters (leaving the maximum lag calculation at its default value of $L = T^{\frac{1}{4}}$):

```
Vw <- function(x) vcovHC(x, method = "white1")
Vcx <- function(x) vcovHC(x, cluster = "group", method = "arellano")
Vct <- function(x) vcovHC(x, cluster = "time", method = "arellano")
Vcxt <- function(x) Vcx(x) + Vct(x) - Vw(x)
Vct.L <- function(x) vcovSCC(x, wj = function(j, maxlag) 1)
Vnw.L <- function(x) vcovNW(x)
Vscc.L <- function(x) vcovSCC(x)
Vcxt.L <- function(x) Vct.L(x) + Vcx(x) - vcovNW(x, wj = function(j, maxlag) 1)
```

then build up a vector of functions on which to loop:

```
vcovs <- c(vcov, Vw, Vcx, Vct, Vcxt, Vct.L, Vnw.L, Vscc.L, Vcxt.L)
names(vcovs) <- c("OLS", "Vw", "Vcx", "Vct", "Vcxt", "Vct.L", "Vnw.L",
                  "Vscc.L", "Vcxt.L")
```

in order to calculate a comprehensive table of p-values from robust estimators. To this end we define a convenience function:

```
cfrtab <- function(mod, vcovs, ...) {
    cfrtab <- matrix(nrow = length(coef(mod)), ncol = 1 + length(vcovs))
    dimnames(cfrtab) <- list(names(coef(mod)),
                             c("Coefficient", paste("s.e.", names(vcovs))))
    cfrtab[,1] <- coef(mod)
    for(i in 1:length(vcovs)) {
        myvcov = vcovs[[i]]
        cfrtab[ , 1 + i] <- sqrt(diag(myvcov(mod)))
        }
    return(t(round(cfrtab, 4)))
}
```

The additive nature of the three basic components V_{WH}, V_{CX}, and V_{CT} allows the researcher to infer on the relative importance of each clustering dimension by looking at the contribution of each to the standard error estimate, so that if, e.g., $V_{CX} \ll V_{CT} \sim V_{CXT}$, then this is evidence of important cross-sectional correlation (Petersen, 2009).

```
cfrtab(plmmod, vcovs)
             (Intercept) log(pcap) log(pc) log(emp)   unemp
Coefficient       1.6433    0.1550  0.3092   0.5939 -0.0067
s.e. OLS          0.0576    0.0172  0.0103   0.0137  0.0014
s.e. Vw           0.0708    0.0185  0.0125   0.0195  0.0013
s.e. Vcx          0.2442    0.0601  0.0462   0.0686  0.0031
s.e. Vct          0.0944    0.0232  0.0063   0.0246  0.0018
s.e. Vcxt         0.2520    0.0617  0.0450   0.0702  0.0033
s.e. Vct.L        0.1875    0.0461  0.0079   0.0480  0.0031
s.e. Vnw.L        0.1144    0.0299  0.0206   0.0316  0.0020
s.e. Vscc.L       0.1503    0.0370  0.0076   0.0387  0.0025
s.e. Vcxt.L       0.2722    0.0657  0.0389   0.0736  0.0036
```

For this pooled OLS model, standard errors estimates assuming group clustering are consistently larger that the rest, including Newey-West and SCC, pointing at non-decaying serial error dependence.

5.1.2 Generic Sandwich Estimators and Panel Models

plm provides a comprehensive set of modular tools: lower-level components, conceptually corresponding to the statistical "objects" involved, (see Zeileis, 2006a,b), and a higher-level set of "wrapper functions" corresponding to standard parameter covariance estimators as they would be used in statistical packages, which work by combining the same, few lower-level components in multiple ways in the spirit of the *Lego principle* of Hothorn et al. (2006).

When estimating regression models, R creates a model object that, together with estimation results, carries on a wealth of useful information, including the original data. Robust testing in R is done retrieving the necessary elements from the model object, using them to calculate a robust covariance matrix for coefficient estimates and then feeding the latter to the actual test function, for example a t-test for significance or a Wald restriction test. This approach to diagnostic testing is more flexible than with standard econometric software packages, where diagnostics usually come with standard output. In our case, for example, one can obtain different estimates of the standard errors under various kinds of dependence without re-estimating the model and present them compactly.

Robust covariance estimators a la White or a la Newey and West for different kinds of regression models are available in package **sandwich** (Lumley and Zeileis, 2007) under form of appropriate methods for the generic functions `vcovHC` and `vcovHAC` (Zeileis, 2004, 2006a). These are designed for data sampled along one dimension; therefore, they cannot generally be used for panel data, yet they provide a uniform and flexible software approach, which has become standard in the R environment. The corresponding `plm` methods described in this chapter have therefore been designed to be sintactically compliant with them.

For example, a `vcovHC.plm` method for the generic `vcovHC` is available, allowing to apply sandwich estimators to panel models in a way that is natural for users of the **sandwich** package. In fact, despite the different structure "under the hood," the user will, e.g., specify a robust covariance for the diagnostics table of a panel model the same way she would for a linear or a generalized linear model, the object-orientation features of R taking care that the right statistical procedure be applied to the model object at hand. What will change, though, are the defaults: the `vcovHC.lm` method defaults to the original White estimator, while `vcovHC.plm` to clustering by groups, both the most obvious choices for the object at hand.

Next to the HC estimator of White (1980), all variants of the panel-specific estimators used in applied practice (Arellano, 1987; Newey and West, 1987; Driscoll and Kraay, 1998; Cameron et al., 2011) are provided; all can be applied to objects representing panel models of different kinds: FE, RE, FD, and, obviously, pooled OLS. The estimate of the parameters' covariance thus obtained can in turn be plugged into diagnostic testing functions, producing either significance tables or hypothesis tests. A function is a regular object type in R, hence compact comparisons of standard errors from different (statistical) methods can be produced by looping on covariance types, as shown in the examples.

Application to Models on Transformed Data

The application of the above estimators to pooled data is always warranted, subject to the relevant assumptions mentioned before. In some, but not all cases, these can also be applied to random or fixed effects panel models, or models estimated on first-differenced data. In all of these cases the estimator is computed as OLS on transformed (partially or totally demeaned, first differenced) data. In general, the same transformation used in estimation is employed. Sandwich estimators can then be computed by applying the usual formula to the transformed data and residuals: $\hat{\tilde{\epsilon}} = \tilde{y} - \tilde{Z}\hat{\gamma}$ (see Arellano (1987) and Wooldridge (2010, Eq. 10.59) for the fixed effects case, Wooldridge (2010, Ch.10) in general).

Under the fixed effects hypothesis, the OLS estimator is biased and FE is required for consistency of parameter estimates in the first place. Similarly, under the hypothesis of a unit root in the errors, first differencing the data is warranted in order to revert to a stationary error term. On the contrary, under the random effects hypothesis, OLS is still consistent, and asymptotically, using RE instead makes no difference. Yet for the sake of parameter covariance estimation, it may be advisable to eliminate time-invariant heterogeneity first, by using one of the above.

One compelling reason for combining a demeaning or a differencing estimator with robust standard errors may be to get rid of persistent individual effects before applying a more parsimonious and efficient kernel-based covariance estimator if cross-serial correlation is suspected or if the sample is simply not big enough to allow double clustering. In fact, as Petersen (2009) shows, the Newey-West- type estimators are biased if effects are persistent, because the kernel smoother unduly downweighs the covariance between faraway observations.

In the following we discuss when it is appropriate to apply clustering estimators to the residuals of demeaned or first-differenced models.

Fixed Effects

The fixed effects estimator requires particular caution. In fact, under the hypothesis of spherical errors in the original model, the time-demeaning of data induces a serial correlation $\text{cor}(\hat{\epsilon}_{nt}, \hat{\epsilon}_{nt-1}) = -1/(T-1)$ in the demeaned residuals (see Wooldridge, 2010, p. 310).

The White-Arellano estimator has originally been devised for this case. By way of symmetry, it can be used for time-clustered data with time fixed effects. The combination of group clustering with time fixed effects and the reverse is inappropriate because of the serial (cross-sectional) correlation induced by the time- (cross-sectional-) demeaning.

By analogy, the Newey-West-type estimators can be safely applied to models with individual fixed effects, while the time and two-way cases require caution. The best policy in both cases, if the degrees of freedom allow, is perhaps to explicitly add dummy variables to account for the fixed effects along the "short" dimension.

Random Effects

In the random effects case, as Wooldridge (2010) notes, the quasi-time demeaning procedure removes the random effects reducing the model on transformed data to a pooled regression, thus preserving the properties of the White-type estimators. By extension of this line of reasoning, all above estimators are applicable to the demeaned data of a random effects model, provided *the transformed errors* meet the relevant assumptions.

First-Differences

First-differencing, like fixed effects estimation, removes time-invariant effects. Roughly speaking, the choice between the two rests on the properties of the error term: if it is assumed to be well behaved in the original data, then FE is the most efficient estimator and is to be preferred; if on the contrary the original errors are believed to behave as a random walk, then first-differencing the data will yield stationary and uncorrelated errors and is therefore advisable (see Wooldridge, 2010, p. 317). Given this, FD estimation is nothing else than OLS on differenced data, and the usual clustering formula applies (see Wooldridge, 2010, p. 318 and Chapter 4 here). As in the RE case, the statistical properties of the different covariance estimators will depend on whether *the transformed errors* meet the relevant assumptions.

Example 5.5 random effects and robust covariances – `Produc` data set
Consider again the comprehensive table of estimators for the Munnell (1990) model in the previous example. The relative magnitude of standard errors under group clustering with respect to the others was hinting at error correlation in time. In the following, the previous table is replicated on a random effects specification:

```
replmmod <- plm(fm, Produc)
cfrtab(replmmod, vcovs)
              log(pcap) log(pc) log(emp)   unemp
Coefficient    -0.0261  0.2920   0.7682 -0.0053
s.e. OLS        0.0290  0.0251   0.0301  0.0010
s.e. Vw         0.0312  0.0305   0.0398  0.0011
s.e. Vcx        0.0603  0.0617   0.0817  0.0025
s.e. Vct        0.0454  0.0480   0.0627  0.0015
s.e. Vcxt       0.0688  0.0720   0.0949  0.0027
s.e. Vct.L      0.0640  0.0644   0.0941  0.0015
s.e. Vnw.L      0.0434  0.0417   0.0562  0.0015
s.e. Vscc.L     0.0575  0.0588   0.0828  0.0015
s.e. Vcxt.L     0.0717  0.0747   0.1054  0.0023
```

The cross-sectional dependence component becomes relatively more important when accounting for time persistence in the model through random individual (country) effects.

5.1.2.1 Panel Corrected Standard Errors

Unconditional covariance estimators are based on the assumption of no error correlation in time (cross-section) and of an unrestricted but invariant correlation structure inside every cross section (time period).[3] They are popular in contexts characterized by relatively small samples, with prevalence of the time dimension. The most common use is on pooled time series, where the assumption of no serial correlation can be accommodated, for example, by adding lagged values of the dependent variable.

Beck and Katz (1995), in the context of political science models with moderate time and cross-sectional dimensions, introduced the so-called panel corrected standard errors (PCSE), which, in the original time-clustering setting, are robust against cross-sectional heteroscedasticity and correlation. The "PCSE" covariance is based on the hypothesis that the covariance matrix of the errors in every group be the same: $\Omega = \Sigma_N \otimes I_T$, with

$$\Sigma_N = \begin{bmatrix} \sigma_1^2 & \sigma_{1,2} & \cdots & \sigma_{1,N-1} & \sigma_{1,N} \\ \sigma_{2,1} & \sigma_2^2 & \cdots & \sigma_{2,N-1} & \sigma_{2,N} \\ \vdots & \vdots & \ddots & \vdots & \vdots \\ \sigma_{N-1,1} & \sigma_{N-1,2} & \cdots & \sigma_{N-1}^2 & \sigma_{N-1,N} \\ \sigma_{N,1} & \sigma_{N,2} & \cdots & \sigma_{N,N-1} & \sigma_N^2 \end{bmatrix} \tag{5.13}$$

so that Σ_N can be estimated by:

$$\hat{\Sigma}_N = \frac{\sum_{t=1}^{T} \hat{\epsilon}_t \hat{\epsilon}_t^\top}{T}$$

from which $\hat{\Omega}$ can be constructed and inserted in the usual "sandwich" formula.

3 A further step in this direction is to use the unconditional estimate of the error covariance in a feasible GLS analysis: see the next section of this chapter.

Example 5.6 time fixed effects model – `agl` data set
Alvarez et al. (1991) estimate a model where economic performance in a panel of 16 countries over 15 years is related to political and labor organization variables: union strength (*central*) and the prevalence of a leftist cabinet (*leftc*). They control for trade openness of countries toward other OECD and for lagged growth, instrumented through an auxiliary regression. They originally use the FGLS estimator of Parks (1967), finding out that economic performance is enhanced where strong unions coexist with an important presence of leftist movements in government or in the opposite situation (rightist governments with weak unions), being less satisfactory for in-between cases. Their original results (see the example in the next section) are very sharp, with narrow standard errors. Beck et al. (1993) attribute the narrow confidence bands to the estimator employed being inappropriate for the sample size at hand; they re-examine the data using OLS estimation of a dynamic model with time fixed effects and time-clustered errors, upholding previous conclusions as regards the effects on growth (although with lower significance) but rendering mixed evidence for inflation and unemployment. The dataset is included in package **pcse** (Bailey and Katz, 2011):

```
library("pcse")
data("agl", package = "pcse")
```

In the following we estimate the model with time fixed effects[4] and produce the diagnostics table with PCSE standard errors:

```
fm <- growth ~ laggl + opengdp + openex + openimp + central * leftc
aglmod <- plm(fm, agl, model = "w", effect = "time")
coeftest(aglmod, vcov=vcovBK)

t test of coefficients:

              Estimate  Std. Error t value  Pr(>|t|)
laggl         0.095085   0.117523     0.81  0.41935
opengdp       0.007256   0.001735     4.18  4.2e-05 ***
openex        0.002373   0.000882     2.69  0.00768 **
openimp      -0.006475   0.002301    -2.81  0.00534 **
leftc        -0.023378   0.008009    -2.92  0.00388 **
central:leftc 0.013172   0.003497     3.77  0.00021 ***
---
Signif. codes:
0 '***' 0.001 '**' 0.01 '*' 0.05 '.' 0.1 ' ' 1
```

5.1.3 Robust Testing of Linear Hypotheses

The main use of robust covariance estimators is together with testing functions from the **lmtest** (Zeileis and Hothorn, 2002) and **car** (Fox and Weisberg, 2011) packages. We have seen the special case of testing single exclusion restrictions through `coeftest`: in order of increasing generality, joint restrictions can be tested through `waldtest`, while `linearHypothesis` from package **car** enables testing a general linear hypothesis on model parameters.

---

4 Notice that time effects, while present in the original Beck et al. (1993), are omitted in the Bailey and Katz (2011) example.

Example 5.7 testing with robust covariance matrices – `Produc` data set
All these functions typically allow passing the `vcov` parameter either as a matrix or as a function (see Zeileis, 2004). If one is happy with the defaults, it is easiest to pass the function itself, as seen in the previous examples; else, one may do the covariance computation inside the call to `coeftest`, thus passing on a matrix:

```
coeftest(plmmod, vcov = vcovHC(plmmod, type = "HC3"))
```

or, rather, define an appropriate function inside the call: in this case, optional parameters are provided as shown below (see also Zeileis, 2004, p. 12):

```
coeftest(plmmod, vcov = function(x) vcovHC(x, type = "HC3"))
```

For some tests, e.g., for multiple model comparisons by `waldtest`, one should always provide a function.

Example 5.8 testing with robust covariance matrices – `Parity` data set
The next example shows how to extend the comparison across models with different kinds of fixed effects, using `linearHypothesis` from package **car**.

Coakley et al. (2006) present a purchasing power parity (PPP) regression on a "long" panel of quarterly data 1973-Q1 to 1998-Q4 for 17 developed countries so that $N = 17$ and $T = 104$. The estimated model is

$$\Delta s_{nt} = \alpha + \beta(\Delta p - \Delta p^*)_{nt} + v_{nt}$$

where s_{nt} is the relative exchange rate against USD and $(\Delta p - \Delta p^*)_{nt}$ is the inflation differential between each country and the US.

```
data("Parity", package = "plm")
fm <- ls ~ ld
pppmod <- plm(fm, data = Parity, effect = "twoways")
```

The hypothesis of interest is $\beta = 1$, meaning that inflation differentials are fully reflected in the exchange rate. We report the corresponding robust Wald test from `linearHypothesis` in package **car** (Fox and Weisberg, 2011), which would be done interactively as follows:

```
library("car")
linearHypothesis(pppmod, "ld = 1", vcov = vcov)
```

(output suppressed), in a compact table supplying different covariance estimators to each of four models: OLS, one-way time or country fixed effects, and two-way fixed effects.

```
vcovs <- c(vcov, Vw, Vcx, Vct, Vcxt, Vct.L, Vnw.L, Vscc.L, Vcxt.L)
names(vcovs) <- c("OLS", "Vw", "Vcx", "Vct", "Vcxt", "Vct.L", "Vnw.L",
                  "Vscc.L", "Vcxt.L")
tttab <- matrix(nrow = 4, ncol = length(vcovs))
dimnames(tttab) <- list(c("Pooled OLS","Time FE","Country FE","Two-way FE"),
                        names(vcovs))
```

```
pppmod.ols <- plm(fm, data = Parity, model = "pooling")
for(i in 1:length(vcovs)) {
    tttab[1, i] <- linearHypothesis(pppmod.ols, "ld = 1",
                                     vcov = vcovs[[i]])[2, 4]
}

pppmod.tfe <- plm(fm, data = Parity, effect = "time")
for(i in 1:length(vcovs)) {
    tttab[2, i] <- linearHypothesis(pppmod.tfe, "ld = 1",
                                     vcov = vcovs[[i]])[2, 4]
}

pppmod.cfe <- plm(fm, data = Parity, effect = "individual")
for(i in 1:length(vcovs)) {
    tttab[3, i] <- linearHypothesis(pppmod.cfe, "ld = 1",
                                     vcov = vcovs[[i]])[2, 4]
}

pppmod.2fe <- plm(fm, data = Parity, effect = "twoways")
for(i in 1:length(vcovs)) {
    tttab[4, i] <- linearHypothesis(pppmod.2fe, "ld = 1",
                                     vcov = vcovs[[i]])[2, 4]
}

print(t(round(tttab, 6)))
        Pooled OLS  Time FE Country FE Two-way FE
OLS       0.000000 0.000000   0.000000   0.000000
Vw        0.000000 0.000000   0.000000   0.000000
Vcx       0.001032 0.000869   0.070773   0.119787
Vct       0.000000 0.000000   0.000000   0.000000
Vcxt      0.000966 0.000842   0.071866   0.121614
Vct.L     0.000000 0.000000   0.001861   0.000748
Vnw.L     0.000000 0.000000   0.000030   0.000000
Vscc.L    0.000000 0.000000   0.000076   0.000013
Vcxt.L    0.000648 0.000672   0.075022   0.129857
```

As is apparent from the results' table, the PPP hypothesis is not rejected any more once one controls for, at a minimum, country fixed effects *and* by-group clustering.

5.1.3.1 An Application: Robust Hausman Testing

Beside the usual quadratic form, Hausman's specification test can be performed in an equivalent form based on testing a linear restriction in an auxiliary linear model. In particular, it can be computed through an artificial regression of the quasi-demeaned response over the quasi-demeaned regressors from the random effects augmented with the fully demeaned regressors from the within model:

$$\tilde{y} = \tilde{Z}\gamma + WX\delta.$$

The Hausman test is then the redundancy test on WX, i.e., the restriction test $\delta = 0$. This artificial regression version of the test can easily be robustified (see Wooldridge, 2010) by using a robust covariance matrix.

Example 5.9 regression-based Hausman test – `Grunfeld` data set
We compare the Hausman test in original and regression-based form for the `Grunfeld` data. The function `phtest` allows for an optional argument `method`, defaulting to `'chisq'` (original form); if `method` is specified as `'aux'`, the test is performed through the auxiliary regression. Below we compare the two versions, using the default estimated covariance matrix in the auxiliary regression.

```
data("Grunfeld", package = "plm")
phtest(inv ~ value + capital, data = Grunfeld)

Hausman Test

data:  inv ~ value + capital
chisq = 2.3, df = 2, p-value = 0.3
alternative hypothesis: one model is inconsistent
phtest(inv ~ value + capital, data = Grunfeld, method = "aux")

Regression-based Hausman test

data:  inv ~ value + capital
chisq = 2.1, df = 2, p-value = 0.3
alternative hypothesis: one model is inconsistent
```

Unsurprisingly, the results from the regression-based and the original Hausman test are consistent: both support the random effects hypothesis.

Example 5.10 robust Hausman test – `RDSpillovers` data set
The `RDSpillovers` data are highly heteroscedastic. In this situation, the original Hausman test is biased toward rejection, as is the alternative regression-based version if not robustified. The latter can nevertheless be computed in robust form, by employing a robust covariance matrix in the restriction test on the auxiliary regression. If `method` is `'aux'`, the `phtest` function admits a further `vcov` argument, possibly allowing to specify the use of a robust estimator for the covariance. As can be seen from the table below, the results change substantially:

```
data("RDSpillovers", package = "pder")
pehs <- pdata.frame(RDSpillovers, index = c("id", "year"))
ehsfm <- lny ~ lnl + lnk + lnrd
phtest(ehsfm, pehs, method = "aux")

Regression-based Hausman test

data:  ehsfm
chisq = 53, df = 3, p-value = 2e-11
alternative hypothesis: one model is inconsistent
```

```
phtest(ehsfm, pehs, method = "aux", vcov = vcovHC)

Regression-based Hausman test, vcov: vcovHC

data:  ehsfm
chisq = 2.3, df = 3, p-value = 0.5
alternative hypothesis: one model is inconsistent
```

The robust version of the Hausman test does not reject the random effects hypothesis any more.

5.2 Unrestricted Generalized Least Squares

If the data-generating process is:

$$y = Z\gamma + \epsilon$$

and $\epsilon \sim (0, \Omega)$ has a general structure, ordinary least squares estimates for β are inefficient, though consistent. By Aitken's theorem (see, e.g., Greene (2003), 10.5), generalized least squares (**GLS**) are the efficient estimator for the model parameters if Ω is known. The estimator is then

$$\hat{\gamma}_{\text{GLS}} = (Z^{\top}\Omega^{-1}Z)^{-1}(Z^{\top}\Omega^{-1}y)$$

Various feasible **GLS** procedures exist drawing on consistent estimators of Ω, which are then plugged into the **GLS** estimator. The key to obtaining a consistent estimate of Ω is, in general, to specify enough structure to faithfully represent its characteristics while keeping the number of parameters to be estimated at a manageable level.

In the standard one-way error components model, as already seen, the disturbance term may be written as $\epsilon_{nt} = \eta_n + \nu_{nt}$ where η_n denotes the (time-invariant) individual-specific effect and ν_{nt} the idiosyncratic error. Observations regarding the same individual n share the same η_n effect, thus the relative errors are autocorrelated. The random effects structure is a very parsimonious way to account for individual heterogeneity, which can be extended in various dimensions, e.g., by specifying an autoregressive process in space and/or time for the idiosyncratic component ν_{nt}.

Under the random effects specification, the variance-covariance matrix of the errors $\Omega = \sigma_\eta{}^2(\mathrm{I}_N \otimes J_T) + \sigma_\nu{}^2(\mathrm{I}_N \otimes \mathrm{I}_T)$ is block-diagonal with $\Omega = \mathrm{I}_N \otimes \Sigma_T$ where

$$\Sigma_T = \sigma_\eta^2 J_T + \sigma_\nu{}^2 \mathrm{I}_T = \begin{bmatrix} \sigma_\nu^2 + \sigma_\eta^2 & \sigma_\eta^2 & \dots & \sigma_\eta^2 \\ \sigma_\eta^2 & \sigma_\nu^2 + \sigma_\eta^2 & \dots & \vdots \\ \dots & & \ddots & \sigma_\eta^2 \\ \sigma_\eta^2 & & & \sigma_\nu^2 + \sigma_\eta^2 \end{bmatrix}$$

The above is the standard specification of random effects panels, described in the previous chapters. It parsimoniously describes the error covariance by means of just two parameters and is, therefore, of very general applicability as far as sample sizes are concerned. In panels with one dimension much larger than the other (typically, large and short panels) a less restrictive approach is possible, termed *general* GLS (Wooldridge, 2010, 10.4.3), which allows for arbitrary *within-individual* heteroscedasticity and serial correlation of errors, *i.e.*, inside the Σ covariance submatrices, provided that these remain the same for every individual.

5.2.1 General Feasible Generalized Least Squares

If one assumes $\Omega = I_N \otimes \Sigma_T$ but leaves the structure of Σ_T completely free except for the obvious requisites of being symmetric and positive definite:

$$\Sigma_T = \begin{bmatrix} \sigma_1^2 & \sigma_{12} & \dots & \sigma_{1T} \\ \sigma_{12} & \sigma_2^2 & \dots & \vdots \\ \dots & & \ddots & \sigma_{T-1,T} \\ \sigma_{1T} & & & \sigma_T^2 \end{bmatrix}$$

individual errors can evolve through time with an unlimited amount of heteroscedasticity and autocorrelation, but they are assumed to be uncorrelated between them in the cross section, and this structure is assumed constant over the different individuals. By this assumption, the components σ_{st} of Σ_T can be estimated drawing on the cross-sectional dimension, using the average over individuals of the outer products of the residuals from a consistent estimator:

$$\hat{\Sigma}_T = \sum_{n=1}^{N} \frac{\hat{\epsilon}_n \hat{\epsilon}_n^\top}{N}$$

where $\hat{\epsilon}_n = (\hat{\epsilon}_{n1}, \dots \hat{\epsilon}_{nT})$ is the subvector of OLS residuals for individual n.

This estimator is called *general feasible* GLS, or GGLS and is also sometimes referred to as the Parks (1967) estimator and is, as observed by Driscoll and Kraay (1998), a variant of the SUR estimator by Zellner (1962). Greene (2003) presents the same estimator in the context of pooled time series, with fixed N and "large" T.

Leaving the intra-group error covariance parameters completely free to vary is an attractive strategy, provided that $N \gg T$ because the number of variance parameters to be estimated with NT data points is $T(T+1)/2$ (Wooldridge, 2010). This is a typical situation in micro-panels such as, e.g., household income surveys, where N is in the thousands but T is typically quite short so that even if estimating an unrestricted covariance, many degrees of freedom are still available.

The original applications have instead been in the field of pooled time series, aimed at accounting for cross-sectional correlation and heteroscedasticity. In this context, Driscoll and Kraay (1998) observe how the lack of degrees of freedom in estimating the error covariance leads to near-singular estimates and hence to downward-biased standard errors, thus overestimating parameter significance. Beck and Katz (1995) also discuss some severe biases of this estimator in small samples. Both start from a pooled time series, T-asymptotic approach, and both are interested in robustness over the cross-sectional dimension. In this light, most of the criticism this estimator has been subject to depends on the peculiar field of application, especially in Beck and Katz (1995) and references therein (Alvarez et al., 1991) where it is applied to political science data with very modest sample sizes; but recent simulations by Chen et al. (2009) show that even in such situations FGLS can be more efficient than the proposed alternatives (OLS with PCSE standard errors).

The **GGLS** principle can be applied to various situations, consistent with different views on heterogeneity (random vs fixed effects hypothesis) or stationarity (e.g., to a model in first differences). That translates into either applying the unrestricted **GLS** estimator directly to the observed data or to a transformation thereof.

This framework allows the error covariance structure inside every group (if `effect` is set to `'individual'`) of observations to be fully unrestricted and is therefore robust against any type of intra-group heteroscedasticity and serial correlation. This structure, by converse, is assumed identical across groups and thus general **FGLS** is inefficient under group-wise heteroscedasticity. Cross-sectional correlation is excluded a priori.

In a pooled time series context (`effect` is set to `'time'`), symmetrically, this estimator is able to account for arbitrary cross-sectional correlation, provided that the latter is time invariant (see Greene, 2003, 13.9.1–2, p. 321–322). In this case, serial correlation has to be assumed away and the estimator is consistent with respect to the time dimension, keeping N fixed.

5.2.1.1 Pooled GGLS

Under the specification described at the beginning of this section, residuals can be consistently estimated by **OLS** and then used to estimate Σ_T as above. Using $\hat{\Omega} = I_N \otimes \hat{\Sigma}_T$, the **FGLS** estimator is:

$$\hat{\gamma}_{\text{GGLS}} = (Z^\top \hat{\Omega}^{-1} Z)^{-1} (Z^\top \hat{\Omega}^{-1} y)$$

The estimated individual submatrix $\hat{\Sigma}_T$ will give an assessment of the structure, if any, of the errors' covariance, which may guide towards more parsimonious specifications like the **RE** one (if all diagonal and, respectively, all off-diagonal elements are of similar magnitude) or possibly an **AR(1)** specification, if covariances between pairs of off-diagonal elements become smaller with distance.

In this small-T, large-N context, one will often want to include time fixed effects to mitigate cross-sectional correlation, which is assumed out of the residuals.

The function `pggls` estimates general **FGLS** models, either with or without fixed effects, or on first-differenced data. In the following we illustrate it on the `EmplUK` data.

Example 5.11 generalized GLS estimator – `EmplUK` data set

The `EmplUK` dataset is a good candidate for **GGLS** estimation as being a relatively big random sample of firms observed over a limited number of years.

The "random effect" equivalent, general **GLS**, is estimated by specifying the `model` argument as `'pooling'`:

```
data("EmplUK", package = "plm")
gglsmod <- pggls(log(emp) ~ log(wage) + log(capital),
                 data = EmplUK, model = "pooling")
summary(gglsmod)

Call:
pggls(formula = log(emp) ~ log(wage) + log(capital), data = EmplUK,
    model = "pooling")
```

```
Unbalanced Panel: n = 140, T = 7-9, N = 1031

Residuals:
   Min. 1st Qu.  Median    Mean 3rd Qu.    Max.
-1.8070 -0.3655  0.0618  0.0323  0.4428  1.5872

Coefficients:
             Estimate Std. Error z-value Pr(>|z|)
(Intercept)    2.0235     0.1585   12.77  < 2e-16 ***
log(wage)     -0.2323     0.0480   -4.84  1.3e-06 ***
log(capital)   0.6105     0.0174   35.02  < 2e-16 ***
---
Signif. codes:
0 '***' 0.001 '**' 0.01 '*' 0.05 '.' 0.1 ' ' 1
Total Sum of Squares: 1850
Residual Sum of Squares: 403
Multiple R-squared: 0.783
```

The `pggls` function is similar to `plm` in many respects. An exception is that the estimate of the group covariance matrix of errors (`sigma`, a matrix) is reported in the model objects instead of the usual estimated variances of the two error components. It can be displayed as follows:

```
round(gglsmod$sigma, 3)
      1976  1977  1978  1979  1980  1981  1982  1983  1984
1976 0.307 0.291 0.277 0.269 0.252 0.254 0.247 0.303 0.362
1977 0.291 0.303 0.296 0.294 0.275 0.259 0.251 0.272 0.428
1978 0.277 0.296 0.299 0.301 0.280 0.264 0.256 0.280 0.433
1979 0.269 0.294 0.301 0.314 0.291 0.273 0.263 0.287 0.452
1980 0.252 0.275 0.280 0.291 0.282 0.265 0.254 0.279 0.426
1981 0.254 0.259 0.264 0.273 0.265 0.266 0.254 0.279 0.447
1982 0.247 0.251 0.256 0.263 0.254 0.254 0.262 0.291 0.473
1983 0.303 0.272 0.280 0.287 0.279 0.279 0.291 0.300 0.486
1984 0.362 0.428 0.433 0.452 0.426 0.447 0.473 0.486 0.505
```

As can be seen, the correlations between pairs of residuals (in time) for the same individual do not die out with the distance in time. The estimated error covariance very much resembles the random effects structure, with a strong prevalence of the individual variance component σ^2_η over σ^2_ν (witness the small difference between values on and outside the diagonal).

5.2.1.2 Fixed Effects GLS

If individual heterogeneity is present but we do not trust the random effects assumption, and moreover the remainder errors are expected to show heteroscedasticity and serial correlation, the **FE** estimator can be employed together with a robust covariance matrix; but if the cross-sectional dimension is sufficient and the assumption of constant covariance matrix across individuals is realistic, then applying the **GGLS** method to time-demeaned data can provide a more efficient alternative, called the fixed effects GLS (FEGLS) estimator (Wooldridge, 2010, 10.5.5).[5]

The errors covariance submatrix for each individual is now:

$$\hat{\Sigma}_T^{\text{FEGLS}} = \sum_{n=1}^{N} \frac{\hat{e}_{\text{w}n}\hat{e}_{\text{w}n}^{\top}}{N}$$

where $\hat{e}_{\text{w}n} = (\hat{e}_{\text{w}n1}, \dots \hat{e}_{\text{w}nT})$ is the subvector of FE (*within*) residuals for individual n. Using $\hat{\Omega}^{(\text{FEGLS})} = \text{I}_N \otimes \hat{\Sigma}_T^{(\text{FEGLS})}$ and the *within* transformed data, the FEGLS estimator is:

$$\hat{\beta}_{\text{FEGLS}} = (X^{\top}W\hat{\Omega}^{(\text{FEGLS})-1}WX)^{-1}(X^{\top}W\hat{\Omega}^{(\text{FEGLS})-1}Wy)$$

This estimator, originally due to Kiefer (1980), takes care of both the serial correlation present in the original errors ϵ_{nt} and, implicitly, of that induced by the demeaning. For this reason, being a combination of both, the estimated $\hat{\Sigma}$ does not give a direct assessment of the original error structure anymore.

Example 5.12 FEGLS estimator – `EmplUK` data set

The fixed effects `pggls` is based on the estimation of a within model in the first step, but this is transparent to the user; estimation follows as above but for the need to specify `model='within'`. For reasons of robustness, as happens with `plm`, this is the default method. It is estimated by:

```
feglsmod <- pggls(log(emp) ~ log(wage) + log(capital), data = EmplUK,
                  model = "within")
summary(feglsmod)
 Within model

Call:
pggls(formula = log(emp) ~ log(wage) + log(capital), data = EmplUK,
    model = "within")

Unbalanced Panel: n = 140, T = 7-9, N = 1031

Residuals:
   Min. 1st Qu.  Median    Mean 3rd Qu.    Max.
-0.5084 -0.0743 -0.0024  0.0000  0.0761  0.6014

Coefficients:
             Estimate Std. Error z-value Pr(>|z|)
log(wage)     -0.6176     0.0308   -20.1   <2e-16 ***
log(capital)   0.5610     0.0172    32.6   <2e-16 ***
---
```

5 Notice that one time period has to be dropped from the data because the empirical covariance matrix of transformed errors has rank $T-1$: see again Wooldrodge (2010, p. 312) and references therein.

```
Signif. codes:
0 '***' 0.001 '**' 0.01 '*' 0.05 '.' 0.1 ' ' 1
Total Sum of Squares: 1850
Residual Sum of Squares: 17.4
Multiple R-squared: 0.991
```

The `phtest` function can be used to assess the need for fixed effects through a Hausman test:

```
phtest(feglsmod, gglsmod)

Hausman Test

data:  log(emp) ~ log(wage) + log(capital)
chisq = 1100, df = 2, p-value <2e-16
alternative hypothesis: one model is inconsistent
```

The Hausman test strongly favours the fixed effects model.

5.2.1.3 First Difference GLS

Analogously, the **GGLS** principle can be applied to data in first differences, in the very same way as for **FEGLS**, giving rise to the first difference **GLS** (**FDGLS**) estimator (Wooldridge, 2010, p. 320).

In this case, the errors covariance submatrix for an individual is:

$$\hat{\Sigma}^{(\text{FDGLS})}_{T-1} = \sum_{n=1}^{N} \frac{\Delta\hat{\epsilon}_n \Delta\hat{\epsilon}_n^\top}{N-1}$$

where $\Delta\hat{\epsilon}_n = (\Delta\hat{\epsilon}_{n1}, \dots \Delta\hat{\epsilon}_{nT})$ is the subvector of **FD** residuals for individual n. Using $\hat{\Omega}^{(\text{FDGLS})} = \text{I}_N \otimes \hat{\Sigma}^{(\text{FDGLS})}_{T-1}$ and the differenced data, the **FDGLS** estimator is:

$$\hat{\gamma}_{\text{FDGLS}} = (\Delta Z^\top \hat{\Omega}^{(\text{FDGLS})-1} \Delta Z)^{-1} (\Delta Z^\top \hat{\Omega}^{(\text{FDGLS})-1} \Delta y)$$

First differencing eliminates time-invariant unobserved heterogeneity, as does the *within* transformation; one difference is that now one time period is lost for each individual. **FD** has to be preferred to **FE** when the original data are likely to be nonstationary, because then the **FD**-transformed residuals will be. Again, elements of $\hat{\Sigma}^{(\text{FDGLS})}$ do not directly represent the correlation structure of residuals because of the induced correlation from first differencing.

To choose which method to use, one can look at the stationarity properties of the residuals. If the residuals of the **FEGLS** estimator are not stationary, then **FDGLS** will be a more appropriate estimator.

Example 5.13 FDGLS estimator – `EmplUK` data set

Specifying `model='fd'`, we obtain the **FDGLS** estimator.

```
fdglsmod <- pggls(log(emp) ~ log(wage) + log(capital), data = EmplUK,
                  model = "fd")
summary(fdglsmod)
 NA
```

```
Call:
pggls(formula = log(emp) ~ log(wage) + log(capital), data = EmplUK,
    model = "fd")

Unbalanced Panel: n = 140, T = 7-9, N = 1031

Residuals:
   Min. 1st Qu.  Median    Mean 3rd Qu.    Max.
-0.7578 -0.0751 -0.0189 -0.0283  0.0260  0.6506

Coefficients:
             Estimate Std. Error z-value Pr(>|z|)
log(wage)     -0.3343     0.0385   -8.67   <2e-16 ***
log(capital)   0.3786     0.0203   18.68   <2e-16 ***
---
Signif. codes:
0 '***' 0.001 '**' 0.01 '*' 0.05 '.' 0.1 ' ' 1
Total Sum of Squares: 1850
Residual Sum of Squares: 11.6
Multiple R-squared: 0.994
```

5.2.2 Applied Examples

Example 5.14 generalized GLS estimator – `RiceFarms` data set
The Rice Farming dataset contains observations on 171 farms over 6 years; therefore, the number of covariance parameters to estimate on 1026 data points is a still manageable 21. Farms come from 6 different regions, each with peculiar characteristics. The random sampling assumption seems to be reasonable within regions, but one might suspect observations from the same region to share some common characteristics, and such characteristics to be possibly related to the regressors: therefore it is advisable to include 5 regional fixed effects, to control for region-related, correlated heterogeneity along the lines of Wooldridge (2010, p. 328). For the reasons given above, we include time effects to control for contemporaneous correlation in the cross section; instead of following the original application of Druska and Horrace (2004) adding one dummy for wet seasons as opposed to dry ones, we simply introduce 5 separate time effects.

```
data("RiceFarms", package = "splm")
RiceFarms <- transform(RiceFarms,
                       phosphate = phosphate / 1000,
                       pesticide = as.numeric(pesticide > 0))

fm <- log(goutput) ~ log(seed) + log(urea) + phosphate +
    log(totlabor) + log(size) + pesticide + varieties +
        + region + time
```

```
gglsmodrice <- pggls(fm, RiceFarms, model = "pooling", index = "id")
summary(gglsmodrice)
 NA
```

```
Call:
pggls(formula = fm, data = RiceFarms, model = "pooling", index = "id")

Balanced Panel: n = 171, T = 6, N = 1026

Residuals:
   Min. 1st Qu.  Median    Mean 3rd Qu.    Max.
-0.9315 -0.2285  0.0151  0.0000  0.2147  1.3740

Coefficients:
                  Estimate Std. Error z-value Pr(>|z|)
(Intercept)         5.3334     0.1788   29.83  < 2e-16 ***
log(seed)           0.1285     0.0241    5.34  9.1e-08 ***
log(urea)           0.1351     0.0151    8.94  < 2e-16 ***
phosphate           0.7040     0.2526    2.79   0.0053 **
log(totlabor)       0.2099     0.0265    7.93  2.1e-15 ***
log(size)           0.5000     0.0281   17.77  < 2e-16 ***
pesticide           0.0355     0.0245    1.45   0.1473
varietieshigh       0.1351     0.0345    3.92  8.9e-05 ***
varietiesmixed      0.1031     0.0446    2.31   0.0209 *
regionlangan       -0.0451     0.0472   -0.96   0.3393
regiongunungwangi   0.0140     0.0532    0.26   0.7926
regionmalausma      0.0200     0.0541    0.37   0.7121
regionsukaambit     0.0671     0.0529    1.27   0.2049
regionciwangi       0.1633     0.0530    3.08   0.0021 **
time2              -0.0328     0.0262   -1.25   0.2102
time3              -0.2049     0.0316   -6.49  8.4e-11 ***
time4              -0.3440     0.0285  -12.08  < 2e-16 ***
time5               0.0576     0.0287    2.01   0.0448 *
time6               0.0441     0.0313    1.41   0.1581
---
Signif. codes:
0 '***' 0.001 '**' 0.01 '*' 0.05 '.' 0.1 ' ' 1
Total Sum of Squares: 1010
Residual Sum of Squares: 101
Multiple R-squared: 0.901
```

Regions do not seem to be so important after all, only Ciwangi being significantly different from the baseline; although a joint restriction test still rejects:

```
library("lmtest")
waldtest(gglsmodrice, "region")
Wald test

Model 1: log(goutput) ~ log(seed) + log(urea) + phosphate + log(totlabor) +
    log(size) + pesticide + varieties + +region + time
Model 2: log(goutput) ~ log(seed) + log(urea) + phosphate + log(totlabor) +
    log(size) + pesticide + varieties + time
  Res.Df Df Chisq Pr(>Chisq)
1   1007
2   1012 -5  28.9    2.5e-05 ***
```

```
---
Signif. codes:
0 '***' 0.001 '**' 0.01 '*' 0.05 '.' 0.1 ' ' 1
```

```
feglsmodrice <- pggls(update(fm, . ~ . - region), RiceFarms, index = "id")
```

Qualitatively, the results do not seem to change much when adding individual fixed effects. The hypothesis that after controlling for the region, all remaining individual heterogeneity be of the random effects type can be tested formally by means of a Hausman test:

```
phtest(gglsmodrice, feglsmodrice)

Hausman Test

data:  fm
chisq = 18, df = 13, p-value = 0.1
alternative hypothesis: one model is inconsistent
```

The Hausman test does in fact not reject. Given the low significance of the regional effects, one might wonder whether a full "random effects" specification can be justified. An updated GGLS specification can readily be compared to the already estimated FEGLS model:

```
phtest(pggls(update(fm, . ~ . - region), RiceFarms,
             model = "pooling", index = "id"),
       feglsmodrice)

Hausman Test

data:  update(fm, . ~ . - region)
chisq = 19, df = 13, p-value = 0.1
alternative hypothesis: one model is inconsistent
```

In fact, even omitting the regional fixed effects, the GGLS specification still passes the Hausman test. The 171 rice farms can actually be seen as random draws from the same population, without the need for either individual or regional fixed effects.

Example 5.15 generalized GLS estimator – `RDSpillovers` data set

The static production function estimation in Eberhardt et al. (2013) is a problematic candidate for GGLS techniques; although it is desirable to allow for a free heteroscedasticity and serial correlation structure across this sample of manufacturing firms observed over a relatively long period of time, care shall be taken with the results exactly because of the relatively big time dimension. As too many covariance parameters, as discussed above, would result in underestimation of standard errors and hence false significance, sharp results should be looked at with suspicion. The example is nevertheless useful for illustration purposes, especially as it can be benchmarked against the thorough specification analysis in the original paper. As it will turn out, the GGLS approach ultimately seems to have satisfactory properties in this setting too.

```
fm <- lny ~ lnl + lnk + lnrd
```

```
gglsmodehs <- pggls(fm, RDSpillovers, model = "pooling")
coeftest(gglsmodehs)

t test of coefficients:

            Estimate Std. Error t value Pr(>|t|)
(Intercept)  1.04589    0.06416    16.3   <2e-16 ***
lnl          0.54825    0.01118    49.0   <2e-16 ***
lnk          0.43762    0.01384    31.6   <2e-16 ***
lnrd         0.08548    0.00548    15.6   <2e-16 ***
---
Signif. codes:
0 '***' 0.001 '**' 0.01 '*' 0.05 '.' 0.1 ' ' 1
```

```
feglsmodehs <- pggls(fm, RDSpillovers, model = "within")
coeftest(feglsmodehs)

t test of coefficients:

     Estimate Std. Error t value Pr(>|t|)
lnl    0.4942     0.0204   24.18  < 2e-16 ***
lnk    0.4922     0.0307   16.01  < 2e-16 ***
lnrd   0.0490     0.0147    3.34  0.00086 ***
---
Signif. codes:
0 '***' 0.001 '**' 0.01 '*' 0.05 '.' 0.1 ' ' 1
```

```
phtest(gglsmodehs, feglsmodehs)

Hausman Test

data:  fm
chisq = 18, df = 3, p-value = 5e-04
alternative hypothesis: one model is inconsistent
```

The Hausman test rejects the "random effects" GGLS specification. Given that correlated heterogeneity seems to be present, an alternative to eliminate it is the first difference transformation:

```
fdglsmodehs <- pggls(fm, RDSpillovers, model = "fd")
```

Which one to choose between FEGLS and FDGLS depends on the properties of transformed residuals. FEGLS residuals show a high level of persistence, as a simple serial correlation test (Wooldridge, 2010, 10.6.3) shows. We make a `data.frame` of the residuals, then estimate a (pooled) autoregressive model:

```
fee <- resid(feglsmodehs)
dbfee <- data.frame(fee=fee, id=attr(fee, "index")[[1]])
coeftest(plm(fee~lag(fee)+lag(fee,2), dbfee, model = "p", index="id"))
```

```
t test of coefficients:

              Estimate Std. Error t value Pr(>|t|)
(Intercept)   0.01096    0.00123     8.89   < 2e-16 ***
lag(fee)      1.07741    0.01926    55.95   < 2e-16 ***
lag(fee, 2) -0.14512    0.01886    -7.69   2.1e-14 ***
---
Signif. codes:
0 '***' 0.001 '**' 0.01 '*' 0.05 '.' 0.1 ' ' 1
```

The **FEGLS** residuals seem close to being nonstationary. The estimated autocorrelation in **FDGLS** residuals is instead much lower:

```
fde <- resid(fdglsmodehs)
dbfde <- data.frame(fde=fde, id=attr(fde, "index")[[1]])
coeftest(plm(fde~lag(fde)+lag(fde,2), dbfde, model = "p", index="id"))

t test of coefficients:

              Estimate Std. Error t value Pr(>|t|)
(Intercept)   0.01392    0.00132    10.50   < 2e-16 ***
lag(fde)      0.10548    0.02085     5.06   4.6e-07 ***
lag(fde, 2)   0.02317    0.01969     1.18      0.24
---
Signif. codes:
0 '***' 0.001 '**' 0.01 '*' 0.05 '.' 0.1 ' ' 1
```

hence it is advisable to resort to the **FDGLS** estimator:

```
coeftest(fdglsmodehs)

t test of coefficients:

       Estimate Std. Error t value Pr(>|t|)
lnl      0.5569     0.0217   25.61   <2e-16 ***
lnk      0.3514     0.0326   10.78   <2e-16 ***
lnrd     0.0611     0.0157    3.89    1e-04 ***
---
Signif. codes:
0 '***' 0.001 '**' 0.01 '*' 0.05 '.' 0.1 ' ' 1
```

The result, despite the limited number of degrees of freedom in estimating Σ_T, is in line with the more sophisticated analyses in the original paper by Eberhardt et al. (2013, Table 7) and with the preferred **FD** specification in Table 5, ibid. Moreover, despite the expected downward bias, standard errors are not too far from those of the above-mentioned **FD** model.

6

Endogeneity

6.1 Introduction

There is an endogeneity problem when the error is correlated with at least one explanatory variable. This phenomenon is very common in econometrics because, compared to experimental sciences, it is not possible (or it is at least difficult) to control the data-generating process. Among the possible causes of endogeneity, the three most important are:

simultaneity. In this case, there is an explanatory variable that is set simultaneously with the response: this is, for example, the case when one seeks to estimate a demand equation for a good, which contains the price of the good itself. The demand and the price are simultaneously set by the condition of equality of supply and demand and, therefore, a variation of the error term of the demand equation will shift the demand curve and therefore induce a variation of the quantity *and* the equilibrium price. The price variable is therefore endogenous.

covariate measured with error. If the "true" model is $y = \alpha + \beta x + \nu$ and what is observed is $x^* = x + \eta$, the estimated model writes: $y = \alpha + \beta(x^* - \eta) + \nu$, or $y = \alpha + \beta x^* + \epsilon$ with $\epsilon = \nu - \beta\eta$. Hence, ϵ is correlated with x^*, which is therefore endogenous.

omitted variable. If the "true" model is $y = \alpha + \beta_x x + \beta_z z + \nu$ and z is unobserved, the estimated model is $y = \alpha + \beta_x x + \epsilon$, with $\epsilon = \beta_z z + \nu$. The error of the estimated model then contains the influence of the omitted variable, and this error is correlated with x if x and z are correlated. Once again, the covariate x is then endogenous.

The OLS estimator is:

$$\hat{\gamma} = (Z^\top Z)^{-1} Z^\top y$$

Replacing y by its expression: $Z\gamma + \epsilon$, we obtain $\hat{\gamma}$ as a function of the errors of the model:

$$\hat{\gamma} = \gamma + (Z^\top Z)^{-1} Z^\top \epsilon$$

We then have, denoting N the sample size:

$$\hat{\gamma} = \gamma + \left(\frac{1}{N} Z^\top Z\right)^{-1} \frac{Z^\top \epsilon}{N}$$

The estimator is consistent ($\text{plim}\ \hat{\gamma} = \gamma$) if $\lim_{N\rightarrow+\infty} \frac{Z^\top\epsilon}{N} = 0$, this expression being the vector of covariances for the population between the covariates and the error. The ordinary least squares model is therefore consistent if the covariates and the error are uncorrelated. When this condition is not met, the method of instrumental variables, which will be presented in detail in this chapter, can be used.

Panel Data Econometrics with R, First Edition. Yves Croissant and Giovanni Millo.

Companion website: www.wiley.com/go/croissant/data-econometrics-with-R

Concerning simultaneity, there is an additional problem as the model is not defined by one equation but by a system of equations. In this case, two strategies can be followed:

- estimating only the equation of interest (limited information estimator),
- estimating simultaneously all the equations (full information estimator).

The latter approach leads to a more efficient estimator, as the correlation of the errors of all the equations is taken into account. But if an equation is wrongly specified, it can contaminate the estimation of the parameters of the other equations of the model.

6.2 The Instrumental Variables Estimator

6.2.1 Generalities about the Instrumental Variables Estimator

Let us consider the following model: $y = Z\gamma + \epsilon$ with $\mathrm{V}(\epsilon) = \sigma^2\mathrm{I}$. if at least one of the covariates is correlated with the errors, the OLS estimator is not consistent. In order to obtain consistency, we use the instrumental variables estimator. The instrumental variables are denoted by L.[1] Denoting by K the number of the covariates and by $M \geq K$ the number of instruments (not including the column of ones), the instrumental variables must verify: $\lim_{N\rightarrow+\infty} \frac{L^\top\epsilon}{N} = 0$. Stated differently, they must not be correlated with the errors.[2] In the simplest case where the number of instruments equals the number of covariates, the instrumental variable estimator is simply obtained by solving the system of equations: $L^\top\epsilon = 0$, which is just identified. Developing this expression, we obtain: $L^\top(y - Z\gamma) = 0$, which can also be written:

$$\hat{\gamma} = (L^\top Z)^{-1}L^\top y \tag{6.1}$$

If there are more instruments than covariates ($M \geq K$), $L^\top\epsilon$ is an over-determined system of linear equations, which, except for very special cases, doesn't have a solution. In this case, two equivalent approaches can be used to obtain the optimal estimator. The first one consists in pre-multiplying the model by $L^\top$:

$$L^\top y = L^\top Z\gamma + L^\top\epsilon \tag{6.2}$$

It is a model that contains $M + 1$ rows and $K + 1$ parameters to estimate γ. If one considers it as a standard regression model, the variance of the errors being $\mathrm{V}(L^\top\epsilon) = \sigma^2 L^\top L$, the best linear estimator is the GLS estimator, and we then obtain the following instrumental variables estimator:

$$\begin{aligned}\hat{\gamma}_{\mathrm{IV}} &= (Z^\top L(L^\top L)^{-1}L^\top Z)^{-1}(Z^\top L(L^\top L)^{-1}L^\top y)\\ &= (Z^\top \mathrm{P}_L Z)^{-1}(Z^\top \mathrm{P}_L y)\end{aligned} \tag{6.3}$$

with $\mathrm{P}_L = L(L^\top L)^{-1}L^\top$.

The second approach is the generalized method of moments. We consider here a vector of $M + 1$ moments: $\mathrm{E}(L^\top\epsilon) = \mathrm{E}(L^\top(y - Z\gamma)) = 0$ for which the variance is $\mathrm{V}(L^\top\epsilon) = \sigma^2 L^\top L$. Using the generalized method of moments, we seek to minimize the quadratic form of the vector of moments, using the inverse of the variance matrix of these moments:

$$\frac{1}{\sigma^2}(y^\top - \gamma^\top Z^\top)L(L^\top L)^{-1}L^\top(y - Z\gamma) = \frac{1}{\sigma^2}(y^\top - \gamma^\top Z^\top)\mathrm{P}_L(y - Z\gamma)$$

The first-order conditions for a minimum are: $-2Z^\top\mathrm{P}_L(y - Z\gamma)/\sigma^2 = 0$, and solving this system of linear equations, we obtain the same estimator as before.

1 As for the Z matrix, the first column of L is a column of ones.

2 It is often the case that some covariates are uncorrelated with the errors and are therefore also used as instruments.

The instrumental variables estimator is also called the two-stage least squares estimator (**2SLS**), as it can be obtained by applying twice the method of ordinary least squares. When we consider the regression of z on L, we obtain the estimator $\hat{\gamma} = (L^\top L)^{-1}L^\top z$ and the fitted values $\hat{z}_L = L\hat{\gamma} = L(L^\top L)^{-1}L^\top z = \mathrm{P}_L z$. The matrix P_L is therefore the projection matrix on the subspace defined by the columns of L. This matrix is symmetric and idempotent, which means that $\mathrm{P}_L\mathrm{P}_L = \mathrm{P}_L$. The instrumental variables estimator (6.3) can also be written, denoting by $\hat{Z}_L = \mathrm{P}_L Z$ the fitted values of the covariates regressed on the instrumental variables:

$$\hat{\gamma}_{\mathrm{2SLS}} = (\hat{Z}_L^\top \hat{Z}_L)^{-1}\hat{Z}_L^\top y = (\hat{Z}_L^\top \hat{Z}_L)^{-1}\hat{Z}_L^\top \hat{y}_L \tag{6.4}$$

and can therefore be obtained by applying **OLS** twice:

- the first time by regressing every covariate on the instruments,
- the second time by regressing the response on the fitted values of the first-stage estimation.

The variance of the instrumental variables estimator is:

$$\mathrm{V}(\hat{\gamma}) = \sigma^2(\hat{Z}_L^\top \hat{Z}_L)^{-1}$$

The estimator is therefore the more efficient the larger the variance of $\hat{Z}_L$, which means that Z and L are highly correlated.

6.2.2 The *within* Instrumental Variables Estimator

The specificity of panel data methods is that the error term is modeled as having two components, an individual effect and an idiosyncratic term. Therefore, the correlation between covariates and instrumental variables, on the one hand, and the errors of the model, on the other hand, must be analyzed separately for each component of the error. In this section, we consider the estimation of the model transformed in deviations from individual means. This transformation wipes out the individual effect; therefore, there is no reason to take care of the correlation between the covariates and the individual effects. The **W2SLS** is obtained by pre-multiplying the model first by W: $Wy = WZ\gamma + W\epsilon$ and then by $L^\top$,

$$L^\top Wy = L^\top WZ\gamma + L^\top W\epsilon \tag{6.5}$$

and applying **GLS** to this transformed data, the variance matrix of the errors of this model being $\sigma^2 L^\top WL$:

$$\hat{\gamma}_{\mathrm{W2SLS}} = (Z^\top WL(L^\top WL)^{-1}L^\top WZ)^{-1}(Z^\top WL(L^\top WL)^{-1}L^\top Wy)$$

or, denoting by: $\mathrm{P}_L^{\mathrm{W}} = WL(L^\top WL)^{-1}L^\top W$ the projection matrix defined by the *within* transformation of the instruments:

$$\hat{\gamma}_{\mathrm{W2SLS}} = (Z^\top \mathrm{P}_L^{\mathrm{W}} Z)^{-1}(Z^\top \mathrm{P}_L^{\mathrm{W}} y) \tag{6.6}$$

A similar reasoning can be followed for the *between* model. We consider the between transformation of the model $By = BZ\gamma + B\epsilon$, with the same transformation applied to the instruments (BL). The instrumental variables estimator is obtained by pre-multiplying the model by $L^\top B$:

$$L^\top By = L^\top BZ\gamma + L^\top B\epsilon \tag{6.7}$$

and applying to this transformed model the **GLS** estimator:

$$\hat{\gamma}_{\mathrm{B2SLS}} = (Z^\top \mathrm{P}_L^{\mathrm{B}} Z)^{-1}(Z^\top \mathrm{P}_L^{\mathrm{B}} y) \tag{6.8}$$

with $\mathrm{P}_L^B = BL(L^\top BL)^{-1}L^\top B$.

The W2SLS is consistent, even if the individual effects are correlated with the covariates. On the contrary, the B2SLS is consistent only if there is no correlation. If this hypothesis is verified, none of them is efficient, as each of them take into account only one component of variability.

Example 6.1 ***within* 2SLS estimator – `SeatBelt` data set**
Cohen and Einav (2003) study the influence of using seat belts on the number of deaths on American roads; they consider occupants of the vehicles involved in accidents (about 35,000 killed a year) and non-occupants (e.g., pedestrians; about 5,000 killed a year). They use panel data for the 50 American states for the 1983-1997 period. This dataset, called `SeatBelt`, is available in the **pder** package. The main covariate is the rate of seat belt `usage`. Two main questions are analyzed:

- the first one concerns the behavior compensation theory developed by Peltzman (1975). According to this theory, using the seat belt makes the driver more confident and leads him to adopt a less prudent driving behavior. Combined with the expected negative effect from seat belts on occupants' deaths, the global effect on mortality may then be insignificant, or even positive if the mortality of non-occupants increases with the use of seat belts,
- the second deals with the problem of endogeneity: if driving conditions get worse, for example for meteorological reasons, other things being equal, road mortality will increase, but seat belt use will also increase, as the drivers are conscious that the probability of having an accident increases. There is therefore in this case a correlation between the error term of the mortality equation and the seat belt use variable. In this case, not taking this endogeneity into account will induce a downward bias on the estimation of the seat belt-use coefficient.

Cohen and Einav (2003) use three estimators. First, the model is estimated using OLS and therefore the endogeneity is not taken into account. The second is the *within* estimator; in this case, the problem of the correlation between the individual effect and the covariate is taken into account as the *within* transformation wipes out the individual effects. On the contrary, the problem of correlation between the idiosyncratic part of the error and the covariate remains. This last problem is solved using the W2SLS estimator. The instruments used are variables that indicate the laws concerning the use of seat belts. These variables (`ds`, `dp` and `dsp`) are correlated with the use of seat belts but not with the errors. Other variables are also used as controls (and are described in the help page of the dataset).

The instrumental variables estimator is computed using the `plm` function. The instruments are specified with a two-part formula, using the **Formula** package (Zeileis and Croissant, 2010). The first part indicates the covariates of the model, while the second part indicates the instruments. Often, a large subset of covariates are used as instruments. In order to avoid repeating almost the same list of variables twice, it is possible to use a more efficient syntax using the `.` operator, constructing the second part of the formula by updating the first part. For example, if the covariates are `x1`, `x2` and `x3`, only `x2` is endogenous, and there is only one external instrument `z`, the model to be estimated can be described by either of the two equivalent following formulas:

```
y ~ x1 + x2 + x3 | x1 + x3 + z
y ~ x1 + x2 + x3 | . - x2 + z
```

The three models estimated by Cohen and Einav (2003), which are reproduced below, include time fixed effects. The response (`occfat`) is the number of vehicle occupants killed on the road.

```
data("SeatBelt", package = "pder")
SeatBelt$occfat <- with(SeatBelt, log(farsocc / (vmtrural + vmturban)))
ols <- plm(occfat ~ log(usage) + log(percapin) + log(unemp) + log(meanage) +
           log(precentb) + log(precenth)+ log(densrur) +
           log(densurb) + log(viopcap) + log(proppcap) +
           log(vmtrural) + log(vmturban) + log(fueltax) +
           lim65 + lim70p + mlda21 + bac08, SeatBelt,
           effect = "time")
fe <- update(ols, effect = "twoways")
ivfe <- update(fe, . ~ . | . - log(usage) + ds + dp +dsp)

rbind(ols = coef(summary(ols))[1,],
      fe = coef(summary(fe))[1, ],
      w2sls = coef(summary(ivfe))[1, ])
      Estimate Std. Error t-value  Pr(>|t|)
ols     0.1140    0.02547   4.478 9.252e-06
fe     -0.0535    0.02252  -2.376 1.790e-02
w2sls  -0.1334    0.04482  -2.975 3.079e-03
```

The results confirm that the endogeneity problem is very important. For the first fitted model, the seat belt-use coefficient is significantly positive. It becomes significantly negative for the fixed effects model, which means that `usage` is strongly correlated with the individual effects. Finally, this coefficient increases importantly (in absolute value) if instrumental variables are used, which indicates that the idiosyncratic error is also correlated with `usage`.

In order to test the behavior compensation theory, the authors estimate the same models, this time using the number of non-occupants killed (`noccfat`) as response.

```
SeatBelt$noccfat <- with(SeatBelt, log(farsnocc / (vmtrural + vmturban)))
nivfe <- update(ivfe, noccfat ~ . | .)
coef(summary(nivfe))[1, ]
  Estimate Std. Error    t-value   Pr(>|t|)
  -0.04237    0.10312   -0.41091    0.68133
```

The results indicate that seat belt use has no influence on out-of -vehicle mortality, in contradiction with Peltzman (1975)'s theory of behavior compensation.

6.3 Error Components Instrumental Variables Estimator

In the previous section, the potential correlation between some covariates and the individual effects has been treated drastically by using the *within* transformation, which wipes out the individual effects. In this section, we present the error component instrumental variables estimator. The two components of the error being present in this model, it is in this case essential to tackle the issue of a potential correlation of some covariates with the two components of the error.

6.3.1 The General Model

Suppose in a first step that the idiosyncratic component of the error is not correlated with the covariates. In this case, if all the covariates are uncorrelated with the individual effects, the unbiased efficient estimator is the GLS estimator. This estimator enables, on the one hand, to take into account part of the inter-individual variation in the sample and, on the other hand, to estimate parameters associated with covariates that don't exhibit temporal variations.

If, on the contrary, all the covariates are correlated with the individual effects, Mundlak (1978) (see subsection 4.2) has shown that the efficient estimator, which is the GLS estimator, is the same as the *within* estimator if the correlation between the individual effects and the covariates (more precisely the individual means of the covariates) is taken into account.

When only some covariates are correlated with the individual effects, none of the two previous estimators is appropriate any more:

- the GLS estimator is not consistent anymore because of the correlation of some covariates with the individual effects,
- the *within* estimator is still consistent but not efficient any more, as it doesn't take into account the fact that some covariates are uncorrelated with the individual effects but wipes out all the inter-individual variation in the sample, especially the covariates that don't exhibit any temporal variation.

The best solution in this case consists then in using an estimator that, on the one hand, uses instrumental variables and, on the other hand, exploits the two sources of variability of the panel in an optimal way. The essential question is then to find good instruments, which is often a difficult task. The richness of panel data allows to overcome this problem. Actually, every covariate can generate two instrumental variables, using the *between* and the *within* transformations. If a rank condition that will be detailed later on is checked, the model can then be estimated without any external instrument. This approach has been used by Hausman and Taylor (1981), Amemiya and MaCurdy (1986), and Breusch et al. (1989).

If, from now, we suspect that some covariates are also correlated with the idiosyncratic part of the error, then none of the estimators we have listed above is consistent. We then use an instrumental variables estimator (*within* or GLS) using external instruments. This strategy has been developed by Baltagi (1981) with his "error component two-stage least squares" EC2SLS estimator and by Balestra and Varadharajan-Krishnakumar (1987) with their "generalized two-stage least squares" G2SLS estimator, which differ by the way the instruments are introduced in the model.

This two branches of the literature have been developed separately, and this dichotomy exists also in most software packages, which usually provide two different functions to estimate these models. We'll follow the approach of Cornwell et al. (1992), who provide a unified view of panel models with instrumental variables. These authors consider three kinds of variables:

- the endogenous variables, which are correlated with the two components of the error,
- the simply exogenous variables, which are correlated with the individual effects but not with the idiosyncratic part of the error,
- the doubly exogenous variables, which are uncorrelated with both components of the error.

Variables from the first category don't provide any usable instrument. For the second one, the *within* transformation is a valid instrument, as it is by construction orthogonal to the individual effects and by hypothesis uncorrelated with the idiosyncratic part. Finally, each covariate of the third category provides two instruments by using the *within* and the *between* transformation.

Consider now the specific case of time-invariant covariates. For these variables, $WX = 0$ and $BX = X$. Therefore, such a variable provides either one instrument, if it is uncorrelated with the individual effects (the covariate itself), or no instrument.

We start with the model to be estimated written in matrix form:

$$y = Z\gamma + \epsilon$$

With the usual hypotheses concerning the error component model, the variance matrix of the error is: $\Omega = \sigma_\nu^2 W + \sigma_\eta^2 B$. We first pre-multiply the model by: $\Omega^{-0.5} = \sigma_\nu^{-0.5} W + \sigma_\eta^{-0.5} B$ and then obtain a transformed model for which the errors are *iid*.

$$\Omega^{-0.5} y = \Omega^{-0.5} Z\gamma + \Omega^{-0.5}\epsilon$$

We then apply to this model the instrumental variables method, using a set of instruments, which, denoting by $L_{(1)}$ the doubly exogenous variables, by $L_{(2)}$ the simply exogenous variables, and by $L = (L_{(1)}, L_{(2)})$ the whole set of instruments, can be written:

$$A = (WL, B\tilde{L})$$

where $\tilde{L}$ is a set of variables that will be defined later. For now, just consider that these variables must provide valid instruments when the *between* transformation is applied.

The instrumental variables estimator is, denoting by $\mathrm{P}_A = A(A^\top A)^{-1}A^\top$ the projection matrix defined by the instruments:

$$\hat{\gamma} = (Z^\top \Omega^{-0.5} \mathrm{P}_A \Omega^{-0.5} Z)^{-1} Z^\top \Omega^{-0.5} \mathrm{P}_A \Omega^{-0.5} y$$

The two matrices W and B being orthogonal, the projection matrix may also be written as the sum of two projection matrices defined by the instruments transformed by the *within* and the *between* matrices:

$$\mathrm{P}_A = WL(L^\top WL)^{-1}L^\top W + B\tilde{L}(\tilde{L}^\top B\tilde{L})^{-1}\tilde{L}^\top B = \mathrm{P}_{WL} + \mathrm{P}_{B\tilde{L}}$$

The estimator is then:

$$\hat{\gamma} = \left(\frac{1}{\sigma_\nu^2} Z^\top \mathrm{P}_{\mathrm{w}L} Z + \frac{1}{\sigma_\iota^2} Z^\top \mathrm{P}_{\mathrm{B}\tilde{L}} Z\right)^{-1} \left(\frac{1}{\sigma_\nu^2} Z^\top \mathrm{P}_{\mathrm{w}L} y + \frac{1}{\sigma_\iota^2} Z^\top \mathrm{P}_{\mathrm{B}\tilde{L}} y\right)$$

or also, denoting $\phi^2 = \sigma_\nu^2/\sigma_\iota^2$:

$$\hat{\gamma} = (Z^\top \mathrm{P}_L^{\mathrm{w}} Z + \phi^2 Z^\top \mathrm{P}_{\mathrm{B}\tilde{L}} Z)^{-1} (Z^\top \mathrm{P}_{\mathrm{w}L} y + \phi^2 Z^\top \mathrm{P}_{\mathrm{B}\tilde{L}} y) \tag{6.9}$$

One can check that, as in the simple error component model, this estimator is a weighted average of the *within* and the *between* estimators: $\hat{\gamma}_{\mathrm{EC2SLS}} = D^{\mathrm{W}}\hat{\gamma}_{\mathrm{W2SLS}} + D^{\mathrm{B}}\hat{\gamma}_{\mathrm{B2SLS}}$, with:

$$\begin{cases} D^{\mathrm{W}} = [Z^\top \mathrm{P}_{\mathrm{w}L} Z + \phi^2 Z^\top \mathrm{P}_{\mathrm{B}\tilde{L}} Z]^{-1} Z^\top \mathrm{P}_{\mathrm{w}L} \\ D^{\mathrm{B}} = \phi^2 [Z^\top \mathrm{P}_{\mathrm{w}L} Z + \phi^2 Z^\top \mathrm{P}_{\mathrm{B}\tilde{L}} Z]^{-1} Z^\top \mathrm{P}_{\mathrm{B}\tilde{L}} \end{cases}$$

Several models proposed in the literature are special cases of this general model.

6.3.2 Special Cases of the General Model

6.3.2.1 The *within* Model

Firstly, if there are no external instruments and if all the covariates are simply exogenous, we have $L = Z$ and $\tilde{L} = 0$, and the *within* estimator results.

Then, if all the covariates are either simply exogenous or endogenous and if the external instruments are simply exogenous, we also have $\tilde{L} = 0$, and L is constituted only by simply exogenous covariates and external instruments. The condition for identification is then that the number of external instruments must be at least equal to the number of endogenous covariates. We then have the *within* instrumental variables estimator:

$$\hat{\gamma} = (Z^\top \mathrm{P}_{\mathrm{w}L} Z)^{-1} Z^\top \mathrm{P}_{\mathrm{w}L} y$$

6.3.2.2 Error Components Two Stage Least Squares

Baltagi (1981)'s estimator is the special case where $L = \tilde{L}$, which means that all the instruments (and potentially some of the covariates) are assumed to be doubly exogenous and are therefore used twice. We start from equations (6.5) and (6.7), which leads respectively to the *within* and *between* estimators. Stacking these two equations, we obtain:

$$\begin{pmatrix} L^\top Wy \\ L^\top By \end{pmatrix} = \begin{pmatrix} L^\top WZ \\ L^\top BZ \end{pmatrix} \gamma + \begin{pmatrix} L^\top W\epsilon \\ L^\top B\epsilon \end{pmatrix}$$

which is justified by the fact that the vector of parameters to be estimated γ is the same in the two equations. In order to apply GLS, we compute the variance of the errors of the stacked model:

$$\mathrm{V}\begin{pmatrix} L^\top W\epsilon \\ L^\top B\epsilon \end{pmatrix} = \mathrm{E}\begin{pmatrix} L^\top W\epsilon\epsilon^\top WL & L^\top W\epsilon\epsilon^\top BL \\ L^\top B\epsilon\epsilon^\top WL & L^\top B\epsilon\epsilon^\top BL \end{pmatrix} = \sigma_\nu^2 \begin{pmatrix} L^\top WL & 0 \\ 0 & \frac{1}{\phi^2} L^\top BL \end{pmatrix}$$

We then apply the formula of the GLS estimator:

$$\begin{aligned}\hat{\gamma} &= \left[\begin{pmatrix} Z^\top WL & Z^\top BL \end{pmatrix} \begin{pmatrix} L^\top WL & 0 \\ 0 & \frac{1}{\phi^2} L^\top BL \end{pmatrix}^{-1} \begin{pmatrix} L^\top WZ \\ L^\top BZ \end{pmatrix} \right]^{-1} \\ &\times \begin{pmatrix} Z^\top WL & Z^\top BL \end{pmatrix} \begin{pmatrix} L^\top WL & 0 \\ 0 & \frac{1}{\phi^2} L^\top BL \end{pmatrix}^{-1} \begin{pmatrix} L^\top Wy \\ L^\top By \end{pmatrix}\end{aligned}$$

$$\begin{aligned}\hat{\gamma} &= [Z^\top WL(L^\top WL)^{-1}L^\top WZ + \phi^2 Z^\top BL(L^\top BL)^{-1}L^\top BZ]^{-1} \\ &\times [Z^\top WL(L^\top WL)^{-1}L^\top Wy + \phi^2 Z^\top BL(L^\top BL)^{-1}L^\top By]\end{aligned}$$

and we finally obtain:

$$\hat{\gamma}_{\text{EC2SLS}} = [Z^\top \mathrm{P}_L^{W} Z + \phi^2 Z^\top \mathrm{P}_L^{B} Z]^{-1}[Z^\top \mathrm{P}_L^{B} y + \phi^2 Z^\top \mathrm{P}_L^{B} y] \tag{6.10}$$

which is the special case of the general model defined by equation (6.9) for which $\tilde{L} = L$.

6.3.2.3 The Hausman and Taylor Model

In the Hausman and Taylor (1981) model, there are no endogenous variables, only simply or doubly exogenous variables. We then have $L_{(1)} = X_{(1)}$, $L_{(2)} = X_{(2)}$ and $L = L_{(1)} + L_{(2)} = X_{(1)} + X_{(2)}$. Moreover, the authors stress the presence of variables with (X^υ) or without (X^c) time variation. The set of instruments they use is:

$$(W(X_{(1)}, X_{(2)}), BX_{(1)}) = (W(X^\upsilon_{(1)}, X^\upsilon_{(2)}), X^c_{(1)}, BX^\upsilon_{(1)})$$

Only covariates that exhibit time variation may be used with their *within* transformation ($WX^c_{(1)} = WX^c_{(2)} = 0$) and doubly exogenous time-invariant variables are used without transformation as instruments ($BX^c_{(1)} = X^c_{(1)}$). Without external instruments, denoting by $K^c_{(1)}, K^c_{(2)}, K^\upsilon_{(1)}, K^\upsilon_{(2)}$ the number of covariates of the 4 categories, the number of instruments is $2K^\upsilon_{(1)} + K^c_{(1)} + K^\upsilon_{(2)}$ as the number of covariates is: $K^c_{(1)} + K^c_{(2)} + K^\upsilon_{(1)} + K^\upsilon_{(2)}$. The model is then identified if $K^\upsilon_{(1)} \geq K^c_{(2)}$, *i.e.*, if the number of doubly exogenous time-varying variables (which provide two instruments) is greater than the number of time-invariant simply exogenous variables, which provide no instrument.

6.3.2.4 The Amemiya-Macurdy Estimator

Hausman and Taylor (1981)'s estimator is consistent if the individual means of the doubly exogenous variables are uncorrelated with the individual effects. Amemiya and MaCurdy (1986) use the stronger hypothesis that the doubly exogenous variables are uncorrelated with the individual effects for each period. We then have: $\mathrm{E}(x_{nt}\eta_n) = 0\ \forall t$ for every doubly exogenous covariate. The corresponding instrument matrix is constructed the following way. Let $X^{\upsilon}_{n(1)}$ be the matrix of doubly exogenous instruments of dimension $T \times K^{\upsilon}_{(1)}$ for individual n. $x^{\upsilon}_{n(1)} = \mathrm{vec}(X^{\upsilon}_{n(1)})$ is a vector of length $T \times K^{\upsilon}_{n(1)}$ obtained by stacking the columns of $X^{\upsilon}_{n(1)}$. The instrument matrix for individual n is then $j_T \otimes x^{\upsilon\,\top}_{n(1)}$, and for the whole sample, we obtain a matrix of dimension $NT \times TK^{\upsilon}_{(1)}$:

$$X^{\upsilon *}_{(1)} = \begin{pmatrix} j_T \otimes x^{\upsilon}_{1(1)} \\ j_T \otimes x^{\upsilon}_{2(1)} \\ \vdots \\ j_T \otimes x^{\upsilon}_{N(1)} \end{pmatrix} \tag{6.11}$$

6.3.2.5 The Breusch, Mizon and Schmidt's Estimator

Breusch et al. (1989) expand the instruments used by Amemiya and MaCurdy (1986) by assuming that the *within* transformations of simply exogenous covariates are valid instruments at every period. Stated differently: $\mathrm{E}((x_{nt(2)} - \bar{x}_{n(2)})\eta_n) = 0$. We then obtain the further matrix of instruments $(WX^{\upsilon}_{(2)})^*$ by applying to $WX^{\upsilon}_{(2)}$ the same transformation than the one used in equation 6.11. The other contribution of Breusch et al. (1989) is to show how the different estimators can be presented in a consistent and nested way. They use the fact that the projection subspace defined by X^* is the same as the one defined by $BX, (WX)^*$:

- Hausman and Taylor (1981): $WX^{\upsilon}_{(1)}, WX^{\upsilon}_{2}, X^{c}_{(1)}, BX^{\upsilon}_{(1)}$,
- Amemiya and MaCurdy (1986): $WX^{\upsilon}_{(1)}, WX^{\upsilon}_{2}, X^{c}_{(1)}, BX^{\upsilon}_{(1)}, (WX^{\upsilon}_{(1)})^*$,
- Breusch et al. (1989): $WX^{\upsilon}_{(1)}, WX^{\upsilon}_{2}, X^{c}_{(1)}, BX^{\upsilon}_{(1)}, (WX^{\upsilon}_{(1)})^*, (WX^{\upsilon}_{(2)})^*$,

As each estimator adds instruments to the previous one, if these instruments are valid, it is necessarily more efficient. Moreover, the validity of extra instruments may be tested by comparing the two models with a Hausman test.

6.3.2.6 Balestra and Varadharajan-Krishnakumar Estimator

This last estimator, proposed by Balestra and Varadharajan-Krishnakumar (1987), is not, contrary to the others, a special case of the general model previously presented. For this model, called the **g2sls** estimator (for "generalized two-stage least squares"), the same transformation is applied to the instruments that is applied also to the covariates and to the response. Therefore, the matrix of instruments is:

$$WL + \phi BL = L - (1 - \phi)BL$$

Baltagi and Li (1992) have shown that the instruments used by Baltagi (1981), $L_{\mathrm{B}} = (WX, BX)$, perform the same projection as $L_{\mathrm{B}} = (WX, WX + \phi BX)$ and $(WX + \phi BX, BX)$. The instruments used by Balestra and Varadharajan-Krishnakumar (1987) are therefore a subset of those used by Baltagi (1981), the supplementary instruments used by Baltagi (1981) being either WX or BX.

Therefore, the estimator of Baltagi (1981) is necessarily not less efficient than the one of Balestra and Varadharajan-Krishnakumar (1987). Baltagi and Li (1992) show, using White (1986), that the supplementary instruments used by Baltagi (1981) are redundant, which means that they don't add any gain in terms of asymptotic efficiency. Consequently, both estimators have the same asymptotic variance.

However, the estimator of Balestra and Varadharajan-Krishnakumar (1987) has an important drawback. A part of the *between* component of every instrumental variable is included in the instruments, and consequently, the estimator of Balestra and Varadharajan-Krishnakumar (1987) is unable to take into account simply exogenous instruments.

With `plm`, the way instruments are introduced is indicated by the `inst.method` argument: `'baltagi'` indicates that instruments are introduced with the *within* and the *between* transformations, `'amc'` uses the set of instruments used by Amemiya and MaCurdy (1986), `'bmsc'` the one used by Breusch et al. (1989), and `'bvk'` indicates that the instrumental variables are transformed the same way as the covariates and the response, as proposed by Balestra and Varadharajan-Krishnakumar (1987).

Example 6.2 EC2SLS estimator – `ForeignTrade` data set

Kinal and Lahiri (1993) studied the determinants of international trade for developing countries and especially the measure of the price and income elasticities. This question is very important because it crucially determines the growth and debt of these countries. The panel dataset used concerns 31 developing countries, for the period 1964-1986. It is available as `ForeignTrade` in the **pder** package.

More precisely, Kinal and Lahiri (1993) estimate three equations: the first one defines the demand for imports, the second one the demand for exports, and the last one the exports supply. The authors suppose that:

- the demand for imports `imports` increases with the domestic income `gnp`, decreases with the price of imports in local currency divided by domestic prices `pmpci`, and rises with the one-period lag of the ratio of reserves to imports `resimp`.
- exports demand `exports` rises with the world income `gnpw` and decreases with the relative price of exports with respect to their foreign substitutes `pxpw`,
- exports supply `exports` increases with the world price in domestic currency divided by the domestic consumer price index `pwpci`, with the potential domestic product `pgnp` (used as a proxy for the capital stock) and also depends positively on a variable that represents the influence of the imports in the supply of exports `importspmpx` (measured by imports in local currency divided by export price).[3]

All the variables are per capita and in logs, in order to avoid heteroscedasticity problems.

In order to take the dynamics of adjustment into account, a one-period lag of the response is introduced as a covariate in every equation.

`gnp`, `exports`, `imports`, and their lags (and therefore `resimp` and `importspmpx`) are assumed to be endogenous, as are the exports price (which induces that `pxpw` is endogenous) and the domestic consumer price index is endogenous (which induces that `pmcpi` and `pwcpi` are also endogenous). Among the covariates, only `gnpw` and `pgnp` are assumed to be exogenous and can therefore be used as instruments. Numerous external instruments are also introduced: a linear trend `trend`, the population `pop`, the exchange rate `exrate`, the consumption `consump`, the disposable income `income`, the reserves `reserves`, money supply `money`, the

3 The authors justify the use of this variable by the fact that, in most developing countries, imports of intermediate and investment goods are very important to be able to produce export goods.

consumer price index cpi, import prices pm, export prices px, and world prices pw, most of the time with a one-period lag.

Kinal and Lahiri (1993) is an extension of the article of Khan and Knight (1988), who estimated a system of equations explaining the determinants of international trade for developing countries using the *within* transformation. They looked for a more efficient estimator, and for this purpose they employed the **EC2SLS** estimator. However, the latter is consistent only if the instruments are uncorrelated with the individual effects. Their strategy is to use the same specification for the *within* and the **EC2SLS** estimators and to test the hypothesis of exogeneity of the instruments through a Hausman test.

We present below the results obtained for the imports demand equation. The *within* and the **EC2SLS** models are estimated. Kinal and Lahiri (1993) use a nonstandard method to estimate the variance of the error components. It is similar to Nerlove (1971), but with a degrees of freedom correction. It is reproduced here by using the random.dfcor argument.

```
data("ForeignTrade", package = "pder")
w1 <- plm(imports~pmcpi + gnp + lag(imports) + lag(resimp)  |
          lag(consump) + lag(cpi) + lag(income) + lag(gnp) + pm +
          lag(invest) + lag(money) + gnpw + pw + lag(reserves) +
          lag(exports) + trend + pgnp + lag(px),
          ForeignTrade, model = "within")
r1 <- update(w1, model = "random", random.method = "nerlove",
             random.dfcor = c(1, 1), inst.method = "baltagi")
```

The hypothesis of no correlation between the instruments and the individual effects implies that the *within* and the **GLS** models are consistent, the latter being more efficient. On the contrary, if this hypothesis is rejected, only the *within* model is consistent. In order to test this hypothesis the authors used the Hausman (1978) test:

```
phtest(r1, w1)

Hausman Test

data:  imports ~ pmcpi + gnp + lag(imports) + lag(resimp) | lag(consump) +  ...
chisq = 11, df = 4, p-value = 0.03
alternative hypothesis: one model is inconsistent
```

The hypothesis of no correlation between the instruments and the individual effects is rejected at the 5% threshold.[4] One solution would be to maintain the *within* estimator, but Kinal and Lahiri (1993), following Cornwell et al. (1992), considered two kinds of instruments:

- those that are not correlated with the individual effects and that therefore can be used twice using the *within* and the *between* transformations,
- those that are correlated with the individual effects and that can therefore only be used in their *within* transformation.

Such a model is defined using a three-part formula:

- the second part indicates the doubly exogenous instruments,
- the third part indicates the simply exogenous instruments.

4 This is also the case for the two other equations: exports supply and exports demand.

Kinal and Lahiri (1993) finally got the following specification:

```
r1b <- plm(imports ~ pmcpi + gnp + lag(imports) + lag(resimp) |
           lag(consump) + lag(cpi) + lag(income) + lag(px) +
           lag(reserves) + lag(exports) | lag(gnp) + pm +
           lag(invest) + lag(money) + gnpw + pw  + trend + pgnp,
           ForeignTrade, model = "random", inst.method = "baltagi",
           random.method = "nerlove", random.dfcor = c(1, 1))

phtest(w1, r1b)

Hausman Test

data:  imports ~ pmcpi + gnp + lag(imports) + lag(resimp) | lag(consump) +  ...
chisq = 7.1, df = 4, p-value = 0.1
alternative hypothesis: one model is inconsistent
```

Based on the Hausman (1978) test, the hypothesis of consistency of the GLS estimator is no longer rejected. Results are presented below; the *within* and GLS estimators give very similar results.

```
rbind(within = coef(w1), ec2sls = coef(r1b)[-1])
          pmcpi     gnp lag(imports) lag(resimp)
within -0.05873 0.02890       0.9512     0.05215
ec2sls -0.05420 0.01361       0.9482     0.04195
```

The short-term elasticity of imports demand is directly given by the price coefficient. The long-term elasticity is obtained by dividing this coefficient by one minus the coefficient of the lagged response. We then have:

```
elast <- sapply(list(w1, r1, r1b),
                function(x) c(coef(x)["pmcpi"],
                              coef(x)["pmcpi"] / (1 - coef(x)["lag(imports)"])))
dimnames(elast) <- list(c("ST", "LT"), c("w1", "r1", "r1b"))
elast
         w1      r1     r1b
ST -0.05873 -0.0552 -0.0542
LT -1.20393 -1.1953 -1.0465
```

The use of this GLS estimator, which efficiently exploits part of the inter-individual variation, has dramatically reduced the standard deviations of the coefficients.

```
rbind(within = coef(summary(w1))[, 2],
      ec2sls = coef(summary(r1b))[-1, 2])
         pmcpi      gnp lag(imports) lag(resimp)
within 0.02915 0.041235      0.03067    0.008257
ec2sls 0.02180 0.006999      0.01289    0.006709
```

Example 6.3 Hausman-Taylor estimator – `TradeEU` data set

The analysis of international trade is often based on the gravity model, inspired by the law of universal gravitation in physics, which indicates that a particle attracts every other particle in the universe using a force that is directly proportional to the product of their masses and inversely proportional to the square of the distance between their centers. By similarity, in international trade the volume of exchange between two countries (imports and exports) is linked to the "masses" of both countries (which can be measured by the population or by their national product) and by the distance between them. Many econometric analyses of the gravity model have drawn on cross sections of countries. The problem of these studies is that they are unable to take into account unobservable heterogeneity at the country level, which leads to biased estimators. In this respect, the use of panel data seems very useful, but the fact that some covariates are correlated with individual effects often leads to employing the *within* estimator. The problem in this case is that that the time-invariant covariates disappear: yet some of these can be of major interest, especially the distance between two countries. The estimator of Hausman and Taylor (1981), which enables, on the one hand, to tackle the problem of correlation between some covariates and the individual effects and on the other hand to estimate the coefficients associated to time-invariant covariates, is very useful in this respect.

Serlenga and Shin (2007) estimate a gravity model for 14 countries of the European Union[5] observed over 42 years (1960-2001). In this panel, the individual unit of observation is not a country but a pair of countries for which the volume of trade is given by the sum of bilateral exports and imports. There are, therefore, $(14 \times 13)/2 = 91$ "individuals".

The response `trade` is the logarithm of the sum of bilateral imports and exports. The covariates are: `gdp`, the sum of the logarithms of the two national products; `dist`, the distance between the capitals of the two countries; `sim`, a measure of the similarity between the pair of countries; `rlf`, the relative factor endowment; and `rer`, the logarithm of the real exchange rate. To this quantitative variables, several qualitative variables are added: mutual adhesion to the European Community, `cee` and to the Euro Zone `emu`; common border; `bor`; and common language, `lan`.

The dataset, called `TradeEU`, is available in the **pder** package.

```
data("TradeEU", package = "pder")
```

Following the authors, we first estimate the OLS and the *within* model:

```
ols <- plm(trade ~ gdp + dist + rer + rlf + sim + cee + emu + bor + lan, TradeEU,
           model = "pooling", index = c("pair", "year"))
fe <- update(ols, model = "within")
fe

Model Formula: trade ~ gdp + dist + rer + rlf + sim + cee + emu + bor + lan

Coefficients:
   gdp    rer    rlf    sim    cee    emu
1.8125 0.0610 0.0325 1.1723 0.3093 0.0852
```

5 Austria, Belgium and Luxemburg (taken as a unique entity), Denmark, Finland, France, Germany, Greece, Ireland, Italy, Netherlands, Portugal, Spain, and the United Kingdom.

As expected, coefficients associated to `dist`, `bor`, and `lan` are not estimated in the *within* model, as these covariates disappear with the *within* transformation. On the contrary, the random effects estimator produces estimates for their coefficients.

```
re <- update(fe, model = "random")
re

Model Formula: trade ~ gdp + dist + rer + rlf + sim + cee + emu + bor + lan

Coefficients:
(Intercept)         gdp        dist         rer         rlf
   -13.9303      1.7949     -0.5909      0.0690      0.0334
        sim         cee         emu         bor         lan
     1.1427      0.3182      0.0927      0.4414      0.4172
```

The results of the random effects model indicate a distance elasticity of bilateral trade of about −0.6 and that having a common border or a common language have a similar effect (an increase of about 40%).

```
phtest(re, fe)

Hausman Test

data:  trade ~ gdp + dist + rer + rlf + sim + cee + emu + bor + lan
chisq = 13, df = 6, p-value = 0.04
alternative hypothesis: one model is inconsistent
```

With the Hausman test, we reject the hypothesis of no correlation at the 5% threshold.

Serlenga and Shin (2007) consider that, among the time-invariant variables, only `lan` is correlated with the individual effects. Two Hausman and Taylor (1981) models are then estimated. In the first one, the only doubly exogenous variable is the real exchange rate `rer`. In this case, the instrumental variables estimator is just identified, as there is only one instrument (the *between* transformation of `rer`) and only one endogenous variable `lan`. In the second one, domestic product `gdp` and relative factor endowment `rlf` are also used as instruments.

```
ht1 <- plm(trade ~ gdp + dist + rer + rlf + sim + cee + emu + bor + lan |
             rer + dist + bor | gdp + rlf + sim + cee + emu + lan ,
             data = TradeEU, model = "random", index = c("pair", "year"),
             inst.method = "baltagi", random.method = "ht")
ht2 <- update(ht1, trade ~ gdp + dist + rer + rlf + sim + cee + emu + bor + lan |
               rer + gdp + rlf + dist + bor| sim + cee + emu + lan)
```

Note than `random.method` is set to `'ht'` so that the *within* residuals used to compute the variance of the components of the error are purged of the influence of the time-invariant covariates.[6] The consistency of either specification is not rejected by the Hausman test.

6 See subsection 2.3.2.

```
phtest(ht1, fe)

Hausman Test

data:  trade ~ gdp + dist + rer + rlf + sim + cee + emu + bor + lan |  ...
chisq = 5e-25, df = 6, p-value = 1
alternative hypothesis: one model is inconsistent
phtest(ht2, fe)

Hausman Test

data:  trade ~ gdp + dist + rer + rlf + sim + cee + emu + bor + lan |  ...
chisq = 2.2, df = 6, p-value = 0.9
alternative hypothesis: one model is inconsistent
```

The last estimated model is suggested by Baltagi (2012). It is similar to the second specification but uses the instruments suggested by Amemiya and MaCurdy (1986) instead. The results are presented in table 6.1 by using the **texreg** package (see Leifeld, 2013).

```
ht2am <- update(ht2, inst.method = "am")
```

```
library("texreg")
texreg(list(ols, fe, re, ht1, ht2, ht2am),
       custom.model.names = c("OLS", "FE", "RE", "HT1", "HT2", "AM2"),
       caption = "Estimations of the gravity model.", label = "table:gravity",
       custom.gof.names  = c("R$^2$", "Adj. R$^2$", "Num. obs.", "s\\_idios",
                             "s\\_id"),
       scriptsize = FALSE)
```

The results of table 6.1 show first that the coefficients of the time-varying covariates are identical for the *within* and the just identified Hausman and Taylor (1981) estimator. This is not the case with the `ht2` model, which is overidentified, as noted by Baltagi (2012). Serlenga and Shin (2007) insist on the fact that the Hausman and Taylor (1981) estimations lead to a great reduction of the influence of the distance and an important increase of the influence of common language and common border. This last conclusion is qualified by Baltagi (2012), which uses the more efficient Amemiya and MaCurdy (1986) estimator. The latter introduces further orthogonality conditions by imposing that doubly exogenous variables be uncorrelated with individual effects at any time, while the Hausman and Taylor (1981) estimator simply requires no correlation between individual effects and the averages of said variables. If these conditions are valid (which can be tested through the Hausman procedure), this estimator is necessarily not less efficient than that of Hausman and Taylor (1981).

```
phtest(ht2am, fe)

Hausman Test

data:  trade ~ gdp + dist + rer + rlf + sim + cee + emu + bor + lan |  ...
chisq = 10, df = 6, p-value = 0.1
alternative hypothesis: one model is inconsistent
```

Table 6.1 Estimations of the gravity model.

	OLS	FE	RE	HT1	HT2	AM2
(Intercept)	-10.95^{***}		-13.93^{***}	-15.76^{***}	-15.66^{***}	-14.00^{***}
	(0.25)		(0.89)	(1.50)	(1.50)	(1.13)
gdp	1.58^{***}	1.81^{***}	1.79^{***}	1.81^{***}	1.81^{***}	1.80^{***}
	(0.01)	(0.02)	(0.02)	(0.02)	(0.02)	(0.02)
dist	-0.65^{***}		-0.59^{***}	-0.38^{*}	-0.38^{*}	-0.59^{***}
	(0.02)		(0.12)	(0.19)	(0.19)	(0.15)
rer	0.10^{***}	0.06^{***}	0.07^{***}	0.06^{***}	0.06^{***}	0.07^{***}
	(0.00)	(0.01)	(0.01)	(0.01)	(0.01)	(0.01)
rlf	0.03^{***}	0.03^{***}	0.03^{***}	0.03^{***}	0.03^{***}	0.03^{***}
	(0.01)	(0.01)	(0.01)	(0.01)	(0.01)	(0.01)
sim	0.88^{***}	1.17^{***}	1.14^{***}	1.17^{***}	1.20^{***}	1.15^{***}
	(0.02)	(0.06)	(0.05)	(0.06)	(0.05)	(0.05)
cee	0.32^{***}	0.31^{***}	0.32^{***}	0.31^{***}	0.31^{***}	0.31^{***}
	(0.02)	(0.02)	(0.02)	(0.02)	(0.02)	(0.02)
emu	0.20^{***}	0.09^{**}	0.09^{***}	0.09^{**}	0.09^{**}	0.09^{***}
	(0.05)	(0.03)	(0.03)	(0.03)	(0.03)	(0.03)
bor	0.52^{***}		0.44^{*}	0.60^{*}	0.61^{*}	0.44
	(0.03)		(0.19)	(0.26)	(0.26)	(0.25)
lan	0.23^{***}		0.42^{*}	1.56^{*}	1.56^{*}	0.43
	(0.03)		(0.18)	(0.71)	(0.68)	(0.24)
R^2	0.90	0.90	0.90	0.90	0.90	0.90
Adj. R^2	0.90	0.90	0.90	0.90	0.90	0.90
Num. obs.	3822	3822	3822	3822	3822	3822
s_idios			0.29	0.29	0.29	0.29
s_id			0.52	0.65	0.67	0.67

$^{***}p < 0.001$, $^{**}p < 0.01$, $^{*}p < 0.05$

The validity of the supplementary instruments used for the Amemiya and MaCurdy (1986) estimator is not rejected by the Hausman test. The standard deviation of the endogenous variable (`lan`) is much lower than in the Hausman and Taylor (1981) estimator (0.24 vs 0.68). The coefficients of the three time-invariant covariates are closer to the **OLS** coefficients than to the Hausman and Taylor (1981) coefficients.

6.4 Estimation of a System of Equations

Instead of estimating only one equation, we can consider a whole system of simultaneous equations, in order to take into account the correlation between the errors of different equations. The estimator obtained is a mix of the **2SLS** estimator described in the previous chapter and the **SUR** estimator (see 3.2.4).

6.4.1 The Three Stage Least Squares Estimator

When there is no correlation between the covariates and the error, the relevant model for the system of equations is the **SUR** model, which is a **GLS** estimator and is described in section 3.2. Denoting by Σ the matrix of covariance of the errors of the L equations, the variance of the errors of the system is $\Omega = \Sigma \otimes \mathrm{I}$, and the **SUR** estimator is:

$$\hat{\gamma} = (Z^\top(\Sigma \otimes \mathrm{I})^{-1}Z)Z^\top(\Sigma \otimes \mathrm{I})^{-1}y$$

This expression involves square matrices of dimensions equal to the sample size. It is therefore not operational for large samples, and it is numerically inefficient anyway. It is therefore preferred, as often happens for **GLS** estimators, to apply **OLS** on transformed data. Denoting by υ_{lm} the elements of the matrix $\Sigma^{-0.5}$, each variable $z^\top = (z_1^\top, z_2^\top, \dots, z_L^\top)$ of the model is transformed by pre-multiplying it by: $\Psi = \Omega^{-0.5} = \Sigma^{-0.5} \otimes \mathrm{I}$. We then have:

$$z^* = (\Sigma^{-0.5} \otimes \mathrm{I})\begin{pmatrix} z_1 \\ z_2 \\ \vdots \\ z_L \end{pmatrix} = \begin{pmatrix} \upsilon_{11}z_1 + \upsilon_{12}z_2 + \dots \upsilon_{1L}z_L \\ \upsilon_{21}z_1 + \upsilon_{22}z_2 + \dots \upsilon_{2L}z_L \\ \vdots \\ \upsilon_{L1}z_1 + \upsilon_{L2}z_2 + \dots \upsilon_{LL}z_L \end{pmatrix}$$

The three-stage least squares estimator is obtained by using the moment conditions: $\mathrm{E}(L^\top\epsilon) = \mathrm{E}(L^\top(y - Z\gamma)) = 0$, for which the variance is: $\mathrm{V}(L^\top\epsilon) = \sigma^2 L^\top\Omega L$. Consistently with the method of moments approach, the estimator is obtained by minimizing a quadratic form of the vector of moments, using the inverse of the variance matrix of these moments:

$$\frac{1}{\sigma^2}(y^\top - \gamma^\top Z^\top)L(L^\top\Omega L)^{-1}L^\top(y - Z\gamma)$$

First order conditions for a minimum are:

$$-2Z^\top L(L^\top\Omega L)^{-1}L^\top(y - Z\gamma)/\sigma^2 = 0$$

Solving this linear system of equations, we obtain the **3SLS** estimator:

$$\hat{\gamma}_{\mathrm{IV}} = (Z^\top L(L^\top(\Sigma \otimes \mathrm{I})L)^{-1}L^\top Z)^{-1}(Z^\top L(L^\top(\Sigma \otimes \mathrm{I})L)^{-1}L^\top y) \quad (6.12)$$

The **3SLS** estimator may be obtained by employing the instrumental variables estimator, pre-multiplying the covariates and the response by $\Psi = \Sigma^{-0.5} \otimes \mathrm{I}$ and the instruments by $(\Psi^{-1})^\top = (\Sigma^{0.5})^\top \otimes \mathrm{I}$. The instruments are then $\tilde{L} = (\Psi^{-1})^\top L$ and define the following projection matrix:

$$\mathrm{P}_{\tilde{L}} = (\Psi^{-1})^\top L(L^\top\Psi^{-1}(\Psi^{-1})^\top L)^{-1}L^\top\Psi^{-1}$$

But:

$$\Psi^{-1}(\Psi^{-1})^\top = \Psi^{-1}(\Psi^\top)^{-1} = (\Psi^\top\Psi)^{-1} = \Omega$$

We then have

$$\mathrm{P}_{\tilde{L}} = (\Psi^{-1})^\top L(L^\top\Omega L)^{-1}L^\top\Psi^{-1}$$

Using this projection matrix in the formula of the instrumental variables estimator (6.3) we finally get:

$$\begin{aligned}\hat{\gamma} &= (Z^\top\Psi^\top(\Psi^{-1})^\top L(L^\top\Omega L)^{-1}L^\top\Psi^{-1}\Psi Z)^{-1} \\ &\times (Z^\top\Psi^\top(\Psi^{-1})^\top L(L^\top\Omega L)^{-1}L^\top\Psi^{-1}\Psi y)\end{aligned} \quad (6.13)$$

or

$$\hat{\gamma} = (Z^\top L(L^\top \Omega L)^{-1} L^\top Z)^{-1} (Z^\top L(L^\top \Omega L)^{-1} L^\top y)$$

which is the formula (6.12) of the **3SLS** estimator. Of course, as in the **GLS** estimator, Ω is in practice unknown and shall be estimated based on the results from a consistent preliminary estimation.

The practical computation of the **3SLS** estimator consists then of the following steps:

- each equation is first estimated independently using the instrumental variables estimator, which leads to a matrix of residuals $\hat{\Xi} = (\hat{\epsilon}_1, \hat{\epsilon}_2, \dots, \hat{\epsilon}_L)$ which is a consistent estimate of the errors of the equations,
- the covariance matrix of the errors of the system is then estimated: $\hat{\Sigma} = \hat{\Xi}^\top \hat{\Xi}/O$
- the Cholesky decomposition of this matrix is computed: $\hat{C} \mid \hat{C}\hat{\Sigma}\hat{C}^\top = \mathrm{I}$,
- the variables are transformed using this matrix: $\tilde{y} = (\hat{C} \otimes \mathrm{I})y$, $\tilde{Z} = (\hat{C} \otimes \mathrm{I})Z$ and $\tilde{L} = ((\hat{C}^{-1})^\top \otimes \mathrm{I})L$.
- and finally the instrumental variables estimator is applied to the transformed system.

The computation of the *within* or *between* **3SLS** estimators is straightforward, as it consists in applying the **3SLS** to *within* or *between* transformed data.

6.4.2 The Error Components Three Stage Least Squares Estimator

Balestra and Varadharajan-Krishnakumar (1987) and Baltagi (1981) have proposed **3SLS** estimators that use the inter- and intra-individual variations of the data in an optimal way.

From now, three indexes must be considered, the individual $n = 1 \dots N$ and time indexes $t = 1 \dots T$ as usual, but also the equation index $l = 1 \dots L$.

$$\epsilon_{lnt} = \eta_{ln} + \nu_{lnt}$$

Denoting by $\epsilon_{ln}^\top = (\epsilon_{ln1}, \dots, \epsilon_{lnT})$, the error vector for individual n and equation l, the error vector for the system of equations is:

$$\epsilon^\top = ((\epsilon_{11}^\top, \epsilon_{12}^\top, \dots, \epsilon_{1N}^\top), (\epsilon_{21}^\top, \epsilon_{22}^\top, \dots, \epsilon_{2N}^\top), \dots, (\epsilon_{L1}^\top, \epsilon_{L2}^\top, \dots, \epsilon_{LN}^\top))$$

The covariance matrix of the errors is then:

$$\Omega = \mathrm{V}(\epsilon) = \Sigma_\eta \otimes (\mathrm{I}_N \otimes J_T) + \Sigma_\nu \otimes (\mathrm{I}_N \otimes \mathrm{I}_T)$$

The presence of individual effects makes this model specific compared to the standard **3SLS** estimator. Compared to the standard error component model, scalars σ_η^2 and σ_ν^2 are replaced by two covariance matrices Σ_η and Σ_ν.

$$\begin{aligned}
\Omega &= (T\Sigma_\nu + \Sigma_\eta) \otimes (\mathrm{I}_N \otimes \bar{J}_T) + \Sigma_\nu \otimes (\mathrm{I}_N \otimes (\mathrm{I}_T - \bar{J}_T)) \\
&= (T\Sigma_\nu + \Sigma_\eta) \otimes B + \Sigma_\nu \otimes W \\
&= \Sigma_\iota \otimes B + \Sigma_\nu \otimes W
\end{aligned}$$

The **3SLS** estimator can then be computed the following way:

- firstly, the different equations are estimated using **2SLS** so that a consistent estimator of the matrix of the errors of the different equations $\hat{\Xi}$ may be computed;
- then, Σ_ν and Σ_ι are estimated by $\hat{\Sigma}_\nu$ and $\hat{\Sigma}_\iota$,
- covariates and responses are transformed by pre-multiplying them by: $\hat{\Psi} = \hat{\Omega}^{-0.5} = \hat{C}_\iota \otimes B + \hat{C}_\nu \otimes W$,

- instrumental variables are transformed by pre-multiplying them by: $(\hat{C}_l^{-1})^\top \otimes B + (\hat{C}_\nu^{-1})^\top \otimes W$,
- the **2SLS** estimator is then applied to the transformed data.

As for the **2SLS** estimator, the difference between the estimators of Baltagi (1981) and Balestra and Varadharajan-Krishnakumar (1987) is that the former uses the *within* and the *between* transformations of the instruments, while the latter uses a quasi-difference transformation.

Example 6.4 error components 3SLS – `ForeignTrade` data set

Kinal and Lahiri (1993) estimate the system composed of the demand for imports and the demand for exports by **3SLS**. To compute this estimator with `plm`, one has to use as first argument a list containing the description of the equations in the system.

```
eqimp <- imports ~ pmcpi + gnp + lag(imports) +
                   lag(resimp) | lag(consump) + lag(cpi) + lag(income) +
                   lag(px) + lag(reserves) + lag(exports) | lag(gnp) + pm +
                   lag(invest) + lag(money) + gnpw + pw  + trend + pgnp
eqexp <- exports ~ pxpw + gnpw + lag(exports) |
                   lag(gnp) + pw + lag(consump) + pm + lag(px) + lag(cpi) |
                   lag(money) + gnpw +  pgnp + pop + lag(invest) +
                   lag(income) + lag(reserves) + exrate
r12 <- plm(list(import.demand = eqimp,
                   export.demand = eqexp),
               data = ForeignTrade, index = 31, model = "random",
               inst.method = "baltagi", random.method = "nerlove",
               random.dfcor = c(1, 1))
summary(r12)
Oneway (individual) effect Random Effect Model
   (Nerlove's transformation)
Call:
plm.list(formula = list(import.demand = eqimp, export.demand = eqexp),
    data = ForeignTrade, model = "random", random.method = "nerlove",
    inst.method = "baltagi", index = 31, ... = pairlist(random.dfcor = c(1,
        1)))

Balanced Panel: n = 31, T = 24, N = 744

Effects:

  Estimated standard deviations of the error
      import.demand export.demand
id           0.0619        0.0782
idios        0.1439        0.1200

  Estimated correlation matrix of the individual effects
                 import.demand export.demand
import.demand            1.000             .
export.demand            0.138             1
```

```
 Estimated correlation matrix of the idiosyncratic effects
                import.demand export.demand
import.demand          1.0000             .
export.demand          0.0975             1

 - import.demand
               Estimate Std. Error t-value Pr(>|t|)
(Intercept)    0.39874    0.11899     3.35  0.00083 ***
pmcpi         -0.05407    0.02170    -2.49  0.01282 *
gnp            0.01103    0.00531     2.08  0.03785 *
lag(imports)   0.95046    0.01187    80.05  < 2e-16 ***
lag(resimp)    0.03948    0.00634     6.22  6.3e-10 ***
---
Signif. codes:
0 '***' 0.001 '**' 0.01 '*' 0.05 '.' 0.1 ' ' 1

 - export.demand
               Estimate Std. Error t-value Pr(>|t|)
(Intercept)     0.1437     0.1395     1.03   0.3032
pxpw           -0.0615     0.0195    -3.16   0.0016 **
gnpw            0.1144     0.0534     2.14   0.0322 *
lag(exports)    0.9465     0.0133    71.11   <2e-16 ***
---
Signif. codes:
0 '***' 0.001 '**' 0.01 '*' 0.05 '.' 0.1 ' ' 1
```

The coefficients for the imports demand equation are very close to those we obtained using the 2SLS estimator. The correlation between the two components of the errors of the two equations is about 10%. Taking into account this correlation slightly reduces the standard errors of the coefficients, as illustrated below.

```
rbind(ec2sls = coef(summary(r1b))[-1, 2],
      ec3sls = coef(summary(r12), "import.demand")[-1, 2])
        pmcpi      gnp lag(imports) lag(resimp) (Intercept)
ec2sls 0.0218 0.006999      0.01289    0.006709      0.0218
ec3sls 0.0217 0.005308      0.01187    0.006342      0.1395
           pxpw    gnpw lag(exports)
ec2sls 0.006999 0.01289     0.006709
ec3sls 0.019467 0.05336     0.013310
```

6.5 More Empirical Examples

Acconcia et al. (2014) seek to estimate the multiplier effect of public spending. This is a difficult task, as public spending can hardly be considered exogenous. They use a panel of 95 Italian administrative regions (provinces) for the years 1990-1999 and take advantage of the implementation of anti-mafia laws, which resulted in the eviction of some elected officials who were replaced by external commissioners. This replacement, which led to a drastic reduction in local public spending, represents an exogenous source of variation in public spending that can be usefully employed as instrument. Using a fixed effects **2SLS** estimator, they estimate the long-term public spending multiplier to be 1.95, a much larger value than the one obtained using the *within* estimator. The `Mafia` dataset is available in the **pder** package.

Egger and Pfaffermayr (2004) studied the determinants of bilateral trade of two countries, Germany and the United States, with their partners, bilateral trade being measured by imports and exports on the one hand, and by foreign direct investment on the other. The authors suspect that the individual effect, which indicates a propensity to trade with a given country for geographical and cultural reasons, is correlated with the distance. In this case, this variable, which is the only time-invariant one, is certainly correlated with the individual effect. The authors use the estimator of Hausman and Taylor (1981) for each equation and also for the system of two equations. The data are provided as `TradeFDI` in the **pder** package.

Hutchison and Noy (2005) study the effects of twin crises, characterized by the simultaneous occurrence of a bank and a currency crisis, on the wealth of countries. The panel consists of 24 developing countries for the 1975-1997 period. The response is the growth rate of the GDP and the two main covariates are the lag of the growth rate and a dummy variable indicating the occurrence of a twin crisis. Employing the lag of the growth rate as a covariate induces an endogeneity problem, which the authors tackle using an error component **2SLS** estimator. The results indicate that the cost of a currency crisis is about 5-8% in terms of growth every year for about 2-4 years, while for the bank crisis this is about 8-10%. The article doesn't provide any evidence of a specific effect of twin crises. The data are provided as `TwinCrises` in the **pder** package.

Cornwell and Trumbull (1994) and Baltagi (2006) estimate a crime economics model for the counties of North Carolina. The response is the criminality rate and, among the covariates, they introduce the probability of being arrested and the number of policemen per inhabitant. These two covariates induce an endogeneity problem: one actually wants to estimate the causal effect of police on crime, but a reverse causality effect is also likely, because more crime will induce the presence of more policemen. Two instrumental variables are used: the offense mix, which is defined as the ratio of crimes involving face-to-face contact to those that do not, and the per capita tax revenue. The first instrument is positively correlated with the probability of being arrested (because the offender may be identified by the victim). The second variable is positively correlated with the number of policemen, more tax income indicating a strong preference for public services and particularly for security. The **2SLS** error component model indicates a much stronger effect of the probability of being arrested than for the other estimators, especially the *within* estimator. The data are provided as `Crime` in the **plm** package.

Baltagi and Khanti-Akom (1990) and Cornwell and Rupert (1988) estimate a wage function using a panel of American individuals, with particular interest in the return to education. A well-known problem of such studies is that unobserved characteristics of individuals, called abilities, are part of the individual effects and may be correlated with education. Using the *within* model, the education covariate disappears: the use of the estimator of Hausman and Taylor (1981) is therefore very relevant in this context. Two time-invariant covariates (being black and being a female) are assumed exogenous, while the level of education is endogenous. Some other time-varying covariates are assumed exogenous and therefore provide two instruments so that the model is identified. The coefficient of education from the Hausman and Taylor (1981) estimator is larger than the one obtained using **GLS** (0.14 vs 0.10). The data are provided as `Wages` in the **plm** package.

7

Estimation of a Dynamic Model

A model is said to be dynamic when one of the regressors is the lagged dependent variable. The usefulness of panel data for estimating dynamic models is self-evident: it is impossible to estimate a dynamic relationship on cross-sectional data while, in the case of time series data, such model cannot be precisely estimated without drawing on long enough a sample. By contrast, with panel data a dynamic model can be estimated over a set of individuals observed over a small number of time periods. The models presented in this chapter are well suited to "micro-panels," i.e., datasets where $N >> T$. For "macro-panels," characterized by a temporal dimension equivalent to, or bigger than, the cross-sectional one, the appropriate models will be based on an adaptation of the methodology employed in unit roots tests and cointegration estimators to the specific issues of panel data.[1]

Among the many applied examples from the literature, one can mention:

- the estimation of per capita income convergence by regressing the growth rate as a function of the initial wealth level or, equivalently, regressing the level of per capita wealth as a function of the lagged wealth level;
- the analysis of the speed of adjustment of the labor force, obtained by regressing employment over different variables, including lagged employment;
- the dynamic analysis of consumption, based on a consumption function including lagged consumption.

The seminal article regarding estimation of dynamic panel models is Balestra and Nerlove (1966). The literature on the subject has become considerable since the 1990s and the papers by Holtz-Eakin, Newey, and Rosen (1988) and Arellano and Bond (1991), who introduced the use of the generalized method of moments for dynamic panels.[2] This one has become the preferred estimation method and the better part of this chapter will be dedicated to presenting it. It shall nevertheless be noted that the field of application of this method to panel data is not limited to dynamic panels and that it can be equally appropriate for static models.

Example 7.1 description of the data – `DemocracyIncome` data set
Along this whole chapter, we will use the paper by Acemoglu, Johnson, Robinson, and Yared (2008) to illustrate results. This study addresses the causal relationship between the level of

1 See section 8.4.
2 Among the many reviews of this literature see in particular Harris et al. (2008), Bond (2002), Roodman (2009a).

Panel Data Econometrics with R, First Edition. Yves Croissant and Giovanni Millo.

Companion website: www.wiley.com/go/croissant/data-econometrics-with-R

wealth and that of democracy in a country. The authors draw on different panel datasets, among which we have considered two:

- the first, made of data observed every 5 years, with 11 observations over the period 1950-2000 for 211 countries;
- the second, corresponding to data observed every 25 years, with 7 observations over the period 1850-2000 for 25 countries.

```
data("DemocracyIncome", package = "pder")
data("DemocracyIncome25", package = "pder")
```

In the cross section, the positive relationship between the degree of democracy and per capita income is apparent and is illustrated in the Figure 7.1, which uses Acemoglu et al. (2008)'s data for the year 2000. However, this contemporaneous correlation does not necessarily imply a causal relationship between the two variables. Using panel data instead allows to investigate causality by specifying a dynamic relationship.

```
library("plm")
pdim(DemocracyIncome)
Balanced Panel: n = 211, T = 11, N = 2321
head(DemocracyIncome, 4)
  country year democracy income sample
1 Andorra 1950        NA     NA      0
2 Andorra 1955        NA     NA      0
3 Andorra 1960        NA     NA      1
4 Andorra 1965        NA     NA      1
```

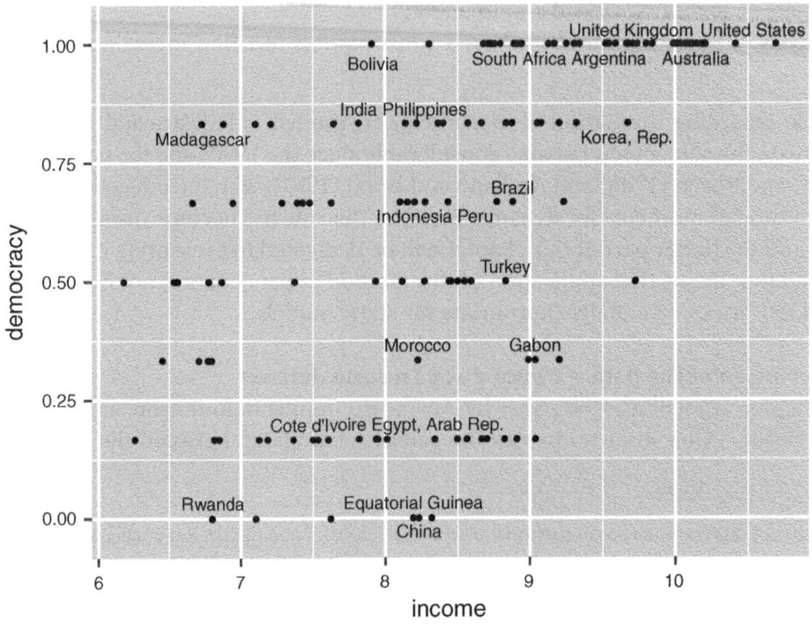

Figure 7.1 Relationship between income and democracy.

The five-year data constitutes a balanced panel of 211 countries observed over 11 periods. However, such balance is artificial because many observations are actually missing, in particular as regards democracy levels. The data comprise the two individual and time indexes (`country` and `year`), the democracy index `democracy`, the log of per capita gross domestic product `income`, and lastly, an indicator allowing to select the subset considered by the authors `sample`.

7.1 Dynamic Model and Endogeneity

The simplest dynamic model is the first order auto-regressive model

$$y_{nt} = \rho y_{n(t-1)} + \eta_n + \nu_{nt}$$

The error of the model is the sum of an individual effect η_n, which is time-invariant, and of an idiosyncratic component ν_{nt}, which would be called the *innovation*.

Throughout this chapter, we'll suppose that the innovations are mutually uncorrelated $\mathrm{E}(\nu_{nt}\nu_{ns}) = 0\ \forall s \neq t$, not correlated with the individual effect $\mathrm{E}(\eta_n \nu_{nt}) = 0$, and that the process is stationary ($\mid \rho \mid < 1$).

For the previous period, the model can be written as: $y_{n(t-1)} = \rho y_{n(t-2)} + \eta_n + \nu_{n(t-1)}$. The error and the covariate $y_{n(t-1)}$ are then correlated because $y_{n(t-1)}$ is correlated with the individual effect η_n.

7.1.1 The Bias of the OLS Estimator

Because of this correlation, the **OLS** estimator (and also the **GLS**) is not consistent. This estimator is:

$$\hat{\rho} = \frac{\sum_{n=1}^{N}\sum_{t=2}^{T} y_{nt}y_{n(t-1)}}{\sum_{n=1}^{N}\sum_{t=2}^{T} y_{n(t-1)}^2} = \rho + \frac{\sum_{n=1}^{N}\sum_{t=2}^{T}(\eta_n + \nu_{nt})y_{n(t-1)}}{\sum_{n=1}^{N}\sum_{t=1}^{T-1} y_{nt}^2}$$

and the numerator of the second term does not converge to 0 because η_n is positively correlated with $y_{n(t-1)}$. The correlation being positive, the **OLS** estimator is biased upward. In order to assess its magnitude, y_{nt} can be rewritten, by recursive substitution and denoting by $-S$ the starting date of the process and 1 that of the first observation:

$$\begin{aligned} y_{nt} &= \rho^{t+S} y_{n(-S)} + \frac{1-\rho^{t+s}}{1-\rho}\eta_n \\ &+ (\nu_{nt} + \rho\nu_{n(t-1)} + \rho^2\nu_{n(t-2)} + \ldots \rho^{t+s-1}\nu_{n(-S+1)}) \end{aligned} \tag{7.1}$$

Supposing that the initial values of y be fixed, for the denominator of the **OLS** estimator, one has then the following limits, first with respect to N and then to T:

$$\begin{aligned} \lim_{N\to+\infty} \frac{1}{N}\sum_{n=1}^{N} y_{nt}^2 &= \left(\frac{1-\rho^{t+S}}{1-\rho}\right)^2 \sigma_\eta^2 + \frac{1-\rho^{2(t+S)}}{1-\rho^2}\sigma_\nu^2 \\ \lim_{T\to+\infty}\lim_{N\to+\infty} \frac{1}{NT}\sum_{n=1}^{N}\sum_{t=1}^{T} y_{nt}^2 &= \frac{\sigma_\eta^2}{(1-\rho)^2} + \frac{\sigma_\nu^2}{1-\rho^2} \end{aligned} \tag{7.2}$$

For the numerator, by the hypothesis of no correlation between the individual effect and the innovations, one has:

$$\lim_{N\to+\infty} \frac{1}{N}\sum_{n}(\eta_n + \nu_{nt})y_{n(t-1)} = \frac{1-\rho^{t+S-1}}{1-\rho}\sigma_\eta^2$$

$$\lim_{N\to+\infty}\lim_{T\to+\infty}\frac{1}{NT}\sum_n\sum_t(\eta_n+\nu_{nt})y_{n(t-1)}=\frac{\sigma_\eta^2}{1-\rho} \tag{7.3}$$

The OLS estimator converges then to:

$$\text{plim}\,\hat{\rho}=\rho+\frac{\frac{\sigma_\eta^2}{1-\rho}}{\frac{\sigma_\eta^2}{(1-\rho)^2}+\frac{\sigma_\nu^2}{1-\rho^2}}=\rho+\frac{(1-\rho^2)\sigma_\eta^2}{(1+\rho)\sigma_\eta^2+(1-\rho)\sigma_\nu^2}$$

In view of this expression, the OLS estimator is biased upward. The bias tends to 0 when σ_η^2 does.

Example 7.2 time effects within model – `DemocracyIncome` data set
In the model of Acemoglu et al. (2008), the dependent variable is the democracy index, and the regressors are the one-period lags of the democracy index itself and of the per capita income. Yearly dummies are also introduced, and estimation is performed on the subsample defined by the `sample` variable. OLS estimation through the R function `lm` is distorted by the presence of lagged values. In fact, the `lag` method used by the program will be the one appropriate for time series, not the one for panel data.[3] For this reason, the function `plm` in the **plm** package will be used instead, setting the `model` argument to `'pooling'`, thus keeping the untransformed data. The `-1` in the formula indicates that we do not want to estimate a general constant but one coefficient for all instances of the `year` variable, which does not affect estimation.

```
ols <- plm(democracy ~ lag(democracy) + lag(income) + year - 1,
           DemocracyIncome, index = c("country", "year"),
           model = "pooling", subset = sample == 1)
```

The same model may be estimated by setting the `model` to `'within'` and the `effect` to `'time'`:

```
ols <- plm(democracy ~ lag(democracy) + lag(income),
           DemocracyIncome, index = c("country", "year"),
           model = "within", effect = "time",
           subset = sample == 1)
coef(summary(ols))
               Estimate Std. Error t-value   Pr(>|t|)
lag(democracy)  0.70637   0.024293  29.077 6.979e-133
lag(income)     0.07232   0.008343   8.668  1.915e-17
```

This first model highlights two results. On one hand, the `democracy` variable shows high persistence, with a coefficient of 0.71. However, we know that the OLS estimator suffers from a positive bias. On the other hand, lagged income seems to exert a significantly positive influence on the democracy index.

7.1.2 The within Estimator

The bias of OLS is due to the correlation between the error term and the lagged endogenous variable due to the presence of an individual effect; hence one may think to solve the problem

3 In particular, this means that the lagged value of the variable for the first observation of the second country will incorrectly be made equal to the last observation of the first country.

through a transformation that eliminates the individual effect. The most obvious choice is the *within* estimator. One has then, denoting $\bar{y}_{n(-1)} = \sum_{t=1}^{T-1} y_{nt}/(T-1)$ and $\bar{y}_n = \sum_{t=2}^{T} y_{nt}/(T-1)$:

$$\hat{\rho} = \frac{\sum_{n=1}^{N}\sum_{t=2}^{T}(y_{nt} - \bar{y}_n)(y_{n(t-1)} - \bar{y}_{n(-1)})}{\sum_{n=1}^{N}\sum_{t=2}^{T}(y_{nt} - \bar{y}_{n(-1)})^2}$$

$$= \rho + \frac{\sum_{n=1}^{N}\sum_{t=2}^{T}(y_{n(t-1)} - \bar{y}_{n(-1)})(v_{nt} - \bar{v}_n)}{\sum_{n=1}^{N}\sum_{t=2}^{T}(y_{nt} - \bar{y}_{n(-1)})^2}$$

The individual effects (and hence the bias) have disappeared from the OLS estimator, but a second source of bias has been introduced. In fact, $y_{n(t-1)} - \frac{1}{T-1}(y_{n1} + \ldots + y_{n(t-1)})$ and $v_t - \frac{1}{T-1}(v_{n2} + \ldots + v_{nt})$ are correlated. For $t > 2$, one has a term in $-\frac{1}{T-1}y_{nt} \times v_{nt}$, one in $y_{n(t-1)} \times -\frac{1}{T-1}v_{n(t-1)}$, and $T-2$ terms in $\frac{1}{(T-1)^2}y_{nt}v_{nt}$. Each term in $v_{nt}y_{nt}$ having an expectation of σ_v^2, one finally has:

$$\sigma_v^2\left(-\frac{1}{T-1} - \frac{1}{T-1} + \frac{T-2}{(T-1)^2}\right) = \sigma_v^2 \times \frac{-T}{(T-1)^2}$$

so that the bias is negative.[4] More precisely, one can show that:[5]

$$\text{plim } \hat{\rho} = \rho - \frac{1+\rho}{T-1}\frac{1 - \frac{1}{T}\frac{1-\rho^T}{1-\rho}}{1 - \frac{2\rho}{(1-\rho)(T-1)}\left(1 - \frac{1-\rho^T}{T(1-\rho)}\right)}$$

This bias differs from that of OLS in two ways. Firstly, it is negative, and secondly, it tends to 0 as T tends to infinity. This bias cannot therefore be ignored in the case of micro-panels, where the time dimension is short. For example, if $t = 10$ (a fairly long time span) and $\rho = 0.5$, the bias is -0.167.

Example 7.3 two-ways within Model – `DemocracyIncome` data set

The *within* model is obtained with `plm` fixing the `model` and `effect` arguments to `'within'` and `'twoways'`, since we want to introduce individual and time effects. The model can be simply estimated by updating the previous OLS model:

```
within <- update(ols, effect = "twoways")
coef(summary(within))
                Estimate Std. Error t-value  Pr(>|t|)
lag(democracy)  0.37863    0.03344 11.3212 1.252e-27
lag(income)     0.01041    0.02640  0.3945 6.933e-01
```

With respect to the OLS model, the autoregressive coefficient is smaller (0.38 vs. 0.71), which was to be expected as the *within* estimator is biased downward while OLS is biased upward. Notice also that after introducing individual effects, the coefficient of income is very close to 0 and not significant any more.

7.1.3 Consistent Estimation Methods for Dynamic Models

The most common estimation methods used for static models being inappropriate, various estimation strategies can be adopted to attain consistency.

4 Nickel (1981).
5 See for example Hsiao (2003) p. 72.

- the first is using the maximum likelihood method. However, it has a major shortcoming of being extremely sensitive to hypotheses made on the starting values of the explanatory variables. Depending on whether one considers the latter to be either fixed or random, and whether it is correlated with the individual effects or not, one obtains rather different models, and the estimation bias can be severe in case of misspecification. For this reason, this estimation method is scarcely used and will not be presented here;[6]
- the second consists in starting with a biased estimator and then correcting the bias. It is the step taken by Kiviet (1995), who proposes a *within* estimator corrected for the previously analyzed bias. However, the interest of this approach is limited by its applicability only to balanced panels and by the fact that it does not consider the possible endogeneity of other regressors;[7]
- the third possibility is using the instrumental variables method, the instruments being lagged levels or differences of the dependent variable. The generalized method of moments, which is an extension of the instrumental variables method, has become more and more popular to this end.

The instrumental variables method is used on a model that was pre-transformed in order to eliminate individual effects. Upon first consideration, the *within* transformation would seem a natural choice; it turns out to be inappropriate instead. In fact, in absence of appropriate external instruments, the only available instrumental variable is often the lagged dependent one, here meaning the dependent variable lagged at least twice. Then, in the *within* model, the error is: $\nu_{nt} - \frac{1}{T-1}\sum_{t=2}^{T} \nu_{nt}$. It contains all realizations of ν_{nt} and is therefore correlated with every lagged value of y_{nt}. Two alternative transformations can be successfully used here: first differences and orthogonal deviations.

For first differences, one simply has $\Delta z_{nt} = z_{nt} - z_{n(t-1)}$, or else, in vector form, $\Delta z_n = Dz_n$ with:

$$D = \begin{pmatrix} 1 & -1 & 0 & \dots & 0 & 0 \\ 0 & 1 & -1 & \dots & 0 & 0 \\ 0 & 0 & 1 & \dots & 0 & 0 \\ \vdots & \vdots & \vdots & \ddots & \vdots & \vdots \\ 0 & 0 & 0 & \dots & -1 & 0 \\ 0 & 0 & 0 & \dots & 1 & -1 \end{pmatrix}$$

The advantage of such transformation is to be simple and intuitive. It has, nevertheless, three drawbacks:

- firstly, one observation, the first one, is necessarily lost;
- secondly, if the original errors are not correlated to begin with, the transformed model's ones are. In fact, one has $\Delta\nu_t\Delta\nu_{t-1} = (\nu_t - \nu_{t-1})(\nu_{t-1} - \nu_{t-2})$ and hence, if the ν are homoscedastic and uncorrelated, the transformed errors are still homoscedastic $\mathrm{E}(\Delta\nu_t^2) = 2\sigma_\nu^2$, but correlated across two successive errors $\mathrm{E}(\Delta\nu_t\Delta\nu_{t-1}) = -\sigma_\nu^2$;
- lastly, for every time period t where one observation is missing, two are lost in differences: t and $t+1$.

---

6 For a detailed presentation of maximum likelihood estimation of dynamic panels, see Hsiao (2003), Chapter 4.

7 See Roodman (2009a), p. 103.

The orthogonal deviations transformation does not suffer from the last two problems, although it is less intuitive, consisting in calculating the difference between each observation and the average of those posterior. Formally, one has:

$$\tilde{z}_{nt} = c_{nt}\left(z_{nt} - \frac{1}{T_{nt}}\sum_{s>t}^{T} z_{ns}\right)$$

where T_{nt} is the number of observations posterior to t for individual n and c_{nt} is a scale factor equal to $\sqrt{\frac{T_{nt}}{T_{nt}+1}}$.

As for the first difference transformation, one observation is lost, but this is usually the last one. In matrix form, for a balanced panel, the transformation is written $\tilde{z}_t = Oz$, with $z = (z_1, z_2, \dots z_t)$ and:

$$O = \begin{pmatrix} \sqrt{\frac{T-1}{T}} & -\frac{1}{\sqrt{T(T-1)}} & -\frac{1}{\sqrt{T(T-1)}} & \dots & -\frac{1}{\sqrt{T(T-1)}} & -\frac{1}{\sqrt{T(T-1)}} \\ 0 & \sqrt{\frac{T-2}{T-1}} & -\frac{1}{\sqrt{(T-1)(T-2)}} & \dots & -\frac{1}{\sqrt{(T-1)(T-2)}} & -\frac{1}{\sqrt{(T-1)(T-2)}} \\ 0 & 0 & \sqrt{\frac{T-3}{T-2}} & \dots & -\frac{1}{\sqrt{(T-2)(T-3)}} & -\frac{1}{\sqrt{(T-2)(T-3)}} \\ \vdots & \vdots & \vdots & \ddots & \vdots & \vdots \\ 0 & 0 & 0 & \dots & \sqrt{\frac{1}{2}} & -\sqrt{\frac{1}{2}} \end{pmatrix}$$

Assuming that the original errors are homoscedastic and uncorrelated, one has, for the transformed errors:

$$\mathrm{V}(\tilde{v}) = \mathrm{E}(\tilde{v}\tilde{v}^\top) = \mathrm{E}(Ovv^\top O^\top) = \sigma_v^2 OO^\top = \sigma_v^2 \mathrm{I}$$

This last result is due to the fact that the rows of O are mutually orthogonal.

Moreover, in case of missing observations for one period, only that observation will be lost for estimation, versus two of them for the model estimated in first differences.

The estimator proposed by Anderson and Hsiao (1982) uses the model written in first difference form in order to eliminate individual effects. The explanatory variable $\Delta y_{n(t-1)} = y_{n(t-1)} - y_{n(t-2)}$ is then correlated with the differenced error $\Delta v_{nt} = v_{nt} - v_{n(t-1)}$. If the innovations are not serially correlated, $\Delta y_{n(t-1)}$ can be instrumented either by $\Delta y_{n(t-2)} = y_{n(t-2)} - y_{n(t-3)}$ or by $y_{n(t-2)}$. In practice, $y_{n(t-2)}$ is often a much better instrument than $\Delta y_{n(t-2)}$.

Example 7.4 Anderson and Hsiao estimator – `DemocracyIncome` data set

To compute the Anderson and Hsiao (1982) estimator, one must specify that both the regressand and regressors are differenced and that the lagged endogenous variable in differences is instrumented with the endogenous in levels lagged two periods. Acemoglu et al. (2008) also chose to instrument per capita income using a second lag. The model is simply described using a two-part formula,[8] the first part indicating the explanatory variables and the second the instruments, the two parts being separated by the sign |.

8 The extended formulas provided in the **Formula** package (Zeileis and Croissant, 2010) are used.

```
ahsiao <- plm(diff(democracy) ~ lag(diff(democracy)) +
              lag(diff(income)) + year - 1 |
              lag(democracy, 2) + lag(income, 2) + year - 1,
              DemocracyIncome, index = c("country", "year"),
              model = "pooling", subset = sample == 1)
coef(summary(ahsiao))[1:2,]
                      Estimate Std. Error t-value   Pr(>|t|)
lag(diff(democracy))    0.4687     0.1182  3.9651 7.971e-05
lag(diff(income))      -0.1036     0.3049 -0.3398 7.341e-01
```

Anderson and Hsiao (1982)'s model being consistent, one expects the estimated autoregressive coefficient to be comprised between that of the *within* model (biased downward) and that of the OLS model (biased upward). This is actually the case here, the obtained value of 0.47 falling between 0.38 and 0.71.

7.2 GMM Estimation of the Differenced Model

The instrumental variables estimator presented in the preceding section is inefficient for two reasons:

- firstly, it does not account for the correlation induced into the errors by first-differencing,
- secondly, there are further valid instruments available.

7.2.1 Instrumental Variables and Generalized Method of Moments

This estimator considers the fact that the number of valid instruments is growing in t. The dynamic character of the model renders the first observation unusable and first-differencing, the second one. Consequently, the first usable observation is the third one, for which the model can be written as:

$$y_{n3} - y_{n2} = \rho(y_{n2} - y_{n1}) + (\nu_{n3} - \nu_{n2})$$

For this observation, y_{n1} is the only valid instrument. For the fourth observation, the error is $\nu_{n4} - \nu_{n3}$, y_{n2} and y_{n1} are valid instruments. Thus, a supplementary instrument is added as t is incremented by 1. For the nth individual, the instruments matrix becomes:

$$L_n = \begin{pmatrix} y_{n1} & 0 & 0 & 0 & 0 & 0 & \dots & 0 & 0 & 0 & 0 \\ 0 & y_{n1} & y_{n2} & 0 & 0 & 0 & \dots & 0 & 0 & 0 & 0 \\ 0 & 0 & 0 & y_{n1} & y_{n2} & y_{n3} & \dots & 0 & 0 & 0 & 0 \\ \vdots & \vdots & \vdots & \vdots & \vdots & \vdots & \ddots & \vdots & \vdots & \ddots & \vdots \\ 0 & 0 & 0 & 0 & \dots & \dots & \dots & y_{n1} & y_{n2} & \dots & y_{n(T-2)} \end{pmatrix} \tag{7.4}$$

The moment conditions correspond to the vector $\mu = L^\top \Delta\nu$. The instruments being by hypothesis uncorrelated with the differenced errors, the expectation of this vector must be 0: $\mathrm{E}(\mu) = 0$.

The generalized method of moments consists in writing the sample equivalent of this vector of theoretical moments, i.e., the arithmetic average of the above expression for the set of individuals in the sample:

$$m = \frac{1}{N}\sum_{n=1}^{N} m_n = \frac{1}{N}\sum_{n=1}^{N} L_n^\top(\Delta y_n - \Delta X_n\beta) \tag{7.5}$$

where, in the simplest case of an autoregressive model, ΔX_n is a column vector containing the endogenous variable differenced and lagged by one period. How to obtain the estimator depends now on the comparison between the number of moments J and that of estimands K. If $J = K$, the method of moments estimator is obtained simply setting (7.5) to 0 and solving for β. One has then:

$$\hat{\beta} = \left(\sum_{n=1}^{N} L_n^{\mathrm{T}} \Delta X_n \right)^{-1} \left(\sum_{n=1}^{N} L_n^{\mathrm{T}} \Delta y_n \right)$$

If $J < K$, the system of linear equations defined by (7.5) is underidentified, and there are infinite combinations of parameter values allowing to equate (7.5) to 0. In the case when $J > K$, the system is overidentified and, apart form very particular cases, there is no combination of parameter values satisfying the equation. In this case, one will look for the parameter combination that minimizes the size of the moment conditions' vector, defined as a quadratic form in the vector of empirical moments:

$$\left(\frac{1}{N} \sum_{n=1}^{N} (\Delta y_n^{\mathrm{T}} - \beta^{\mathrm{T}} \Delta X_n^{\mathrm{T}}) L_n \right) A \left(\frac{1}{N} \sum_{n=1}^{N} L_n^{\mathrm{T}} (\Delta y_n - \Delta X_n \beta) \right) \tag{7.6}$$

where A is the weighting matrix of the moments. Setting to 0 the derivatives of (7.6) with respect to β, and solving with respect to β, one obtains the generalized method of moments estimator:

$$\begin{aligned} \hat{\beta} = & \left[\left(\sum\nolimits_n \Delta X_n^{\mathrm{T}} L_n \right) A \left(\sum\nolimits_n L_n^{\mathrm{T}} \Delta X_n \right) \right]^{-1} \\ & \times \left[\left(\sum\nolimits_n \Delta X_n^{\mathrm{T}} L_n \right) A \left(\sum\nolimits_n L_n^{\mathrm{T}} \Delta y_n \right) \right] \end{aligned} \tag{7.7}$$

7.2.2 One-step Estimator

In order to make this estimator computable, a weighting matrix has to be chosen. The simplest choice for A is identity. In this case, the function to minimize is simply the sum of squares of the elements in the vector. This solution is inefficient if the variances of these elements are different. In this case, intuitively, it is more efficient to assign a correspondingly higher weight to elements of the vector that have lower variance. The weighting matrix is then a diagonal one containing the inverse of the variance of each element. Moreover, if any elements in the vector are correlated, their joint weight will have to be reduced because they carry similar information. In general, the optimal weighting matrix is the inverse of the variance-covariance matrix of the vector of moments.[9] One has therefore:

$$A^{-1} = \mathrm{V}(\bar{m}) = \mathrm{V}\left(\frac{1}{N} \sum_{n=1}^{N} m_n \right) = \frac{1}{N^2} \sum_{n=1}^{N} \mathrm{V}(m_n)$$

If the errors in levels are homoscedastic and uncorrelated, $\mathrm{V}(m_n)$ has a very simple expression. In fact, one has:

$$\mathrm{V}(m_n) = \mathrm{E}(L_n^{\mathrm{T}} \Delta v_n \Delta v_n^{\mathrm{T}} L_n) = L_n^{\mathrm{T}} \mathrm{E}(D v_n v_n^{\mathrm{T}} D^{\mathrm{T}}) L_n = \sigma_v^2 L_n^{\mathrm{T}} h L_n$$

with:

$$h = DD^{\mathrm{T}} = \begin{pmatrix} 2 & -1 & 0 & \dots & 0 \\ -1 & 2 & -1 & \dots & 0 \\ 0 & -1 & 2 & \dots & 0 \\ \vdots & \vdots & \vdots & \vdots & \vdots \\ 0 & 0 & 0 & -1 & 2 \end{pmatrix} \tag{7.8}$$

9 See Hansen (1982).

In fact, the model errors are the differenced innovations $v_{nt} - v_{n(t-1)}$. Supposing that these errors are homoscedastic and uncorrelated, one has:

- $\mathrm{E}(\Delta v_{nt}^2) = 2\sigma_v^2$;
- $\mathrm{E}(\Delta v_{nt}\Delta v_{n(t-1)}) = -\sigma_v^2$;
- $\mathrm{E}(\Delta v_{nt}\Delta v_{ns}) = 0$ si $\mid t - s \mid > 1$.

The inverse of the weighting matrix can then be written as:

$$A^{(1)-1} = \mathrm{V}(\bar{m}) = \frac{1}{N^2}\sum_{n=1}^{N}\mathrm{V}(m_n) = \frac{\sigma_v^2}{N^2}\sum_{n=1}^{N}L_n^\top hL_n \tag{7.9}$$

σ_v^2 is an unknown scalar that does not play any role in estimation and which can therefore be ignored. The estimator using this weighting matrix is called the one-step estimator. It can be obtained simply by substituting $\left(\sum_{n=1}^{N}L_n^\top hL_n\right)^{-1}$ for A in the equation (7.7). To calculate its variance, one starts with replacing Δy_n in (7.7) by $\Delta X_n\beta + \Delta v_n$. One then has:

$$\begin{aligned}\hat{\beta}^{(1)} - \beta = &\left[\left(\sum_n \Delta X_n^\top L_n\right)\left(\sum_n L_n^\top hL_n\right)^{-1}\left(\sum_n L_n^\top \Delta X_n\right)\right]^{-1}\\ &\times\left[\left(\sum_n \Delta X_n^\top L_n\right)\left(\sum_n L_n^\top hL_n\right)^{-1}\left(\sum_n L_n^\top \Delta v_n\right)\right]\end{aligned} \tag{7.10}$$

which allows to obtain the variance of $\hat{\beta}^{(1)}$, denoted $\mathrm{V}^{(1)}$:

$$\begin{aligned}\mathrm{V}^{(1)} &= \mathrm{E}((\hat{\beta}-\beta)(\hat{\beta}-\beta)^\top)\\ &= \left[\left(\sum_n \Delta X_n^\top L_n\right)\left(\sum_n L_n^\top hL_n\right)^{-1}\left(\sum_n L_n^\top \Delta X_n\right)\right]^{-1}\\ &\times\left[\left(\sum_n \Delta X_n^\top L_n\right)\left(\sum_n L_n^\top hL_n\right)^{-1}\right.\\ &\times\mathrm{E}\left[\left(\sum_n L_n^\top \Delta v_n\right)\left(\sum_n \Delta v_n^\top L_n\right)\right]\\ &\times\left.\left(\sum_n L_n^\top hL_n\right)^{-1}\left(\sum_n L_n^\top \Delta X_n\right)\right]\\ &\times\left[\left(\sum_n \Delta X_n^\top L_n\right)\left(\sum_n L_n^\top hL_n\right)^{-1}\left(\sum_n L_n^\top \Delta X_n\right)\right]^{-1}\end{aligned} \tag{7.11}$$

If the hypotheses on errors are verified, one has:

$$\mathrm{E}\left[\left(\sum_n L_n^\top \Delta v_n\right)\left(\sum_n \Delta v_n^\top L_n\right)\right] = \sigma_v^2\sum_n L_n^\top hL_n$$

and the expression for the variance simplifies to:

$$\hat{\mathrm{V}}^{(1)} = \sigma_v^2\left[\left(\sum_n \Delta X_n^\top L_n\right)\left(\sum_n L_n^\top hL_n\right)^{-1}\left(\sum_n L_n^\top \Delta X_n\right)\right]^{-1} \tag{7.12}$$

The generalized moments estimator and its variance can be expressed more compactly using the following matrix notation: $\Delta X^\top = (\Delta X_1^\top, \Delta X_2^\top, \dots, \Delta X_n^\top)$, $\Delta y^\top = (\Delta y_1^\top, \Delta y_2^\top, \dots, \Delta y_n^\top)$, $L^\top = (L_1^\top, L_2^\top, \dots, L_n^\top)$ and H a block-diagonal matrix obtained by repeating h N times.

$$\hat{\beta}^{(1)} = [(\Delta X^\top L)(L^\top HL)^{-1}(L^\top \Delta X)]^{-1}[(\Delta X^\top L)(L^\top HL)^{-1}(L^\top \Delta y)] \tag{7.13}$$

$$\hat{\mathrm{V}}^{(1)} = \sigma_v^2[(\Delta X^\top L)(L^\top HL)^{-1}(L^\top \Delta X)]^{-1} \tag{7.14}$$

If, contrary to the assumptions made, the errors are actually heteroscedastic and/or autocorrelated, the one-step estimator remains consistent, but two classic problems arise:

- on the one hand, the weighting matrix employed isn't a consistent estimate of the "true" weighting matrix any more, which leads to an efficiency loss;
- on the other hand, equation (7.14) is an inconsistent estimator of the variance. As a consequence, any test statistic based upon it will be biased.

7.2.3 Two-steps Estimator

In order to partly resolve the first problem, one can use a two-step estimator, consisting in recovering the residuals from the one-step model $\Delta\hat{v}_n^{(1)}$ and estimating $\mathrm{E}\left[\left(\sum_n L_n^\top \Delta v_n\right)\left(\sum_n \Delta v_n^\top L_n\right)\right]$ through $\sum_n L_n^\top \Delta\hat{v}_n^{(1)} \Delta\hat{v}_n^{(1)\top} L_n$, this estimator being robust to the presence of heteroscedasticity and/or autocorrelation. In this case, the inverse of the weighting matrix of moments used is written as:

$$
\begin{aligned}
A^{(2)-1} &= \hat{\mathrm{V}}(\bar{m}) = \frac{1}{N^2}\sum_n \hat{\mathrm{V}}(m_n) \\
&= \frac{1}{N^2}\sum_n L_n^\top \Delta\hat{v}_n^{(1)} \Delta\hat{v}_n^{(1)\top} L_n = \frac{1}{N^2} L^\top \hat{\Omega}_{\hat{\beta}^{(1)}} L
\end{aligned}
\tag{7.15}
$$

with $\hat{\Omega}_{\hat{\beta}^{(1)}}$ a block diagonal matrix with blocks: $\Delta\hat{v}_n^{(1)} \Delta\hat{v}_n^{(1)\top}$ for $n = 1 \dots N$. The two-step GMM estimator is then obtained substituting (7.15) for A in equation (7.7):

$$
\begin{aligned}
\hat{\beta}^{(2)} &= [(\Delta X^\top L)(L^\top \hat{\Omega}_{\hat{\beta}^{(1)}} L)^{-1}(L^\top \Delta X)]^{-1} \\
&\times [(\Delta X^\top L)(L^\top \hat{\Omega}_{\hat{\beta}^{(1)}} L)^{-1}(L^\top \Delta y)]
\end{aligned}
\tag{7.16}
$$

Regarding the estimator's variance, by similar reasoning to that of equations (7.11 and 7.12), one has:

$$
\hat{\mathrm{V}}^{(2)} = [(\Delta X^\top L)(L^\top \hat{\Omega}_{\hat{\beta}^{(1)}} L)^{-1}(L^\top \Delta X)]^{-1}
\tag{7.17}
$$

The problem of this estimator is that it contains $\hat{\Omega}_{\hat{\beta}^{(1)}}$, which depends on the residuals of the one-step model and therefore on $\hat{\beta}^{(1)}$ and on y. This estimator is therefore biased and the derivation of a robust estimator for the variance will be presented in section 7.2.3.

Example 7.5 difference GMM estimator – `DemocracyIncome` data set

GMM estimation of a panel model is performed through the pgmm function in the **plm** library, The arguments of this function are the same as for the `plm` function, plus some specific ones:

- `formula`: the formula is peculiar, as it has three parts: the first, as usual, contains the explanatory variables, the second the "GMM" instruments, the third the "normal" instruments,
- `model`: the model to be estimated, either in one step: `'onestep'`, or two: `'twosteps'`,
- `effect`: the effects are either `'individual'` (they are then eliminated by differencing), or `'twoways'`, in which case indicator variables for each period are added as "normal" instruments.

We first compute the one-step estimator:

```
diff1 <- pgmm(democracy ~ lag(democracy) + lag(income) |
           lag(democracy, 2:99)| lag(income, 2),
           DemocracyIncome, index=c("country", "year"),
           model="onestep", effect="twoways", subset = sample == 1)
```

```
coef(summary(diff1))
                 Estimate Std. Error z-value  Pr(>|z|)
lag(democracy)  0.50499    0.09049    5.581 2.396e-08
lag(income)    -0.09011    0.08029   -1.122 2.617e-01
```

The two-step model is obtained by setting the `model` argument to `'twosteps'`:

```
diff2 <- update(diff1, model = "twosteps")
coef(summary(diff2))
                 Estimate Std. Error z-value  Pr(>|z|)
lag(democracy) 0.554007    0.10783 5.13777 2.780e-07
lag(income)    0.001844    0.06054 0.03045 9.757e-01
```

All available lags having been used, the number of instruments is sizable. One actually has: $0.5 \times (11-1) \times (11-2) = 45$ **GMM** instruments; plus the 9 indicator variables for time periods and the second lag of income, $J = 55$.

Notice also how the results are near those of the Anderson and Hsiao (1982) model.

7.2.4 The Proliferation of Instruments in the Generalized Method of Moments Difference Estimator

For the generalized method of moments estimator, the number of instruments grows with the time dimension of the sample. For the difference **GMM** model, considering only the levels of y which instrument Δy, one has: an instrument y_1 for the third observation (the first usable one); two instruments y_1, y_2 for the fourth; and $T-2$ instruments for the last observation $y_1, y_2, \ldots, y_{T-2}$, for a total $J = 1 + 2 + \ldots + (T-2) = 0.5(T-1)(T-2)$ instruments. For example, if $t = 10$, one has 36 instruments. The number of instruments grows quadratically with T. The weighting matrices of moments (7.9 and 7.15) are of dimension $J \times J$. Because of their symmetry, they contain $J \times (J+1)/2$ unique elements. The number of estimands of the matrix is therefore given by a polynomial in T whose dominant element is $T^4/8$. Every element of this matrix having to be estimated through an empirical average calculated over the N individuals in the sample, the precision in estimating the matrix elements is not guaranteed unless N is "big" with respect to J. Else, it can frequently happen that (7.9 et 7.15) be singular. The generalized moments estimator cannot then be calculated through the formula (7.7) because it uses the inverse of said matrix. One can then resort to generalized inversion methods, but this is clearly the symptom of too many instruments for the given number of individuals in the sample.

To understand the consequences of too many instruments, it is simplest to consider the case of the instrumental variables estimator. This estimator can be obtained by applying least squares twice: the first time regressing each column of the explanatory variables' matrix X on that of instrumental variables L, the second time regressing the dependent variable y on the predicted values of the previous regression $\hat{X}$. The bigger the number of instruments J, the better the first stage fit, i.e., the closest $\hat{X}$ will be to X. Should J become equal or greater than the number of observations, one will have $\hat{X} = X$ and the instrumental variables estimator will be identical to the ordinary least squares one. This is referred to as the "overfitting" problem.[10]

Different solutions are possible in order to limit the number of instruments. The first one consists in limiting the number of lags considered. For example, for $t = 10$, if limiting the number of

10 See Roodman (2009a), pp. 98-99.

lags to 3 one gets 1 instrument for $t = 3$, 2 for $t = 4$ and 3 for $t = 5 \dots 10$: a total of 21 instruments versus 36 if using all lags.

The second possibility is to "collapse" the moment conditions.[11] In this case, the matrix of instruments (7.4) is replaced by the following one:

$$L_n = \begin{pmatrix} y_{n1} & 0 & 0 & 0 & \dots & 0 & 0 & 0 \\ y_{n2} & y_{n1} & 0 & 0 & \dots & 0 & 0 & 0 \\ y_{n3} & y_{n2} & y_{n1} & 0 & \dots & 0 & 0 & 0 \\ \vdots & \vdots & \vdots & \vdots & \ddots & \vdots & \vdots & \vdots \\ y_{n(T-3)} & y_{n(T-4)} & y_{n(T-5)} & y_{n(T-6)} & \cdots & y_{n2} & y_{n1} & 0 \\ y_{n(T-2)} & y_{n(T-3)} & y_{n(T-4)} & y_{n(T-5)} & \cdots & y_{n3} & y_{n2} & y_{n1} \end{pmatrix} \tag{7.18}$$

The vector of $(T-2)$ empirical moments is then: $\bar{m} = \frac{1}{N}\sum_n L_n^\top \Delta v_n$ with:

$$(L_n^\top \Delta v_n)^\top = \left(\sum_{t=3}^{T} y_{n(t-2)}\Delta v_{nt}, \sum_{t=4}^{T} y_{n(t-3)}\Delta v_{nt}, \sum_{t=5}^{T} y_{n(t-4)}\Delta v_{nt}, \dots, \right.$$
$$\left. \sum_{t=T-1}^{T} y_{n(t-T+2)}\Delta v_{nt}, y_{n1}\Delta v_{nt} \right)$$

Example 7.6 instruments proliferation – `DemocracyIncome25` data set
To illustrate the problem of instrument proliferation, we consider the second dataset, where the frequency is 25 years.

```
data("DemocracyIncome25", package = "pder")
pdim(DemocracyIncome25)
Balanced Panel: n = 25, T = 7, N = 175
```

We estimate the GMM model in differences, using the two variables `democracy` and `income` as GMM instruments with all the available lags.

```
diff25 <- pgmm(democracy ~ lag(democracy) + lag(income) |
            lag(democracy, 2:99) + lag(income, 2:99),
            DemocracyIncome25, model = "twosteps")
```

For each GMM instrument, there are $0.5 \times 6 \times 5 = 15$ moment conditions and hence a total of 30 GMM instruments plus the 5 time dummies, i.e. $J = 35$, when the number of individuals is $N = 25$.

```
diff25lim <- pgmm(democracy ~ lag(democracy) + lag(income) |
               lag(democracy, 2:4)+ lag(income, 2:4),
               DemocracyIncome, index=c("country", "year"),
               model="twosteps", effect="twoways", subset = sample == 1)
diff25coll <- pgmm(democracy ~ lag(democracy) + lag(income) |
                lag(democracy, 2:99)+ lag(income, 2:99),
                DemocracyIncome, index=c("country", "year"),
                model="twosteps", effect="twoways", subset = sample == 1,
                collapse = TRUE)
```

11 See Roodman (2009b), p. 148.

```
sapply(list(diff25, diff25lim, diff25coll), function(x) coef(x)[1:2])
                    [,1]     [,2]      [,3]
lag(democracy)  0.4066  0.4678  0.50273
lag(income)    -0.1713 -0.1258 -0.04221
```

As can be readily seen, the results of the three models are quite similar, which seems to indicate that the proliferation of instruments is not an important issue in this particular context.

7.3 Generalized Method of Moments Estimator in Differences and Levels

The main drawback of the difference GMM estimator is that lagged levels of the dependent variable are often very weakly correlated with its lagged first difference. To solve this weak instruments problem, one can add moment conditions on the model in levels.

7.3.1 Weak Instruments

One can clearly see the weakness of the correlation between the instruments of the difference model and the regressor Δy_{t-1} in the case of a simple autoregressive model with $T = 3$.[12] In this case, the difference model for the third observation (the only usable one) can be written:

$$\Delta y_{n3} = \rho \Delta y_{n2} + \Delta v_{n3}$$

The only available instrument for this observation is y_{n1}. The GMM estimator reverts then to the instrumental variables one, Δy_{n2} being instrumented by y_{n1}. Applying two-stage least squares, one first estimates Δy_{n2} as a function of y_{n1}, then in a second step Δy_{n3} as a function of $\Delta \hat{y}_{n2}$. T

The structural model being $y_{nt} = \rho y_{n(t-1)} + \eta_n + v_{nt}$, the equation to be estimated in the first step can equivalently be written:

$$\Delta y_{n2} = (\rho - 1) y_{n1} + \eta_n + v_{n2}$$

The OLS estimator is then:

$$\hat{\pi} = (\rho - 1) + \frac{1/N \sum_n y_{n1}(\eta_n + v_{n2})}{1/N \sum_n y_{n1}^2}$$

Supposing that the process began many periods ago, one can calculate the limit of $\hat{\pi}$ observing that the numerator tends to $\sigma_\eta^2/(1-\rho)$ (see 7.3) and the denominator to $\sigma_\eta^2/(1-\rho)^2 + \sigma_v^2/(1-\rho^2)$ (see 7.2). One has then, denoting $k = (1-\rho)^2/(1-\rho^2)$:

$$\text{plim } \hat{\pi} = (\rho - 1)\frac{k}{\sigma_\eta^2/\sigma_v^2 + k} \quad (7.19)$$

Observing that $\lim_{\rho=1} k = 0$, one can clearly see how, if the process is close to having a unit root, $\hat{\pi}$ will be close to 0. Figure 7.2, representing plim $\hat{\pi}$ and $\rho - 1$ as a function of ρ illustrates the fact that, even for values of ρ well below 1, plim $\hat{\pi}$ is very close to 0. The instruments are therefore weak, and the quality of the second step in the two-stage least squares estimator will be low (erratic estimate, high standard error).

The instruments will be equally weak if the variance of the individual effect is much larger than that of the innovation.

12 See Blundell and Bond (1998) p. 120.

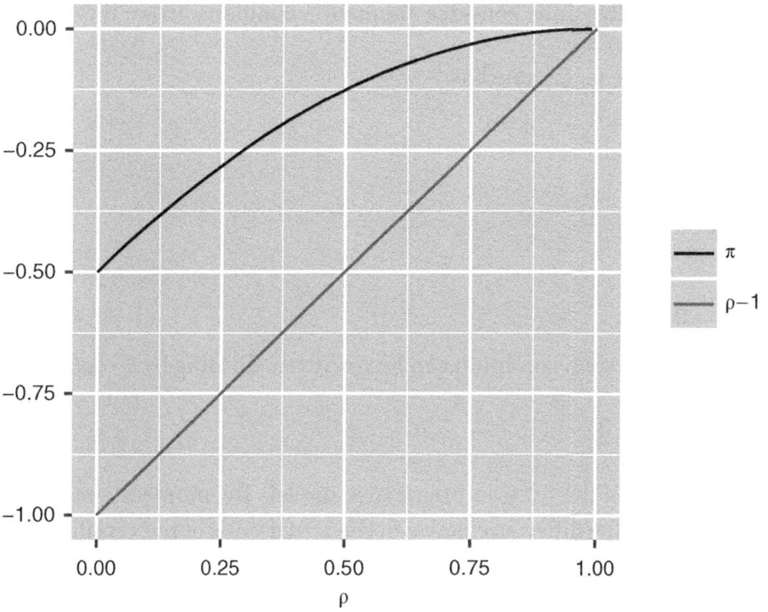

Figure 7.2 First step coefficient as a function of ρ.

7.3.2 Moment Conditions on the Levels Model

Arellano and Bover (1995) and Blundell and Bond (1998) show that under weak hypotheses on the data-generating process, another moment condition exists for the levels equation, which can be written as:

$$y_{nt} = \rho y_{n(t-1)} + \eta_n + \nu_{nt}$$

The supplementary moment conditions can be written:

$$\mathrm{E}(\Delta y_{n(t-s)}(\eta_n + \nu_{nt})) = 0 \quad s = 1 \ldots t-1$$

These show how $\Delta y_{n(t-s)}$ are valid instruments for $y_{n(t-1)}$ in the levels equation. If one takes into account the moment conditions for the differenced model too, only the condition corresponding to $s = 1$ is appropriate, the others being redundant. For example, for $T = 4$ there are 3 moment conditions for the levels equation:[13]

$$(\eta + \nu_3)\Delta y_2 \tag{7.20}$$

$$(\eta + \nu_4)\Delta y_3 \tag{7.21}$$

$$(\eta + \nu_4)\Delta y_2 \tag{7.22}$$

and 3 conditions for the differenced model:

$$(\nu_3 - \nu_2)y_1 \tag{7.23}$$

$$(\nu_4 - \nu_3)y_2 \tag{7.24}$$

$$(\nu_4 - \nu_3)y_1 \tag{7.25}$$

Subtracting (7.20) from (7.22) or subtracting (7.25) from (7.24), one has in both cases: $(\nu_4 - \nu_3)\Delta y_2$. Consequently, one moment condition is redundant. One can omit the

13 The individual index is temporarily omitted.

condition (7.22) and, more generally, only consider moment conditions from the levels model of the type: $\mathrm{E}(\Delta y_{n(t-1)}(\eta_n + v_{nt})) = 0$.

Replacing $y_{n(t-1)}$ with $\rho y_{n(t-2)} + \eta_n + v_{n(t-1)}$, one has:

$$\mathrm{E}[(\eta_n + v_{nt})((\rho - 1)y_{n(t-2)} + \eta_n + v_{n(t-1)})] = 0$$

As the v are uncorrelated, one has:

$$\mathrm{E}[\eta_n((\rho - 1)y_{n(t-2)} + \eta_n)] = 0$$

Or else, for period t:

$$\mathrm{E}[\eta_n((\rho - 1)y_{nt} + \eta_n)] = 0$$

For $|\rho| < 1$ (stationarity hypothesis), such condition can be rewritten, dividing by $1 - \rho$:

$$m_{nt} = \mathrm{E}\left[\eta_n\left(y_{nt} - \frac{\eta_n}{1-\rho}\right)\right] = 0$$

Now, $\frac{\eta_n}{1-\rho}$ is the steady state of y_{nt} in the pure autoregressive model. The moment condition indicates therefore that, for period t, the difference between the actual value of the variable and the steady state must be uncorrelated with the individual effect.

Replacing y_{nt} with $\rho y_{n(t-1)} + \eta_n + v_{nt}$, one has:

$$\begin{aligned} m_{nt} &= \mathrm{E}\left[\eta_n\left(\rho y_{n(t-1)} + \eta_n + v_{nt} - \frac{\eta_n}{1-\rho}\right)\right] \\ &= \mathrm{E}\left[\eta_n\left(\rho y_{n(t-1)} - \frac{\rho\eta_n}{1-\rho}\right)\right] = \rho m_{n(t-1)} \end{aligned}$$

Therefore: $m_{n(t-1)} = 0 \Rightarrow m_{nt} = 0$. This equation indicates that the moment condition is verified either for all periods or for none. This situation is illustrated in the first panel of Figure 7.3.[14]

A more pragmatic interpretation of this equation is that m_n decreases in time at a rate ρ. If the process has begun a long time ago, y is near its steady state value and the moment condition is acceptable, even if it is not exactly verified. This situation is illustrated in the second panel of Figure 7.3.

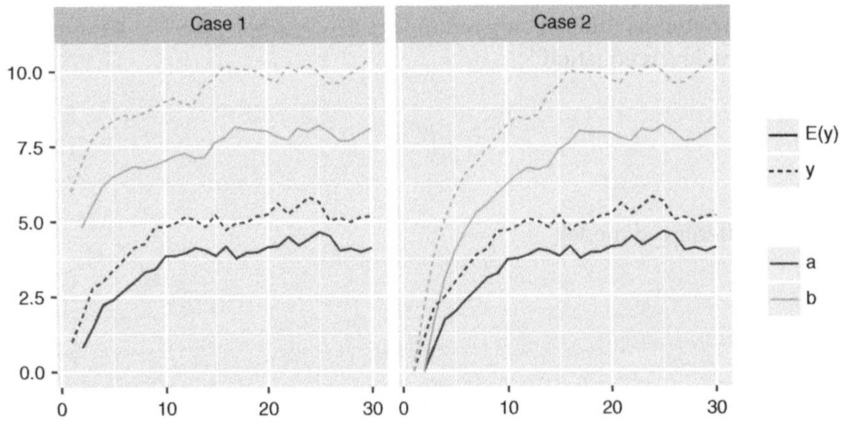

Figure 7.3 The supplementary condition of th system-GMM estimator.

14 This figure is inspired by Roodman (2009b) p. 145 et 147.

7.3.3 The System GMM Estimator

The estimator adding moment conditions from the levels estimator to the difference model is called generalized method of moments system estimator, or "sys-**GMM**". This estimator is obtained using the vector of errors in differences and in levels:

$$\epsilon_n^+ = (\Delta\epsilon_n, \epsilon_n) = (\Delta v_n, \epsilon_n)$$

and the corresponding matrix of moments:

$$L_n^+ = \begin{pmatrix} L_n & 0 & 0 & \dots & 0 \\ 0 & \Delta y_{n2} & 0 & \dots & 0 \\ 0 & 0 & \Delta y_{n3} & \dots & 0 \\ 0 & 0 & 0 & \dots & \Delta y_{n(t-1)} \end{pmatrix}$$

The moment conditions are then:

$$\left(\sum_n L_n^{+\top}\begin{pmatrix}\Delta v_n(\beta)\\ \epsilon_n(\gamma)\end{pmatrix}\right)^\top = \left(\sum_n y_{n1}\Delta v_{n3}, \sum_n y_{n1}\Delta v_{n4}, \sum_n y_{n2}\Delta v_{n4}, \dots,\right.$$
$$\sum_n y_{n1}\Delta v_{nt}, \sum_n y_{n2}\Delta v_{nt}, \dots, \sum_n y_{n(T-2)}\Delta v_{nt},$$
$$\left.\sum_n \epsilon_{n3}\Delta y_{n2}, \sum_n \epsilon_{n4}\Delta y_{n3}, \dots, \sum_n \epsilon_{nt}\Delta y_{n(t-1)}\right)^\top$$

where, as usual, $\gamma^\top = (\alpha, \gamma^\top)$. There is an intercept to be estimated for the sys-**GMM** model, which is not the case for the diff-**GMM** one.

The choice of an initial weighting matrix is less obvious than in the case of the difference model. In fact, there, only the vector of differenced errors is used, and consequently the variance of said vector, under the hypotheses of homoscedasticity and no serial correlation of the innovations, is proportional to a known matrix according to the coefficient σ_ν^2: hence estimation is unnecessary (see 7.8). By contrast, here the augmented vector of errors includes the errors in levels and hence the individual effects. In this case, the variance matrix depends on σ_ν^2 and on σ_η^2. To solve this problem and allow for a known starting matrix, one can assume $\sigma_\eta^2 = 0$. In this case:

$$V(\epsilon_n^+) = E\left(\begin{pmatrix}\Delta v_n\\ v_n\end{pmatrix}(\Delta v_n^\top, v_n^\top)\right) = E\begin{pmatrix} Dv_n v_n^\top D^\top & Dv_n v_n^\top \\ v_n v_n^\top D^\top & v_n v_n^\top\end{pmatrix} = \sigma_\nu^2\begin{pmatrix} h & D \\ D^\top & I\end{pmatrix}$$

Example 7.7 system GMM – `DemocracyIncome` data set

The system **GMM** model is obtained in a way similar to that in differences, the only change being the argument `transformation`, which defaults to `'d'` for *difference*, must be set to `'ld'` (for *level* and *difference*).

```
sys2 <- pgmm(democracy ~ lag(democracy) + lag(income) |
             lag(democracy, 2:99)| lag(income, 2),
             DemocracyIncome, index = c("country", "year"),
             model = "twosteps", effect = "twoways",
             transformation = "ld")
coef(summary(sys2))
                Estimate Std. Error z-value  Pr(>|z|)
lag(democracy)    0.6176    0.05714  10.809 3.134e-27
lag(income)       0.1200    0.01792   6.696 2.142e-11
```

The autoregressive coefficient obtained with the difference and the system models are close. The income coefficient is now significantly positive and much larger than previously.

7.4 Inference

Generalized method of moments estimation poses two types of problems in terms of inference:

- the first is that, even if the estimator is consistent, the same is not necessarily true for the variance-covariance matrix of coefficients if applying the classical formula. One can then use robust estimators of said matrix;
- the second is that estimation is consistent only under certain hypotheses: in particular those of no error correlation and of moments' validity.

7.4.1 Robust Estimation of the Coefficients' Covariance

The variance of the one-step estimator is given by equation (7.11). If the innovations are heteroscedastic and/or correlated, $L^{\top}HL$ is an inconsistent estimator of $\mathrm{E}\left[\left(\sum_n L_n^{\top}\Delta v_n\right)\left(\sum_n \Delta_n v_n^{\top} L_n\right)\right]$ and the variance estimator given by (7.14) is not robust. By contrast, $L^{\top}\hat{\Omega}_{\hat{\beta}^{(1)}}L$ is a consistent estimator of the moments' variance, which allows, plugging this expression into (7.11), to obtain a robust estimator of the coefficients' variance for the one-step estimator:

$$\begin{aligned}\hat{\mathrm{V}}^{(1)} &= [\Delta X^{\top}L(L^{\top}HL)^{-1}L^{\top}\Delta X]^{-1}\\ &\times \Delta X^{\top}L(L^{\top}HL)^{-1}(L^{\top}\hat{\Omega}_{\hat{\beta}^{(1)}}L)(L^{\top}HL)^{-1}L^{\top}\Delta X \\ &\times [\Delta X^{\top}L(L^{\top}HL)^{-1}L^{\top}\Delta X]^{-1}\end{aligned} \tag{7.26}$$

The expression of the two-step estimator is given by (7.16). The problem is its dependence on $\hat{\Omega}_{\hat{\beta}^{(1)}}$, in turn depending on $\hat{\beta}^{(1)}$ and hence on Δy. Consequently, $\hat{\beta}^{(2)}$ is a nonlinear function of Δy and the usual variance formula is inappropriate.

Estimation of the variance $\hat{\Omega}_{\hat{\beta}^{(1)}}$ of the J moments' vector is typically very imprecise for two reasons. The first is that the number of parameters is large ($J \times (J+1)/2$). The second is that these parameters are second-order moments of second-order moments, hence fourth-order moments of the original data.[15]

The solution proposed by Windmeijer (2005) allows to obtain a consistent estimator of the two-step estimators' variance. To begin with, one replaces in (7.16) Δy by $\Delta X\beta + \Delta v$. One has then:

$$\begin{aligned}\hat{\beta}^{(2)} - \beta &= [(\Delta X^{\top}L)(L^{\top}\hat{\Omega}_{\hat{\beta}^{(1)}}L)^{-1}(L^{\top}\Delta X)]^{-1}\\ &\times [(\Delta X^{\top}L)(L^{\top}\hat{\Omega}_{\hat{\beta}^{(1)}}L)^{-1}(L^{\top}\Delta v)]\end{aligned} \tag{7.27}$$

In general, define:

$$\begin{aligned}g(\Delta y, \hat{\Omega}) &= [(\Delta X^{\top}L)(L^{\top}\hat{\Omega}L)^{-1}(L^{\top}\Delta X)]^{-1}\\ &\times [(\Delta X^{\top}L)(L^{\top}\hat{\Omega}L)^{-1}(L^{\top}\Delta v)]\end{aligned} \tag{7.28}$$

implying that $\hat{\beta}^{(2)} - \beta = g(\Delta y, \hat{\Omega}_{\hat{\beta}^{(1)}})$. The variance of $\hat{\beta}^{(2)}$ is then that of $g(\Delta y, \hat{\Omega}_{\hat{\beta}^{(1)}})$. One subsequently approximates g around the true value of parameters β. Denote by D the gradient of g evaluated at the true parameter values:

$$D = \frac{\partial}{\partial \hat{\beta}} g(\Delta y, \hat{\Omega}_{\hat{\beta}})|_{\hat{\beta}=\beta}$$

15 See Roodman (2009b) p. 140.

The first-order approximation can then be written:

$$g(\Delta y, \hat{\Omega}_{\hat{\beta}^{(1)}}) \approx g(\Delta y, \hat{\Omega}_{\beta}) + D(\hat{\beta}^{(1)} - \beta)$$

Or, $(\hat{\beta}^{(1)} - \beta) = g(\Delta y, H)$. Consequently, the approximation becomes:

$$g(\Delta y, \hat{\Omega}_{\hat{\beta}^{(1)}}) \approx g(\Delta y, \hat{\Omega}_{\beta}) + Dg(\Delta y, H)$$

The variance of $\hat{\beta}^{(2)}$ is then approximated by:

$$\hat{\hat{V}}^{(2)} \approx [g(\Delta y, \hat{\Omega}_{\beta}) + Dg(\Delta y, H)][g(\Delta y, \hat{\Omega}_{\beta}) + Dg(\Delta y, H)]^{\top}$$

or:

$$\begin{aligned} \hat{\hat{V}}^{(2)} \approx\; & g(\Delta y, \hat{\Omega}_{\beta})g(\Delta y, H)^{\top}D^{\top} \\ & + Dg(\Delta y, H)g(\Delta y, \hat{\Omega}_{\beta})^{\top} \\ & + g(\Delta y, \hat{\Omega}_{\beta})g(\Delta y, \hat{\Omega}_{\beta})^{\top} \\ & + Dg(\Delta y, H)g(\Delta y, H)^{\top}D^{\top} \end{aligned} \tag{7.29}$$

Replacing Δv by $\Delta\hat{v}^{(1)}$ and $\hat{\Omega}_{\beta}$ by $\hat{\Omega}_{\hat{\beta}^{(1)}}$, $g(\Delta y, \hat{\Omega}_{\beta})g(\Delta y, \hat{\Omega}_{\beta})^{\top}$ and $g(\Delta y, \hat{\Omega}_{\beta})g(\Delta y, H)^{\top}$ are both approximated by $\hat{V}^{(2)} = [\Delta X^{\top}L(L^{\top}\hat{\Omega}_{\hat{\beta}^{(1)}}L)^{-1}L^{\top}\Delta X]^{-1}$. Moreover, $g(\Delta y, H)g(\Delta y, H)^{\top} = [\Delta X^{\top}L(L^{\top}HL)^{-1}L^{\top}\Delta X]^{-1} = \hat{V}^{(1)}$. One has then, finally, the expression for the robust covariance of the two-step estimator:

$$\hat{\hat{V}}^{(2)} = \hat{V}^{(2)}D^{\top} + D\hat{V}^{(1)}D^{\top} + \hat{V}^{(2)} + D\hat{V}^{(2)}$$

The expression of D is to be found in Windmeijer (2005).

Example 7.8 robust estimation of the covariance matrix – `DemocracyIncome` data set

The function `vcov` computes the "classical" (and inconsistent) version of the variance, and `vcovHC` the robust version (equations 7.26 for the one-step model, and 7.29 for the two-step one). Below we extract the standard errors of the first two coefficients for the two-step difference model.

```
sqrt(diag(vcov(diff2)))[1:2]
lag(democracy)      lag(income)
       0.04795          0.04646
sqrt(diag(vcovHC(diff2)))[1:2]
lag(democracy)      lag(income)
       0.10783          0.06054
```

One can actually see that in this example the classical variance formula seems to be biased downward. In fact, "robust" standard errors are clearly superior.

7.4.2 Overidentification Tests

If the moment conditions are valid, the empirical moments' vector $\bar{m} = \frac{1}{N}\sum_n L_n^{\top}\Delta v_n$ has expectation zero. If this hypothesis is verified, the Wald statistic:

$$\bar{m}^{\top}V(\bar{m})^{-1}\bar{m}$$

is distributed as a χ^2 with $J - K$ degrees of freedom. This test has been proposed by Sargan (1958) and applied to GMM models by Hansen (1982). Various versions of this test can be obtained, depending on whether:

- the one-step or the two-step residuals are used to approximate $\bar{m}$;
- the simple estimator ($\frac{\sigma_\nu^2}{N^2}L^\top HL$) or the robust one ($\frac{1}{N^2}L^\top\hat{\Omega}_{\hat{\beta}^{(1)}}L$) is used for the moments' variance matrix.

For example, the test on the two-step model using the robust estimator of the moments' matrix is based on the following statistic:

$$\left(\frac{1}{N}\Delta\hat{\nu}^{(2)\top}L\right)\left(\frac{1}{N^2}L^\top\hat{\Omega}_{\hat{\beta}^{(1)}}L\right)^{-1}\left(\frac{1}{N}L^\top\Delta\hat{\nu}^{(2)}\right)$$
$$= (\Delta\hat{\nu}^{(2)\top}L)(L^\top\hat{\Omega}_{\hat{\beta}^{(1)}}L)^{-1}(L^\top\Delta\hat{\nu}^{(2)})$$

which is the value of the objective function of the two-step GMM model evaluated in $\hat{\beta}^{(2)}$.

It is recommended, in the case of "sys-GMM" models, to perform a Sargan-Hansen test on the subset of moment conditions concerning the levels model, in order to separately test the validity of the supplementary hypotheses imposed by that model.

Example 7.9 Sargan-Hansen test – `DemocracyIncome` data set

The Sargan-Hansen test can be performed through function `sargan`. For example, for the one-step difference model, one has:

```
sargan(diff2)

Sargan test

data:  democracy ~ lag(democracy) + lag(income) | lag(democracy, 2:99) | ...
chisq = 50, df = 44, p-value = 0.3
sargan(sys2)

Sargan test

data:  democracy ~ lag(democracy) + lag(income) | lag(democracy, 2:99) | ...
chisq = 56, df = 54, p-value = 0.4
```

For the difference model, one has $J = 55$ (the 45 "GMM" instruments, the income variable and nine time dummies) and $K = 11$ (the lagged endogenous variable, income and the nine time dummies). The number of degrees of freedom is then $J - K = 44$. In this case, the hypothesis of moments' validity is not rejected.

For the system model, the number of periods used is 10 (one more than in the difference model). There are therefore one more coefficient and one more instrument (the coefficient associated to the added time dummy) and 10 supplementary instruments corresponding to the 10 moment conditions for the 10 observations of the model in levels. One therefore has $J = 55 + 1 + 10 = 66$ and $K = 11 + 1 = 12$. Hence, the number of degrees of freedom is $J - K = 66 - 12 = 54$, and again, the hypothesis of validity of the moment conditions for the system GMM model is not rejected.

The Hansen-Sargan test is particularly sensitive to the problem of instrument proliferation. Roodman (2009b), using the studies by Levine et al. (2000) and Forbes (2000), shows that the p-value of this test tends to be very high, leading to non-rejecting the validity of moment conditions, when the same test performed on models more parsimonious in terms of instruments may lead to the opposite conclusions. To illustrate this result, we compute Sargan's test on the model previously estimated on the dataset with 7 observations of 25 countries.

```
sapply(list(diff25, diff25lim, diff25coll),
       function(x) sargan(x)[["p.value"]])
  chisq   chisq   chisq
0.91890 0.07105 0.21531
```

The p-value for the model using all moment conditions is near 1, while those of the other models are much lower; in particular, for the model limiting the number of lags to 3, the hypothesis of instruments validity is rejected at the 10% significance level.

7.4.3 Error Serial Correlation Test

The generalized method of moments is not consistent unless the moment conditions are verified, which, in particular, implies that the innovations are serially uncorrelated. Arellano and Bond (1991) proposed an appropriate test, based on the following statistic:

$$a_l = \frac{1}{\sqrt{N}} \Delta\hat{v}^\top \Delta\hat{v}^{-l}$$

where Δv^{-l} is the l-th lag of Δv. Using the expression of the theoretical model and of the estimated one: $\Delta y = \Delta X\beta + \Delta v = \Delta X\hat{\beta} + \Delta\hat{v}$, one gets:

$$\Delta\hat{v} = \Delta v - \Delta X(\hat{\beta} - \beta)$$

Inserting this expression into the test statistic, one obtains:

$$\begin{aligned}
a_l &= \frac{1}{\sqrt{N}}(\Delta v^\top - (\hat{\beta} - \beta)^\top \Delta X^\top)(\Delta v^{-l} - \Delta X^{-l}(\hat{\beta} - \beta)) \\
&= \frac{1}{\sqrt{N}} \Delta v^\top \Delta v^{-l} \\
&\quad - \frac{1}{N} \Delta v^\top \Delta X^{-l} \sqrt{N}(\hat{\beta} - \beta) \\
&\quad - \sqrt{N}(\hat{\beta} - \beta)^\top \frac{1}{N} \Delta X^\top \Delta v^{-l} \\
&\quad + \sqrt{N}(\hat{\beta} - \beta)^\top \frac{1}{\sqrt{N}} \frac{1}{N} \Delta X^\top \Delta X^{-l} \sqrt{N}(\hat{\beta} - \beta)
\end{aligned}$$

This expression simplifies if $N \rightarrow +\infty$ observing that:

- $\hat{\beta}$ converges at a rate $\sqrt{N}$, $\sqrt{N}(\hat{\beta} - \beta)$ does neither diverge nor converge to 0;
- if the explanatory variables are not post-determined, they are not correlated with posterior values of v. One has then: $\frac{1}{N} v^\top \Delta X^{-l} \rightarrow 0$;
- $\frac{1}{N} \Delta X^\top \Delta X^{-l}$ does not diverge.

which implies that the second and fourth terms converge to 0. A consistent estimator of the variance of a_l can therefore be based on:

$$b_l = \frac{1}{\sqrt{N}}(\Delta v^\top \Delta v^{-l} - (\hat{\beta} - \beta)^\top \Delta X^\top \Delta v^{-l})$$

A consistent estimator of b_l (and hence of a_l) is:

$$\frac{1}{N}(\Delta\hat{v}^{-l\top}\hat{V}(\Delta\hat{v})\Delta\hat{v}^{-l} + \Delta\hat{v}^{-l\top}\Delta X\hat{V}(\hat{\beta})\Delta X^\top\hat{v}^{-l}$$
$$- 2\Delta\hat{v}^{-l\top}\Delta X(\Delta X^\top LAL^\top\Delta X)^{-1}\Delta XLAL^\top\hat{V}(\Delta\hat{v})\Delta\hat{v}^{-l})$$

The test statistic is thus obtained by dividing a_l by the square root of the above expression and it is normally distributed under the hypothesis of no serial correlation. The model being expressed in first differences, the first-order serial correlation test is inappropriate because $\Delta v_{nt} = v_{nt} - v_{n(t-1)}$ is correlated with $\Delta v_{n(t-1)} = v_{n(t-1)} - v_{n(t-2)}$ because of the presence of $v_{n(t-1)}$ in the two successive differences. On the converse, the second-order serial correlation test is appropriate, as it consists in testing correlation between $\Delta v_{nt} = v_{nt} - v_{n(t-1)}$ and $\Delta v_{n(t-2)} = v_{n(t-2)} - v_{n(t-3)}$, which will be only present if $v_{n(t-1)}$ is correlated with $v_{n(t-2)}$, i.e., if the innovations in levels are serially correlated of order 1.

Example 7.10 autocorrelation test – `DemocracyIncome` data set
The error serial correlation test of Arellano and Bond (1991) is obtained through the function `mtest`. The argument `order` is here set to 2 according to the preceding remark.

```
mtest(diff2, order = 2)

Autocorrelation test of degree 2

data:  democracy ~ lag(democracy) + lag(income) | lag(democracy, 2:99) | ...
normal = 0.88, p-value = 0.4
```

The hypothesis of no serial correlation is not rejected.

7.5 More Empirical Examples

There are many articles using GMM on panel data. We will here limit ourselves to a short description of those for whose the data are available in the **plm** and **pder** packages.

The study by Levine et al. (2000) tests for a causal relationship between the quality of the financial system (which limits information asymmetries and facilitates transactions) and economic growth. To this end, they estimate a model where economic growth is a function of a number of control variables and of exogenous characteristics of the financial system. They draw on a panel of 74 countries with 7 observations of 5-year periods from 1960 to 1995. The log of the growth rate is regressed on the log of initial wealth and of three indicators of financial system quality: the degree of liquidity of the financial system, the ratio of commercial banks' to central bank deposits, and the ratio of outstanding credit to GDP. The two GMM models - difference and system - are estimated, and the three indicators turn out having a positive and significant influence on growth, especially in the system case. Roodman (2009b) returns on this study elaborating on the instruments proliferation problem, potentially leading to incorrect acceptance of the validity of moment conditions. In particular, in the original study the p-value of the Hansen test of overidentifying restrictions is 0.97. Different specifications, more parsimonious as far as the number of instruments is concerned, used by Roodman (2009b) yield much different results. In fact, the p-value is now 0.001 and the validity of the supplementary restrictions of the system model is rejected. The data allowing to reproduce these results are available as `FinanceGrowth` in the **pder** package.

Forbes (2000) is concerned with the effect of income inequality on economic growth. To this end, a panel of 45 countries over 6 5-year periods from 1960 to 1995 is analyzed. Growth is estimated as a function of the log of income per capita lagged by one period, of the Gini coefficient lagged by one period, of the education level of male and female, and of the price level of investments. Various estimation methods are used, in particular the difference GMM of Arellano and Bond (1991). The main result of the study is that the sign of the Gini coefficient is positive and significant at the 5% level. This result is against those of many cross-sectional studies suggesting a negative relationship between inequality and economic growth. This study has been reconsidered by Roodman (2009b) in order to illustrate the pitfalls of using many weak instruments. In fact, the autoregressive coefficient is near 1, and the number of instruments is very high (89, against only 138 observations). Roodman (2009b) employs various other specifications with a limited number of instruments, and in these cases the Gini coefficient is not significant any more. These data are provided as `IneqGrowth` in the **pder** package

Caselli et al. (1996) address the issue of countries' economic growth, and in particular the phenomenon of convergence. They start from the results obtained in many cross-sectional studies, for the most part coming to the conclusion that countries converge tho their steady state at a very low rate, generally in the region of 2-3%. Their point is that such studies suffer from two specification problems: the first is neglecting the dynamic nature of the model, and the second is not considering the possible endogeneity of explanatory variables. The authors apply the estimator of Arellano and Bond (1991) to a panel of 93 countries and 6 5-year periods, 1965 to 1985. They find a much higher convergence rate, in the order of 10%. Bond et al. (2001) indicate that these results must be taken with caution in so far as the dependent variable is close to having a unit root, and hence the instruments used are weak. They reestimate the same model using the Blundell and Bond (1998) estimator and thus obtain a much lower convergence rate of about 2-4%. These data are provided as `Solow` in the **pder** package

In their seminal paper, Arellano and Bond (1991) used data on 140 British firms from 1976 to 1984 in order to estimate a labor demand equation. The covariates are two lags of the dependent variable and, also including two lags, the salary rate, the capital stock, and the production level. These data have been used in many further articles, in particular Blundell and Bond (1998), Windmeijer (2005) and Roodman (2009a). They are available as `EmplUK` in the **plm** package.

Alonso-Borrego and Arellano (1999) perform a study on similar data concerning 738 Spanish firms over the period 1983-1990. A VAR model is used for employment and the salary rate. These data are provided as `Snmesp` in the **plm** package.

Mairesse and Hall (1996), Blundell and Bond (2000) and Bond (2002) have estimated a Cobb-Douglas production function over a panel of 509 American firms over the period 1982-1989. The explanatory variables, taken in logs, are the lagged dependent variable and the two production factors (labor and capital) contemporaneous and lagged by one period. The results by Mairesse and Hall (1996), obtained using the Arellano and Bond (1991) estimator, are surprising: the hypothesis of constant returns to scale is rejected and the coefficient on capital is small and not significant. Blundell and Bond (2000) show how these unsatisfactory results are due to the variables used being near to having a unit root. In such cases, we know that the difference GMM estimator yields bad results because the instruments are weak. The system estimator instead yields more plausible results (hypothesis of constant returns to scale not rejected and significant coefficient on capital). These data are available as `RDPerfComp` in the **pder** package.

Kessler et al. (2011) address the influence of inter-regional transfers within a federal State on regional inequality. Their theoretical model predicts that, counterintuitively, such transfers may aggravate inter-regional inequalities. They use data on 17 OECD countries over the period 1982-1999 and the Arellano and Bond (1991) estimator. The results actually point at an aggravating effect of an increase in transfers on inter-regional inequality. These data are provided as `RegIneq` in the **pder** package.

8

Panel Time Series

8.1 Introduction

Panel time series methods were born to address the issues of "long" panels of possibly non-stationary series, usually of macroeconomic nature. Such datasets, pooling together a sizable number of time series from different countries (regions, firms) have become increasingly common and are the main object of empirical research in many fields: development economics, regional or political science to name a few; the most typical unit of observation being a country or region within a reasonably large set of similar units and over at least two decades of either yearly or quarterly data.

Unlike "large" panels, the emphasis is therefore not only on N-asymptotics but on both N and T tending to infinity, either sequentially or jointly (a seminal paper in this respect is Phillips and Moon, 1999). Specifying the order with which N and T diverge is essential for the properties of estimators.

The dynamics holds a more important, often prominent place (see e.g. Pesaran and Smith, 1995; Eberhardt et al., 2013). Under cointegration, error correction specifications are often of interest (see e.g. Holly et al., 2010). The assumption of parameter homogeneity is also often questioned in this field, often leading to relaxing it in favor of heterogeneous specifications where the coefficients of individual units are free to vary over the cross section. The parameter of interest can then be either the whole population of individual ones or the cross-sectional average thereof.

Lastly, the issue of cross-sectional correlation, which is assumed away in the case of dynamic GMM estimators a la Arellano and Bond (1991), takes a central role in panel time series methods. In fact, observations coming from countries of the world, or regions within one country or continent, are more likely than not to be correlated in the cross section either by some spatial process, whereby shocks spread to neighboring units because of proximity, or by the effect of common factors.

For example, consider a dynamic error component model:

$$y_{nt} = \eta_n + \rho y_{nt-1} + \beta x_{nt} + \nu_{nt}$$

where η is allowed to be correlated with x; for $N \to \infty$ and fixed T, the OLS estimator of (ρ, β) is inconsistent because of the presence of the unobserved correlated effects η. From Chapter 7, we know that the within estimator for this model is in turn biased downward, the bias being inversely proportional to T so that it becomes less severe as the available time dimension gets

Panel Data Econometrics with R, First Edition. Yves Croissant and Giovanni Millo.

Companion website: www.wiley.com/go/croissant/data-econometrics-with-R

longer. If N and T both diverge, then for consistency T is needed to grow "fast enough" relative to N, i.e., at a rate such that the limit of N/T is finite.

From a different viewpoint, if each time series in the panel is considered separately, as $T \to \infty$, OLS are a consistent estimator for the individual parameters (ρ_n, β_n) so that separately estimating, and then either averaging or pooling, the coefficients becomes a feasible strategy.

More generally, the abundance of data along both dimensions in large N, large T panels opens up possibilities and issues, other than the familiar ones of large, short panels: heterogeneity can be considered, where coefficients are not fixed across individuals but are allowed to vary, either freely or randomly around an average; nonstationarity, where the long time dimension allows to address unit roots and cointegration; and cross-sectional dependence across individual units, possibly due to common factors to which individual units react idiosyncratically.

8.2 Heterogeneous Coefficients

Long panels allow to estimate separate regressions for each unit. Hence it is natural to question the assumption of parameter homogeneity ($\beta_n = \beta \; \forall n$, also called the *pooling* assumption) as opposed to various kinds of *heterogeneous* specifications. This is a vast subject, which we will keep as simple as possible here; in general it can be said that imposing the pooling restriction reduces the variance of the pooled estimator but may introduce bias if these restrictions are false (Baltagi et al., 2008). Moreover, the heterogeneous model is usually a generalization of the homogeneous one so that estimating it may allow to test for the validity of the pooling restriction.

The panel data model with individual heterogeneity:

$$y_{nt} = \alpha + \beta_n x_{nt} + \eta_n + \nu_{nt}$$

generalizes the familiar individual effects model: here, all parameters vary across units, while in the former only the intercept did. The decision "to pool or not to pool" spans a vast literature; it is analyzed thoroughly by Baltagi et al. (2000) (see also Baltagi and Griffin, 1997; Baltagi et al., 2003a) in a forecasting perspective. Summing up the results of a number of studies, Baltagi et al. (2008) conclude that for forecasting purposes, the simplicity and stability of the pooled estimators dominate the flexibility of the heterogeneous ones, but seen from other perspectives, conclusions may reverse. It can be safely stated that data rich environments favor the latter, while the appeal of pooling restrictions becomes higher the smaller the dataset.

8.2.1 Fixed Coefficients

The heterogeneous panel model is:

$$y_{nt} = \alpha + \beta_n x_{nt} + \eta_n + \nu_{nt} \tag{8.1}$$

where β_n are individual-specific parameters and x_{nt} is a vector of K explanatory variables.

If the pooling assumption is relaxed and one does not want to make any other assumption about how the β_n are generated, and if the T dimension permits, one can simply estimate a separate vector of coefficients for each regression.

Individual slope parameters β_n can be estimated (T-consistently) by least squares as:

$$\hat{\beta}_{\text{OLS},n} = (X_n^\top X_n)^{-1} X_n^\top y_n \tag{8.2}$$

This can be accomplished by subsetting the data and running OLS; more efficient functionality is provided in **plm** through the function `pvcm`, leaving the `model` argument at the default value of `'within'`.

8.2.2 Random Coefficients

Estimating separate regressions negates the advantages of panel datasets in that degrees of freedom are greatly reduced with respect to the pooled data. If β_ns are treated as fixed, there will be NK parameters to estimate with NT observations. Random coefficients specifications allow instead for cross-sectional variability while still reaping the benefits of pooling.

8.2.2.1 The Swamy Estimator

Swamy (1970) proposed a model with all individual-specific coefficients. In this case, we have:

$$y_{nt} = \gamma_n^\top z_{nt} + v_{nt}$$

where homoscedasticity of v is not assumed and $\gamma_n \sim N(\gamma, \Delta)$, or $\delta_n = \gamma_n - \gamma \sim N(0, \Delta)$. The model is then rewritten as:

$$y_{nt} = \gamma^\top z_{nt} + \epsilon_{nt}$$

with $\epsilon_{nt} = v_{nt} + \delta_n^\top z_{nt}$. The model errors can be heteroscedastic (in particular because we did not impose homoscedasticity of v) and the errors of each individual are correlated as containing the same parameter vector δ_n. For the n–th individual, the error covariance is then:

$$\Omega_n = \mathrm{E}(\epsilon_n \epsilon_n^\top) = \mathrm{E}[(v_n + Z_n\delta_n)(v_n^\top + \delta_n^\top Z_n^\top)]$$

v and δ being uncorrelated by hypothesis, we have:

$$\Omega_n = \mathrm{E}(\epsilon_n \epsilon_n^\top) = \sigma_n^2 \mathrm{I}_T + Z_n \Delta Z_n^\top$$

For the whole sample, $\Omega = \mathrm{E}(\epsilon\epsilon^\top)$ is a block diagonal matrix, each block being equal to Ω_n.

OLS estimation of this model is inefficient, not taking into account the heteroscedasticity and the correlation of errors. The model can be efficiently estimated by generalized least squares by computing $\Omega^{-0.5}$ and then applying OLS to the variables transformed by pre-multiplying them by $\Omega^{-0.5}$. Given that the latter is a block diagonal matrix, the same result is obtained by pre-multiplying each individual's data by the corresponding block $\Omega_n^{-0.5}$. The generalized least squares method is clearly infeasible because Ω_n is unknown, but it can be made operational by employing an estimate thereof from a consistent model. This amounts to estimating N σ_n^2 and the elements of the Δ matrix, or in total $N + K(K+1)/2$ parameters.

To this end, we start by estimating each individual model by OLS. We then have:

$$\hat{\gamma}_n = (Z_n^\top Z_n)^{-1} Z_n^\top y_n = \gamma_n + (Z_n^\top Z_n)^{-1} Z_n v_n$$

A natural estimator of σ_n^2 is then:

$$\hat{\sigma}_n^2 = \sum_t^T \hat{\epsilon}_{nt}^2 / (T - K - 1)$$

The estimates are then averaged:

$$\bar{\hat{\gamma}} = \frac{1}{N} \sum_{n=1}^{N} \hat{\gamma}_n$$

The estimation of Δ is based on the expression $\hat{\gamma}_n - \bar{\hat{\gamma}}$, which, developing and regrouping terms, can be written:

$$\begin{aligned}\hat{\gamma}_n - \bar{\hat{\gamma}} &= \gamma_n + (Z_n^\top Z_n)^{-1} Z_n^\top v_n - \frac{1}{N}\sum_{n=1}^{N}(\gamma_n + (Z_n^\top Z_n)^{-1}Z_n^\top v_n) \\ &= \frac{N-1}{N}\gamma_n + \frac{N-1}{N}(Z_n^\top Z_n)^{-1}Z_n^\top v_n - \frac{1}{N}\sum_{m\neq n}\gamma_m \\ &- \frac{1}{N}\sum_{m\neq n}(Z_m^\top Z_m)^{-1}Z_m^\top v_m\end{aligned}$$

The usefulness of this expression is in writing $\hat{\gamma}_n - \bar{\hat{\gamma}}$ as a linear combination of uncorrelated random variates, which considerably simplifies the computation of the variance of $\hat{\gamma}_n$ as all covariances are zero. We then have:

$$\begin{aligned}\mathrm{E}((\hat{\gamma}_n - \bar{\hat{\gamma}})^2) &= \left(\frac{N-1}{N}\right)^2 \Delta + \left(\frac{N-1}{N}\right)^2 \sigma_n^2 (Z_n^\top Z_n)^{-1} \\ &+ \frac{N-1}{N^2}\Delta + \frac{1}{N^2}\sum_{m\neq n}\sigma_m^2(Z_m^\top Z_m)^{-1}\end{aligned}$$

Finally, regrouping terms:

$$\mathrm{E}((\hat{\gamma}_n - \bar{\hat{\gamma}})^2) = \frac{N-1}{N}\Delta + \frac{N-2}{N}\sigma_n^2(Z_n^\top Z_n)^{-1} + \frac{1}{N^2}\sum_n \sigma_n^2(Z_n^\top Z_n)^{-1}$$

We then have:

$$\begin{aligned}\mathrm{E}\left(\sum_n(\hat{\gamma}_n - \bar{\hat{\gamma}})^2\right) &= (N-1)\Delta + \frac{N-2}{N}\sum_n \sigma_n^2(Z_n^\top Z_n)^{-1} + \frac{1}{N}\sum_n \sigma_n^2 (Z_n^\top Z_n)^{-1} \\ &= (N-1)\Delta + \frac{N-1}{N}\sum_n \sigma_n^2(Z_n^\top Z_n)^{-1}\end{aligned}$$

$$\mathrm{E}\left(\frac{1}{N-1}\sum_n(\hat{\gamma}_n - \bar{\hat{\gamma}})^2\right) = \Delta + \frac{1}{N}\sum_n \sigma_n^2(Z_n^\top Z_n)^{-1}$$

which gives the estimator of Δ:

$$\hat{\Delta} = \frac{1}{N-1}\sum_n(\hat{\gamma}_n - \gamma)^2 - \frac{1}{N}\sum_n \sigma_n^2(Z_n^\top Z_n)^{-1}$$

Example 8.1 Random coefficient model – `Dialysis` data set

Caudill et al. (1995) examine the effect that certificate-of-need regulation by state health planning organizations has on the speed of diffusion of a medical technology, hemodialysis. More specifically, they test the hypothesis that this regulation has slowed the rate of adoption of this technology. They use a panel of 50 American states for 14 years (from 1977 to 1990). The degree of adoption of the technology `diffusion` is measured as the ratio of the number of dialysis machines in a particular state for a year divided by the number of machines for the last period of observation. A logistic diffusion function is used for the response. Two covariates are used: a time trend and a dummy variable that equals one for observations for which certificate-of-need regulation is in effect, interacted with the time trend.

The Swamy (1970) model can be estimated with the `pvcm` function, setting the `model` argument to `'random'`.

```
data("Dialysis", package = "pder")
rndcoef <- pvcm(log(diffusion / (1 - diffusion)) ~ trend + trend:regulation,
                Dialysis, model="random")
summary(rndcoef)
Oneway (individual) effect Random coefficients model

Call:
pvcm(formula = log(diffusion/(1 - diffusion)) ~ trend + trend:regulation,
    data = Dialysis, model = "random")

Balanced Panel: n = 50, T = 14, N = 700

Residuals:
total sum of squares: 629.5
    id   time
0.4685 0.2659

Estimated mean of the coefficients:
                 Estimate Std. Error z-value Pr(>|z|)
(Intercept)       -1.4266     0.1284  -11.11   <2e-16 ***
trend              0.3416     0.0260   13.15   <2e-16 ***
trend:regulation  -0.0581     0.0237   -2.45    0.014 *
---
Signif. codes:
0 '***' 0.001 '**' 0.01 '*' 0.05 '.' 0.1 ' ' 1

Estimated variance of the coefficients:
                 (Intercept)   trend trend:regulation
(Intercept)           0.6617 -0.0736           0.0398
trend                -0.0736  0.0288          -0.0205
trend:regulation      0.0398 -0.0205           0.0179

Total Sum of Squares: 33900
Residual Sum of Squares: 642
Multiple R-Squared: 0.981
```

The results indicate that certificate-of-need regulation has slowed the diffusion of hemodialyis technology, as the coefficient is significantly (at the 5% level) negative. The estimated covariance matrix of the random coefficients is an element of the fitted model called `"Delta"`; the following command extracts the mean values of the three coefficients and their standard deviations.

```
cbind(coef(rndcoef), stdev = sqrt(diag(rndcoef$Delta)))
                        y  stdev
(Intercept)      -1.42656 0.8135
trend             0.34161 0.1697
trend:regulation -0.05806 0.1339
```

The random coefficients have large standard deviations: about half the mean for the trend coefficient and about two times the mean for the regulation coefficients. These large values justify the use of the random coefficient model.

8.2.2.2 The Mean Groups Estimator

Under less restrictive parametric assumptions than those of the Swamy model, assuming only exogeneity of the regressors and independently sampled errors, the average γ can be estimated by the simpler mean groups (MG) method

$$\hat{\gamma}_{\mathbf{MG}} = \frac{1}{N}\sum_{n=1}^{N} \hat{\gamma}_{\mathbf{OLS},n} \tag{8.3}$$

and its dispersion, in a nonparametric fashion, through the empirical covariance of the individual $\hat{\gamma}_n$:

$$\mathrm{V}(\hat{\gamma}_{\mathbf{MG}}) = \frac{1}{N(N-1)}\sum_{n=1}^{N}(\hat{\gamma}_{\mathbf{OLS},n} - \hat{\gamma}_{\mathbf{MG}})(\hat{\gamma}_{\mathbf{OLS},n} - \hat{\gamma}_{\mathbf{MG}})^{\top} \tag{8.4}$$

which is in fact the simplified version of the Swamy covariance seen above. In the context of the Swamy model, it is biased but T-consistent and, differently from the original, always non-negative definite; as such, it has been suggested by Swamy (1970) himself as an alternative for cases when his parametric covariance is not. In general, it can be shown that the MG estimator is a special case with equal GLS weighting of the Swamy estimator, to which it converges as T grows sufficiently large (Hsiao and Pesaran, 2008). The function pmg performs mean groups estimation by default (`model='mg'`).

Example 8.2 Heterogeneous coefficients – `HousePricesUS` data set

Holly et al. (2010) analyze the long-run relationship between house prices and economic fundamentals (per capita income, net borrowing cost and population growth) in a sample of 49 US states over 29 years. The hypothesis of interest is whether house prices have an income elasticity of one. Their specification allows for variable coefficients in the random sense, as discussed above. The core of their model is the relationship between the logs of the nonstationary variables house prices `price` and income `income`. Their initial approach is to estimate a static specification by mean groups (MG). In the following we compare the coefficients from the asymptotically equivalent Swamy and MG estimators:

```
data("HousePricesUS", package = "pder")
swmod <- pvcm(log(price) ~ log(income), data = HousePricesUS, model= "random")
mgmod <- pmg(log(price) ~ log(income), data = HousePricesUS, model = "mg")
coefs <- cbind(coef(swmod), coef(mgmod))
dimnames(coefs)[[2]] <- c("Swamy", "MG")
coefs
             Swamy     MG
(Intercept) 3.8914 3.8498
log(income) 0.2867 0.3018
```

One can see that for $T = 29$, the efficient Swamy estimator and the simpler MG are already very close; moreover, both are statistically very far from one.

Dynamic Mean Groups Importantly, Pesaran and Smith (1995) consider the MG estimator in dynamic models of the type

$$y_{nt} = \rho_n y_{nt-1} + \gamma_n^{\top} z_{nt} + \nu_{nt} \tag{8.5}$$

and show that, unlike aggregated or pooled regressions, it provides consistent estimates of both coefficients and standard errors. Considering the full parameter vector $\theta_n = (\rho_n, \gamma_n)$, they observe that, while for fixed T the estimator $\hat{\theta}_n$ is biased of order $1/T$, the individual regressions (8.2) become consistent estimators of θ_n as T diverges. Hence the **MG** estimator of the average parameter vector $\bar{\theta}$ is consistent for both N and $T \to \infty$ (see the discussion in Hsiao and Pesaran, 2008). Explicit calculation of the individual parameters' covariance as in (8.4) in turn provides a consistent estimate of $V(\hat{\gamma})$.

Example 8.3 dynamic MG estimation – `RDSpillovers` data set

In their analysis of the returns of own vs general R&D, Eberhardt et al. (2013) consider both static and dynamic heterogeneous specifications in the production function of European firms. In doing so, every country-industry is allowed to follow its own production function; individual parameters are then averaged for the purpose of general inference. Static and dynamic specifications alike are considered.

In the following, we estimate both the static **MG** model (see their Table 7) and the dynamic **MG**. As in the original paper, we include individual trends by specifying `trend =TRUE`:

```
library("texreg")
data("RDSpillovers", package = "pder")
fm.rds <- lny ~ lnl + lnk + lnrd
mg.rds <- pmg(fm.rds, RDSpillovers, trend = TRUE)
dmg.rds <- update(mg.rds, . ~ lag(lny) + .)
screenreg(list('Static MG' = mg.rds, 'Dynamic MG'= dmg.rds), digits = 3)
```

```
=========================================
              Static MG     Dynamic MG
-----------------------------------------
(Intercept)      4.550 ***     4.038 ***
                (0.841)       (0.778)
lnl              0.568 ***     0.507 ***
                (0.086)       (0.059)
lnk              0.117         0.020
                (0.122)       (0.085)
lnrd            -0.058        -0.092
                (0.079)       (0.071)
trend            0.022 **      0.023 ***
                (0.008)       (0.004)
lag(lny)                       0.223 ***
                              (0.034)
-----------------------------------------
Num. obs.     2637          2518
=========================================
*** p < 0.001, ** p < 0.01, * p < 0.05
```

The lagged dependent variable turns out significant, although the autoregressive parameter's magnitude is modest. On the basis of the dynamic model, the authors proceed to calculate the long-run coefficients with or without common factor restrictions (see comment to their Table 8). Here we only reproduce the computation of the long- run elasticity of production to own R&D (which is the ratio of the coefficient of R&D to one minus the autoregressive coefficient), and the estimation of its standard error, through a Taylor approximation, by the delta method. With reference to a vector of K random variates, the function `deltamethod` from package **msm** (Jackson, 2011) requires: a formula describing the transformation (here, `x5/(1-x2)` as the coefficients on `lag(lny)` and `lnrd` are respectively 2nd and 5th); a vector of K estimates for the means; and a $K \times K$ matrix of covariance estimates. For the latter two, here we provide the `coef.panelmodel` and `vcov.panelmodel` of the dynamic model:

```
library("msm")
b.lr <- coef(dmg.rds)["lnrd"]/(1 - coef(dmg.rds)["lag(lny)"])
SEb.lr <- deltamethod(~ x5 / (1 - x2),
                      mean = coef(dmg.rds), cov = vcov(dmg.rds))
z.lr <- b.lr / SEb.lr
pval.lr <- 2 * pnorm(abs(z.lr), lower.tail = FALSE)
lr.lnrd <- matrix(c(b.lr, SEb.lr, z.lr, pval.lr), nrow=1)
dimnames(lr.lnrd) <- list("lnrd (long run)", c("Est.", "SE", "z", "p.val"))
round(lr.lnrd, 3)
                  Est.    SE      z p.val
lnrd (long run) -0.118 0.091 -1.301 0.193
```

After obtaining the point estimate and standard error of the long-run coefficient, we compute the t-statistic and the corresponding asymptotic p-value for the two-tailed test. The long-run elasticity of production to own R&D from the dynamic MG model is not significant at any conventional confidence level.[1]

8.2.3 Testing for Poolability

Heterogeneous estimators relax the assumption made in the error components model, which imposes homogeneity of all model parameters (but the intercept) across individuals. Under this assumption, one can estimate a single model for the whole sample, at most including individual-specific constant terms. This restriction, which is usually called *poolability*, can be tested by comparing the estimation results from the different approaches. Furthermore, one can impose the further restriction of no individual-specific intercepts.

In the variable coefficients framework, unrestricted estimation consists in estimating by OLS one different model for each individual. The sum of squared residuals is then: $\hat{\epsilon}_{np}^\top \hat{\epsilon}_{np}$. For this model, degrees of freedom are: $N(T-K-1)$. The restricted model to compare to can be either pooled OLS ($\hat{\epsilon}_{\text{OLS}}^\top \hat{\epsilon}_{\text{OLS}}$ with $NT-K-1$ degrees of freedom) or the *within* model ($\hat{\epsilon}_{\text{w}}^\top \hat{\epsilon}_{\text{w}}$ with $N(T-1)-K$ degrees of freedom), depending on whether the absence of individual effects is imposed or not. The test statistic is then (taking the *within* specification as the restricted model):

$$\frac{\hat{\epsilon}_{\text{w}}^\top \hat{\epsilon}_{\text{w}} - \hat{\epsilon}_{np}^\top \hat{\epsilon}_{np}}{\hat{\epsilon}_{np}^\top \hat{\epsilon}_{np}} \frac{N(T-K-1)}{(N-1)K}$$

This takes the form of a well-known stability test (known as the Chow test) distributed under H_0 as an F with $(N-1)K$ and $N(T-K-1)$ degrees of freedom.

The function performing this kind of test is called `pooltest`. One possible usage is to provide two models, one estimated separately for each individual, and either an OLS or a *within* model. In the first case, *all* parameters are supposed constant under H_0, including the constant terms. The unrestricted model is estimated by the function `pvcm`. As seen above,

1 The authors report instead the results from a common factor-restricted model, reaching qualitatively similar conclusions.

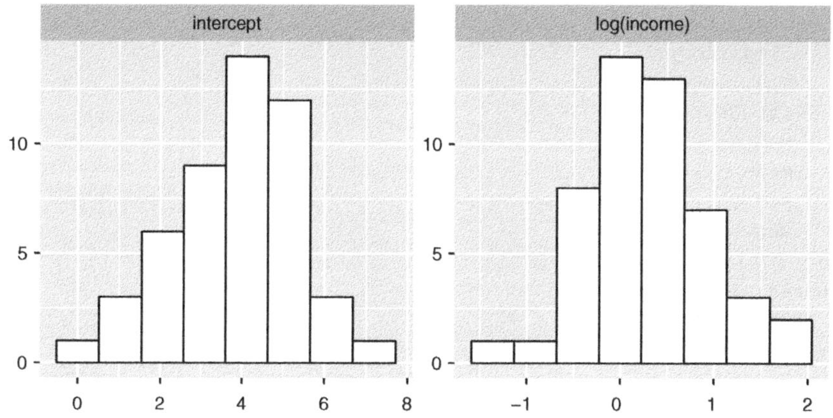

Figure 8.1 Individual coefficients, HousePriceUS.

this function allows to estimate two different models, depending on the parameter `model`; here, the appropriate value for this argument is `'within'` (the other possible choice being illustrated in the next section).

Example 8.4 Poolability test – `HousePricesUS` data set
Estimating the competing models for the `HousePricesUS` data, we have:

```
housep.np <- pvcm(log(price) ~ log(income), data = HousePricesUS,
          model = "within")
housep.pool <- plm(log(price) ~ log(income), data = HousePricesUS,
          model = "pooling")
housep.within <- plm(log(price) ~ log(income), data = HousePricesUS,
          model = "within")
```

As usual, the `pvcm` function provides a `coef.pvcm` method to retrieve individual coefficients. As a first assessment of their dispersion, in Figure 8.1 we display a histogram of the distribution of either coefficient.

The `summary.pvcm` method instead returns, for each coefficient, the synthetic statistics usually produced by `summary` for a generic numeric vector:

```
summary(housep.np)
Oneway (individual) effect No-pooling model

Call:
pvcm(formula = log(price) ~ log(income), data = HousePricesUS,
    model = "within")

Balanced Panel: n = 49, T = 29, N = 1421

Residuals:
   Min. 1st Qu.  Median    Mean 3rd Qu.    Max.
-0.2790 -0.0699 -0.0058  0.0000  0.0647  0.3524
```

```
Coefficients:
  (Intercept)         log(income)
 Min.    :-0.295   Min.    :-1.141
 1st Qu.: 3.152   1st Qu.:-0.138
 Median : 4.146   Median : 0.228
 Mean    : 3.850   Mean    : 0.302
 3rd Qu.: 4.777   3rd Qu.: 0.661
 Max.    : 6.911   Max.    : 2.037

Total Sum of Squares: 3870
Residual Sum of Squares: 13.7
Multiple R-Squared: 0.996
```

The stability test can then be performed supplying `housep.np` and either `housep.pool` or `housep.within` to the test function, depending on whether we want to assume absence of individual effects or not. Notice the different degrees of freedom.

```
pooltest(housep.pool, housep.np)

F statistic

data:  log(price) ~ log(income)
F = 26, df1 = 96, df2 = 1300, p-value <2e-16
alternative hypothesis: unstability
pooltest(housep.within, housep.np)

F statistic

data:  log(price) ~ log(income)
F = 16, df1 = 48, df2 = 1300, p-value <2e-16
alternative hypothesis: unstability
```

Coefficient stability is very strongly rejected, even in its weakest form (specific constants). The same tests can be performed using a `formula-data` syntax, specifying the nature of the restricted model through the `model` argument.

8.3 Cross-sectional Dependence and Common Factors

Dependence across individual units, or cross-sectional dependence, can take two main forms. Either it depends on the relative position of units in (some) space, so that – according to the so-called Tobler law – nearby units are "more related" than far away ones; or it depends on being observed at the same time and thus being subject to the same set of common, global factors that affect each unit to an extent that does not depend on distance.

The former kind of dependence is called *spatial* and is more appropriate to describe phenomena that *spill over* from one unit to nearby ones through vicinity, such as the diffusion of a disease or of know-how in the labor force or the alteration in cigarette sales from cross-border smuggling. In this case, one does therefore often speak of *local* dependence; although in many spatial models, effects do actually carry over across all spatial units, they in fact always do so in a *distance-decaying* fashion, whereby influence is strongest between the closest units. In the characterization of Pesaran and Tosetti (2011), this kind of dependence is also dubbed "cross-sectional weak dependence."

The latter kind of dependence does instead not need units to be referenced in any space: the relative position does not matter because correlation is assumed to stem from being exposed to the same, cross-sectionally invariant *common factors* (the world interest rate, the price of oil, the rate of technological progress, the stock market booms or busts, the price of homes in some reference market). Common factors can well originate from one or more main locations (think of a primary stock exchange, such as New York or London, setting prices that affect all other peers worldwide) but the effect will not depend on distance. Because factor-related dependence does typically not decrease with the distance between units, it is also called *global* dependence. In the characterization of Pesaran and Tosetti (2011), it is named "cross-sectional strong dependence."

As can be seen from the examples, common factors can be observable or not: the case when they are unobservable is of course the most interesting one. Most importantly, they can also be correlated with the regressors included in the model so that if they are omitted because they are unobservable, they will be a source of endogeneity and hence of inconsistency for estimators, unless they are appropriately accounted for (for an assessment of the properties of panel time series estimators under different omitted factors scenarios, see Coakley et al., 2006).

The first kind of dependence will be the subject of the chapter on spatial panels. In the following, common factor induced correlation will be our primary concern; nevertheless, the methods presented here are generally robust to spatial correlation as well.

8.3.1 The Common Factor Model

Consider the factor-augmented panel model

$$y_{nt} = \gamma_n^\top z_{nt} + \delta_n^\top f_t + \epsilon_{nt}$$

where $n = 1, \dots, N$ is the cross-sectional index and $t = 1, \dots, T$ the time index. z_{nt} is a $K + 1$ vector of observed, strictly exogenous regressors including a 1 and f_t is a vector of unobserved, cross-sectionally invariant common factors.

Such structure is capable of generating cross-sectional correlation in case of a similar, albeit not identical, response across countries to modifications in the common factors, measured by the factor loadings δ_n. The common factors are allowed to be correlated with the regressors, as is most likely to be the case, so their effect comes both through factor loadings and through the indirect effect on the observed regressors. The common factors are also allowed to be nonstationary. Moreover, the remainder error term ϵ is allowed to be spatially correlated as in

$$\epsilon_{nt} = \rho \sum_{m=1}^{N} w_{nm} \epsilon_{mn} + \nu_{nt}$$

where w_{nm} is the generic element of an $N \times N$ spatial weights matrix W in which nonzero elements correspond to pairs of spatially close observation units (e.g., regions sharing a common border, or below a given distance threshold); so that each error is correlated with a weighted average of the errors in close-by observations according to the parameter ρ.[2]

The two kinds of error dependence induced by omitted common factors and by spatial error correlation have serious consequences on the properties of estimators if they are neglected.

2 This is known in the spatial econometrics literature as the spatial autoregressive model and will be covered in Chapter 10.

The former induces cross-sectional correlation of a pervasive type, not dying out with distance, characterized by Pesaran and Tosetti (2011) as *strong*; moreover, if the omitted common factors are correlated with the regressors, the latter become endogenous and estimators become inconsistent. The latter type of dependence, dubbed *weak* because it dies out with distance, has less serious consequences on estimation but can still cause inefficiency (and hence inconsistent standard errors and invalid inference); moreover, as discussed in the next section, it weakens consistency in the particular case of spurious panel regression. Estimators able to control for the strong kind of dependence, as it turns out, are consistent in the presence of weak dependence as well.

In the special case of only one factor with uniform factor loadings $\delta_n = \delta \; \forall n$, the common factor model becomes a time fixed effects model, which can be estimated either by OLS with time dummies or by the appropriate *within* estimator, i.e., OLS on cross-sectionally demeaned data.

8.3.2 Common Correlated Effects Augmentation

The principle of common correlated effects (CCE) augmentation of Pesaran (2006) is based on the idea that, for large N, the factors f_t can be approximated by cross-sectional averages of the response and regressors. Following the original paper (see also Holly et al., 2010), consider the model:

$$y_{nt} = \alpha_n + \beta_n z_{nt} + \epsilon_{nt} \tag{8.6}$$

where both the (composite) error ϵ and the regressors z are generated by linear combinations of the unobserved, cross-sectionally invariant factors f:

$$\epsilon_{nt} = \gamma_n^\top f_t + \epsilon_{nt} \tag{8.7}$$

$$z_{nt} = a_n + \Gamma_n^\top f_t + v_{nt} \tag{8.8}$$

Substituting (8.7) in (8.6) and combining the result with (8.8), we get:

$$z_{nt} = d_n + C_n^\top f_t + v_{nt} \tag{8.9}$$

where $z_{nt} = (y_{nt}, z_{nt})^\top$ and

$$v_{nt} = \begin{pmatrix} \epsilon_{nt} + \beta_n^\top v_{nt} \\ v_{nt} \end{pmatrix}$$

$$d_n = \begin{pmatrix} 1 & \beta_n^\top \\ 0 & I_k \end{pmatrix} \begin{pmatrix} \alpha_n \\ a_n \end{pmatrix}$$

$$C_n = \begin{pmatrix} \gamma_n & \Gamma_n \end{pmatrix} \begin{pmatrix} 1 & 0 \\ \beta_n & I_k \end{pmatrix}$$

Taking cross-section averages of (8.9),

$$\bar{z}_t = \bar{d} + \bar{C}^\top f_t + \bar{v}_t$$

so that, if $(\bar{C}\bar{C}^\top)^{-1}$ is invertible, the common factors can be written as:

$$f_t = (\bar{C}\bar{C}^\top)^{-1}\bar{C}(\bar{z}_t - \bar{d} - \bar{v}_t)$$

If as $N \to \inf \bar{v}_t \to 0$ and $\bar{C} \xrightarrow{p} C$, then

$$f_t - (CC^\top)^{-1}C(\bar{z}_t - \bar{d}) \xrightarrow{p} 0$$

Following this line of reasoning, Pesaran (2006) shows that the cross-sectional averages of the response ($\bar{y}_t$) and regressors ($\bar{X}_t$) are N-consistent estimators of the unobserved common factors and can therefore be used as observable proxies thereof. Augmenting the regression with these averages is known as the common correlated effects (CCE) principle. CCE estimators can be used to consistently estimate the individual slope parameters β_n by applying least squares to the augmented regression

$$y_{nt} = \alpha_n + d_{nt} + \beta_n^\top Z_{nt} + g_n^\top W_t + \epsilon_{nt}$$

where $W_t = (\bar{y}_t, \bar{X}_t)^\top$.

The estimator for each individual slope coefficient can then be written compactly as

$$\hat{\gamma}_{\text{CCE},n} = (Z_n^\top \bar{M} Z_n)^{-1} Z_n^\top \bar{M} y_n$$

with $\bar{M} = \mathrm{I}_T - \bar{H}(\bar{H}^\top \bar{H})^{-1}\bar{H}^\top$, $\bar{H}$ contains: the $T \times (K+1)$ matrix of cross-sectional averages $W_t, t = 1, \dots T$; and a deterministic component comprising individual intercept and time trend (Pesaran, 2006, p. 974). The average is then estimated by the MG method,

$$\hat{\gamma}_{\text{CCEMG}} = \frac{1}{N}\sum_{n=1}^{N} \hat{\gamma}_{\text{CCE},n}$$

This estimator is known as CCEMG, for "common correlated effects mean groups."

The covariance matrix is estimated nonparametrically, on the basis of the empirical covariance of the individual coefficients, just like in the MG case:

$$\mathrm{V}(\hat{\gamma}_{\text{CCEMG}}) = \frac{1}{N(N-1)}\sum_{n=1}^{N}(\hat{\gamma}_{\text{CCE},n} - \hat{\gamma}_{\text{CCEMG}})(\hat{\gamma}_{\text{CCE},n} - \hat{\gamma}_{\text{CCEMG}})^\top \qquad (8.10)$$

Unlike other estimators, the CCE is (N-) consistent for any fixed, unknown number of possibly nonstationary common factors. Being robust to strong forms of cross-sectional dependence, the CCE estimator is also robust to weak ones such as spatial correlation (see Pesaran and Tosetti, 2011). Moreover, the CCE strategy has proved most effective in a number of simulation studies, e.g., Coakley et al. (2006), Pesaran and Tosetti (2011), Kapetanios et al. (2011).

Example 8.5 Common correlated effects MG – `HousePricesUS` data set

The function pmg will perform CCE augmentation in the context of the MG model, if the argument `model` is set to `'cmg'`. In their article, Holly et al. (2010) augment their model with the cross-section averages in order to obtain a consistent estimate of the income elasticity of house prices in the presence of common factors. Below we reproduce and compare their MG and CCEMG results. The MG and CCEMG coefficients are substantially different; with CCE the income elasticity turns out much higher and not significantly different from 1 any more, in line with economic theory. `summary.pmg` explicitly outputs the coefficients and significance diagnostics for the added cross-sectional averages, denoted with the suffix `.bar`. The coefficients on the latter are not meaningful *per se*, but their joint significance can be seen as an informal test for the presence of common factors.

```
library("texreg")
cmgmod <- pmg(log(price) ~ log(income), data = HousePricesUS, model = "cmg")
screenreg(list(mg = mgmod, ccemg = cmgmod), digits = 3)
```

```
=============================================
                  mg               ccemg
---------------------------------------------
(Intercept)          3.850 ***      -0.115
                    (0.204)        (0.256)
log(income)          0.302 **        1.135 ***
                    (0.093)        (0.195)
y.bar                                1.047 ***
                                   (0.058)
log(income).bar                     -1.195 ***
                                   (0.199)
---------------------------------------------
Num. obs.          1421           1421
=============================================
*** p < 0.001, ** p < 0.01, * p < 0.05
```

8.3.2.1 CCE Mean Groups vs. CCE Pooled

Estimation by the **CCE** principle can be performed either leaving parameters β_n free to vary, as above, or imposing parameter homogeneity (but maintaining heterogeneity in intercepts, factor loadings, and possibly time trends), which leads to the **CCEP** (pooled) estimator

$$\hat{\beta}_{\text{CCEP}} = \left(\sum_{n=1}^{N} Z_n^{\top} \bar{M} Z_n \right)^{-1} \sum_{n=1}^{N} Z_n^{\top} \bar{M} y_n \tag{8.11}$$

and is to be preferred on efficiency grounds when the underlying assumption that $\beta_n = \beta$ is reasonable. It must be observed that the **CCEP** estimator, although imposing $\beta_n = \beta$, still allows individual factor loadings δ_n to differ.

The standard pooled or heterogeneous estimators can be seen an special cases of this more general formulation where augmentation is eliminated or reduced: pooled **OLS** as **CCEP** with $\bar{M} = I_T$, individual fixed effects as **CCEP** with $\bar{H}$ containing only individual dummies. The mean groups (**MG**) estimator can in turn be seen as **CCEMG** where $\bar{M} = I_T$.

Example 8.6 CCEMG and CCEP – `HousePricesUS` data set

The function `pcce` estimates **CCE** models of either type by projection of the original regressors on the matrix $\bar{\mathbf{M}}$; by default (`model='mg'`) one gets the **CCEMG**, if `model='p'` the **CCEP**. This is the only way to perform **CCEP** estimation, while **CCEMG** results from `pcce` will be equivalent to those obtained through explicit augmentation with pmg, the only difference being that here one cannot see the significance diagnostics for the added cross-sectional averages:

```
ccemgmod <- pcce(log(price) ~ log(income), data=HousePricesUS, model="mg")
summary(ccemgmod)
Common Correlated Effects model
Call:
pcce(formula = log(price) ~ log(income), data = HousePricesUS,
    model = "mg")

Balanced Panel: n = 49, T = 29, N = 1421

Residuals:
    Min.   1st Qu.    Median      Mean   3rd Qu.      Max.
-0.23744 -0.03549  0.00027  0.00000  0.03639  0.22423
```

```
Coefficients:
             Estimate Std. Error z-value Pr(>|z|)
log(income)     1.135      0.195    5.81  6.3e-09 ***
---
Signif. codes:
0 '***' 0.001 '**' 0.01 '*' 0.05 '.' 0.1 ' ' 1
Total Sum of Squares: 47.2
Residual Sum of Squares: 5.66
HPY R-squared: 0.74
```

Holly et al. (2010) are interested in estimating the relationship between house prices and income net of the influence of common factors under the pooled specification as well. To this end, they estimate a homogeneous CCEP version of the baseline model:

```
ccepmod <- pcce(log(price) ~ log(income), data=HousePricesUS, model="p")
summary(ccepmod)
Common Correlated Effects model
Call:
pcce(formula = log(price) ~ log(income), data = HousePricesUS,
    model = "p")

Balanced Panel: n = 49, T = 29, N = 1421

Residuals:
     Min.   1st Qu.    Median      Mean   3rd Qu.      Max.
-0.27883 -0.03928 -0.00209  0.00000  0.03927  0.29993

Coefficients:
             Estimate Std. Error z-value Pr(>|z|)
log(income)     1.199      0.207    5.79  7.2e-09 ***
---
Signif. codes:
0 '***' 0.001 '**' 0.01 '*' 0.05 '.' 0.1 ' ' 1
Total Sum of Squares: 47.2
Residual Sum of Squares: 6.89
HPY R-squared: 0.696
```

The results from the two specifications are very close, as regards both the coefficients and the standard errors thereof, which speaks in favor of imposing the pooling restriction.

8.3.2.2 Computing the CCEP Variance

According to Pesaran (2006, 5.2), the variance of the CCEP estimator can be computed in two different ways, depending on whether the assumption of parameter homogeneity is imposed here as well (homogeneous estimator) or not (heterogeneous, or nonparametric, estimator).

The heterogeneous version (Pesaran, 2006, Th. 3) is based again on the nonparametric estimate of the individual coefficients' covariance. Defining

$$\Psi = \frac{1}{N}\sum_{n=1}^{N}\frac{Z_n^\top \bar{M} Z_n}{T}$$

and

$$\hat{R} = \frac{1}{N-1}\sum_{n=1}^{N} \frac{Z_n^{\mathsf{T}}\bar{M}Z_n}{T}(\hat{\gamma}_{\text{CCE},n} - \hat{\gamma}_{\text{CCEMG}})(\hat{\gamma}_{\text{CCE},n} - \hat{\gamma}_{\text{CCEMG}})^{\mathsf{T}}\frac{Z_n^{\mathsf{T}}\bar{M}Z_n}{T}$$

the estimator is

$$\mathrm{V}(\hat{\gamma}_{\text{CCEP}}) = \frac{1}{N}\Psi^{-1}\hat{R}\Psi^{-1} \tag{8.12}$$

This estimator is consistent under quite general conditions as regards the rate of growth of N vs T and the distribution of individual parameters; it is the one that fares best in the original papers' simulation study and the one the author recommends to use. It is therefore the default method in the `pcce` function.

Nevertheless, strictly speaking, (8.12) is not appropriate under complete homogeneity. Pesaran (2006, Th. 4) presents an alternative, which is appropriate for large panels (i.e., if $T/N \to 0$ as $(N,T) \xrightarrow{j} \infty$). The latter, which is presented in detail in Pesaran (2006, p. 988), is based on the nonparametric kernel-smoothed estimator of Newey and West (see 5.1.1.3) and can be calculated using standard methods. Analogously, and again in large-N settings, the familiar clustering estimator can be applied. In fact, $\bar{M}$ being idempotent, the CCEP estimator in (8.11) can be seen as OLS on the transformed variables $\bar{M}Z$; hence methods for robust covariances can be applied to `pcce` objects the same way they are to `plm` ones – *e.g.*, those representing a *within* model. From a software viewpoint, the `pcce` function is compliant with both `vcovNW` and `vcovHC`.

Example 8.7 variance of the CCEP estimator – `RDSpillovers` data set

The main point in Eberhardt et al. (2013) is to control for cross-sectional R&D spillovers in estimating the productivity of own R&D of any observation unit. To this end, they employ CCE augmentation both in heterogeneous and in homogeneous flavors. Below we present the CCEP estimates from their Table 5 with three alternative estimators for the standard errors:

```
ccep.rds <- pcce(fm.rds, RDSpillovers, model="p")
library(lmtest)
ccep.tab <- cbind(coeftest(ccep.rds)[, 1:2],
                  coeftest(ccep.rds, vcov = vcovNW)[, 2],
                  coeftest(ccep.rds, vcov = vcovHC)[, 2])
dimnames(ccep.tab)[[2]][2:4] <- c("Nonparam.", "vcovNW", "vcovHC")
round(ccep.tab, 3)
     Estimate Nonparam. vcovNW vcovHC
lnl     0.562     0.088  0.031  0.045
lnk     0.289     0.161  0.045  0.077
lnrd    0.084     0.068  0.020  0.033
```

A priori, homogeneous variance estimators are relatively well-suited to this comparatively large and short dataset, provided that the homogeneity assumption holds. From the results we can instead see that the nonparametric standard errors are much more conservative, hinting at pooling assumptions being too restrictive.

8.4 Nonstationarity and Cointegration

The time series dimension of "long" panel datasets raises the issue of possible nonstationarity and cointegration. From an econometric viewpoint, if two (single) nonstationary time series are cointegrated, then the least squares estimator of the regression parameter characterizing the relationship is superconsistent and converges to the true value faster than its stationary counterpart (Stock, 1987). If on the contrary they are nonstationary but not cointegrated, the statistical relationship is spurious, and least squares estimates do not converge to their true values at all, while fit and significance diagnostics yield the false positive results famously discussed by Granger and Newbold (1974).

In a panel time series context, there is one more dimension available for inference: the cross section. Assuming cross-sectional independence, Phillips and Moon (1999) show that a spurious panel data regression can still deliver a consistent estimate of long-run parameters. Yet its convergence properties will be weaker than those of a cointegrating one: in particular, the coefficients of a spurious panel regression will still converge to their true values, although at a much slower rate $\sqrt{N}$ than that of a cointegrating panel, which is $T\sqrt{N}$.

This result depends on an assumption of cross-sectional independence. It is weakened if the errors are cross-sectionally weakly correlated, for example if they follow a spatial process, and can be expected to fail in presence of strong cross-sectional dependence, as would arise when omitting to control for common factors (Phillips and Moon, 1999, pages 1091–1092). Both pooled OLS (Phillips and Sul, 2003) and mean groups estimators (Coakley et al., 2006) lose their advantage in precision from pooling when cross-sectional dependence is present.

8.4.1 Unit Root Testing: Generalities

Detecting unit roots has become a central subject in macroeconometrics. The techniques employed are adaptations from the time series literature to the panel case. We will begin by reviewing the main results regarding time series.

Consider a variable y_t generated by an autoregressive process of order one:

$$y_t = \rho y_{t-1} + z_t^{\top}\gamma + \epsilon_t$$

The vector of explanatory variables may contain an intercept, a linear trend, and different explanatory variables. To keep things simple, in the following we will assume $\gamma = 0$, so that y follows a "pure" autoregressive process. As regards the error (which in this context is often called the *innovation*), we will assume that it has mean zero and standard deviation σ. By recursive substitution, one has:

$$y_t = \rho^t y_0 + \rho^{t-1}\epsilon_1 + \rho^{t-2} + \ldots + \rho\epsilon_{t-1} + \epsilon_t$$

If y_0 is deterministic and the ϵ are not correlated, the variance of y can be written:

$$\mathrm{V}(y_t) = (\rho^{t-1} + \rho^{t-2} + \ldots + \rho + 1)\sigma^2$$

If $\rho \neq 1$, we have:

$$\mathrm{V}(y_t) = \frac{1-\rho^t}{1-\rho}\sigma^2 \rightarrow \frac{1}{1-\rho}\sigma^2$$

On the other hand, if $\rho = 1$, $\mathrm{V}(y_t) = t\sigma^2$ so that the variance grows to infinity with t; the series is then nonstationary and is said to have a *unit root*. The presence of unit roots poses various problems, first and foremost that of *spurious regressions*. In the presence of a unit root, a series presents a peculiar sort of trend that is not deterministic but stochastic, and the presence of such trends in two series containing unit roots may induce an artificial correlation between them. In Figure 8.2 we present two autoregressive series with respectively $\rho = 0.2$ and $\rho = 1$. We see how in the former case the autoregressive process translates into correlation between successive values of y_t; in particular, if $y_{t-1} < 0$ then y_t is more likely to be negative than positive. However, the curve representing the realization of the process crosses the horizontal axis frequently. On

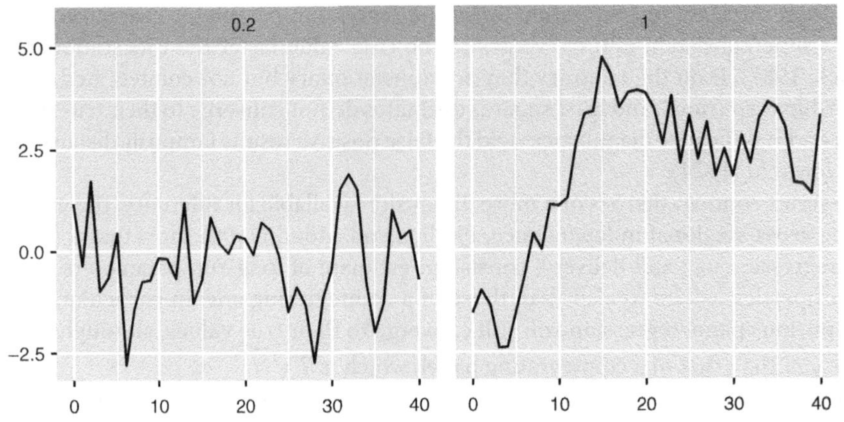

Figure 8.2 Autoregressive processes with different ρ parameters.

the other hand, in the case of a unit root, one can clearly detect the presence of a stochastic trend (in this case, on the rise): y_t only changes sign once, and most of its realizations are positive.

To illustrate the importance of the spurious regression problem, we perform a short simulation exercise; we draw two autoregressive series independently, regress one on the other, and recover the t-statistic corresponding to the null hypothesis $H_0 : \beta = 0$. This hypothesis is true by construction; therefore, in a normal context the t-statistic should not reject (i.e., be roughly less than 2) in 95% of cases. Let us begin by illustrating this result for $\rho = 0.2$. To this end, we employ two functions: code generates an autoregressive series, `tstat` performs OLS estimation, and recovers the t-statistic:

```
autoreg <- function(rho = 0.1, T = 100){
  e <- rnorm(T)
  for (t in 2:(T)) e[t] <- e[t] + rho *e[t-1]
  e
}
tstat <- function(rho = 0.1, T = 100){
  y <- autoreg(rho, T)
  x <- autoreg(rho, T)
  z <- lm(y ~ x)
  coef(z)[2] / sqrt(diag(vcov(z))[2])
}
result <- c()
R <- 1000
for (i in 1:R) result <- c(result, tstat(rho = 0.2, T = 40))
quantile(result, c(0.025, 0.975))
  2.5%   97.5%
-2.114  1.990
prop.table(table(abs(result) > 2))

FALSE  TRUE
0.943 0.057
```

We can see how the empirical quantiles are very close to their expected values and the share of false positives is in the region of 5%. Let us now do the same with two series, each containing a unit root:

```
result <- c()
R <- 1000
for (i in 1:R) result <- c(result, tstat(rho = 1, T = 40))
quantile(result, c(0.025, 0.975))
  2.5%   97.5%
-9.158  8.227
prop.table(table(abs(result) > 2))

FALSE  TRUE
0.379 0.621
```

Judging by the usual t-statistic, in two thirds of cases one would conclude in favor of a significant relationship between our two independently generated variables.

It is therefore crucial to detect the presence of unit roots in time series data; otherwise, there are considerable chances to obtain falsely significant results. To this end, it is simplest to write the equation of the autoregressive process subtracting y_{t-1} to both sides. One has then:

$$\Delta y_t = (\rho - 1)y_{t-1} + \epsilon_t$$

The unit root test then becomes a zero restriction test for the coefficient associated to y_{t-1} in the model where the regressand is Δy_t. One might want to use a classic t-statistic, obtained dividing $\hat{\rho} - 1$ by its standard error. Setting $H_0 : \rho = 1$ vs $H_1 : \rho < 1$, one will then reject the unit root hypothesis at the 5% level if the statistic is less than -1.64.

```
R <- 1000
T <- 100
result <- c()
for (i in 1:R){
  y <- autoreg(rho=1, T=100)
  Dy <- y[2:T] - y[1:(T-1)]
  Ly <- y[1:(T-1)]
  z <- lm(Dy ~ Ly)
  result <- c(result, coef(z)[2] / sqrt(diag(vcov(z))[2]))
}
```

In Figure 8.3 we depict a histogram of the realizations of the t-statistic, superposing a normal density curve:

One can easily see that employing classic inference procedures to detect the presence of unit roots is unwarranted, as the t-statistic follows a distribution that is very far from the normal. Employing the usual critical value of -1.64, one has here:

```
prop.table(table(result < -1.64))

FALSE  TRUE
0.542 0.458
```

which leads to reject the true hypothesis of a unit root one half of the times. To perform the Dickey-Fuller test, one needs specific critical values that are not those of the normal (or the t) distribution. The test can be performed augmenting the auxiliary model with a constant

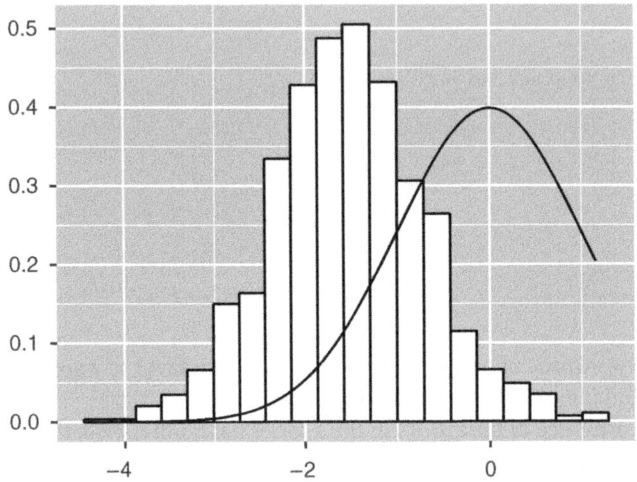

Figure 8.3 Histogram of the Student statistic in case of a unit root.

and/or a deterministic trend; lags of Δy can also be added in order to clean out any possible autocorrelation of ϵ.

The regression between two series both containing a unit root is only appropriate if they present a long-term structural relationship. One speaks then of *co-integration*. More precisely, we will say that two variables x and y are cointegrated if there exists β such that:

$$y = \alpha + \beta x + \epsilon$$

where ϵ is stationary, i.e., it does not have unit roots. A simple cointegration test can then be performed as follows:

1) verify whether y and x have unit roots with a Dickey-Fuller test,
2) if they both do, then estimate a model of y on x and recover the residuals $\hat{\epsilon}$,
3) do a Dickey-Fuller test on $\hat{\epsilon}$: if the unit root hypothesis is rejected, then x and y are cointegrated and the regression of y on x is meaningful; otherwise, x and y are integrated but not cointegrated, and the regression of y on x will be spurious.

8.4.2 First Generation Unit Root Testing

The classical test for unit roots is usually called ADF for "augmented Dickey-Fuller". Many extensions of this test have been proposed to adapt it to a panel data setting.

8.4.2.1 Preliminary Results

Some of these tests are obtained by applying separate ADF tests to every individual in the sample. To perform these preliminary tests, one shall choose the number of lags and the relevant set of deterministic variables d_{mt}, which can be either $d_{1t} = \emptyset$, $d_{2t} = 1$ (an intercept), or $d_{3t} = 1, t$ (an intercept and a time trend).

$$\Delta y_{nt} = (\rho_n - 1)y_{n(t-1)} + \sum_{s=1}^{L_n} \theta_{ns}\Delta y_{n(t-s)} + \alpha_{mn}^{\top} d_{mt} \quad m = 1, 2, 3 \tag{8.13}$$

This choice can be based on a number of criteria:

- the Schwarz information criterion (SIC),

- the Akaike information criterion (AIC),
- the Hall method, consisting in adding as many lags as there are significant ones.

The regression is performed on $T - L_n - 1$ observations for each individual, which leads to $N \times \tilde{T}$ in total, with $\tilde{T} = T - (\bar{L} - 1)$, $\bar{L}$ being the average number of lags. The variance of the residuals for individual n is estimated by:

$$\hat{\sigma}^2_{\epsilon_n} = \frac{\sum_{t=L_n+1}^{T} \hat{\epsilon}^2_{nt}}{df_n} \tag{8.14}$$

with df_n the degrees of freedom of the regression.

8.4.2.2 Levin-Lin-Chu Test

Levin et al. (2002) proposed the first panel unit root test. In order to perform it, one must run two preliminary regressions: respectively, of Δy_{nt} and of y_{nt-1} as functions of Δy_{nt-s} $s = 1, \dots L_n$ and d_{mt}, obtaining two residual vectors denoted respectively by z_{nt} and $v_{n(t-1)}$.

These two residuals are then normalized dividing them by the estimated standard error (equation 8.14). The estimator of ρ is obtained by regressing $z_{nt}/\hat{\sigma}_n$ on $v_{nt}/\hat{\sigma}_n$ for the whole sample. Its standard deviation and t-statistic are denoted respectively by $\hat{\sigma}(\hat{\rho})$ and $t_{\hat{\rho}} = \hat{\rho}/\hat{\sigma}(\hat{\rho})$.

The long-term variance of y_n is estimated by:

$$\hat{\sigma}^2_{y_n} = \frac{1}{T-1} \sum_{t=2}^{T} \Delta y^2_{nt} + 2 \sum_{s=1}^{\bar{K}} \omega_{\bar{K}s} \left[\frac{1}{t-1} \sum_{t=2+s}^{T} \Delta y_{nt} \Delta y_{nt-s} \right]$$

where $\bar{K}$ is the truncation lag parameter and $\omega_{\bar{K}s}$ are the sample covariance weights, which depend on the choice of kernel.

Calling $s_n = \frac{\hat{\sigma}_{yn}}{\hat{\sigma}_{\epsilon_n}}$ the ratio between the long-term and the short-term variance for the n-th individual and $\bar{s} = \frac{1}{N}\sum_n s_n$ the sample average thereof, Levin et al. (2002) show that the statistic:

$$t^*_\rho = \frac{t_\rho - N\tilde{T}\bar{s}\hat{\sigma}_{\tilde{\epsilon}}^{-2}\hat{\sigma}(\hat{\rho})\mu^*_{m\tilde{T}}}{\sigma^*_{m\tilde{T}}}$$

is normally distributed under the null hypothesis of a unit root. $\mu^*_{m\tilde{T}}$ and $\sigma^*_{m\tilde{T}}$ can be found in the original paper.

8.4.2.3 Im, Pesaran and Shin Test

One of the drawbacks of the Levin et al. (2002) test is that the alternative hypothesis holds that $\rho \neq 1$, but at the same time it is the same for all individuals. The test proposed by Im et al. (2003) (IPS) overtakes this limitation: the null hypothesis is still $\rho = 1$ for all individuals, but the alternative is now that ρ can be different across individuals, provided that $\rho_n < 1$ at least for some of them. The IPS test takes the form of a simple average of the t-statistics for $H_0 : (\rho - 1) = 0$ from the individual ADF regressions (8.13):

$$\bar{t} = \frac{1}{N} \sum_{n=1}^{n} t_{\rho n}$$

The IPS statistic follows a nonstandard distribution, and must be therefore compared with values tabulated *ad hoc*. Alternatively, it can be standardized with mean and variance $E(\bar{t})$ and $V(\bar{t})$ given in the Im et al. (2003) paper. The test statistic is then $\frac{\sqrt{N}(\bar{t} - E(\bar{t}))}{\sqrt{V(\bar{t})}}$:
which, under the null of a unit root, is normally distributed.

8.4.2.4 The Maddala and Wu Test

Maddala and Wu (1999) proposed a similar test, again not imposing homogeneity of ρ under the alternative. Instead of the t-statistics, it is based on combining the N critical values p-value$_n$ obtained from the individual ADF tests. The test statistic is then simply:

$$P = -2 \sum_{n=1}^{N} \ln \text{p-value}_n$$

and, under the null of a unit root for all N individuals, it is distributed as a χ^2 with $2N$ degrees of freedom.

Example 8.8 First generation unit root testing – `HousePricesUS` data set

The first-generation unit root tests can be computed using the `purtest` function. A `formula-data` can be used to describe the variable for which the test has to be computed and the deterministic covariates (`0`, `1` for an intercept, and `trend` for an intercept and a time trend). The same description of the test to be computed can be performed using a `pseries` and specifying the deterministic covariates using the `exo` argument.

We set below the `lags` argument to 2 for comparability across procedures, instead of leaving the choice to one of the flexible procedures described above (e.g., by setting the `lags` argument to `'Hall'` to select the lags using Hall (1994)'s method). We apply the test to the `price` variable of the `HousePricesUS` data set.

```
data("HousePricesUS", package = "pder")
price <- pdata.frame(HousePricesUS)$price
purtest(log(price), test = "levinlin", lags = 2, exo = "trend")

Levin-Lin-Chu Unit-Root Test (ex. var.: Individual
Intercepts and Trend)

data:  log(price)
z = -1.3, p-value = 0.1
alternative hypothesis: stationarity
purtest(log(price), test = "madwu", lags = 2, exo = "trend")

Maddala-Wu Unit-Root Test (ex. var.: Individual
Intercepts and Trend)

data:  log(price)
chisq = 100, df = 98, p-value = 0.4
alternative hypothesis: stationarity
purtest(log(price), test = "ips", lags = 2, exo = "trend")

Im-Pesaran-Shin Unit-Root Test (ex. var.: Individual
Intercepts and Trend)

data:  log(price)
z = 0.77, p-value = 0.8
alternative hypothesis: stationarity
```

The three tests strongly don't reject the null hypothesis of unit root.

8.4.3 Second Generation Unit Root Testing

The above panel unit root tests do all rest on the hypothesis of absence of cross-sectional correlation. When, after the turn of the millennium, the panel data literature started recognizing how pervasive cross-sectional correlation is in applications and progressed toward the development of consistent methods in its presence, the above assumption started to be seen as too restrictive. The tests assuming no cross-sectional correlation became known under the collective name of "first-generation" panel unit root tests, to distinguish them from the new breed of testing procedures that was emerging. These new panel unit root tests, sharing the quality of being consistent in the face of cross-sectional correlation, were dubbed "second generation" to distinguish them from the former and are currently most often employed in applications.

The reference framework for cross-sectionally correlated panels is, as discussed above, the common factor model. A number of cross-correlation-compliant panel unit root procedures have been devised in this framework based on various defactoring procedures. One of the most popular second-generation tests, due to Pesaran (2007), takes the approach of controlling for the common factors, instead of trying to eliminate them; it does so in the CCE framework, by augmenting the auxiliary regressions through cross-sectional averages of the response and regressors. The individual ADF regressions are augmented with the cross-sectional averages of lagged levels and differences of the individual series:

$$\Delta y_{nt} = (\rho_n - 1)y_{n(t-1)} + \sum_{s=1}^{L_n} \theta_{ns}\Delta y_{n(t-s)} + \theta_{(s+1)}\Delta\bar{y}_t + \theta_{(s+2)}\bar{y}_{(t-1)}$$
$$+ \sum_{s=L_n+3}^{2L_n+2} \theta_s \Delta\bar{y}_{(t-s)} + \alpha_{mn}^{\top} d_{mt} \quad (m = 1, 2, 3) \tag{8.15}$$

The individual ADF regressions are therefore denoted "cross-sectionally augmented ADF" (CADF) regressions; the resulting individual CADF statistics can in principle be combined as described above, forming the basis for either a "cross-sectionally augmented IPS" (CIPS) or a Maddala-Wu test. However, the limiting distributions for the latter do not apply anymore in the absence of cross-sectional independence; for this reason, Pesaran (2007) tabulated critical values for the CIPS test for the three different cases where the auxiliary CADF regressions contain an intercept, a deterministic trend, or none of the above.

Example 8.9 IPS and CIPS tests – `HousePricesUS` data set

Holly et al. (2010) analyze the stationarity of their target variable, the house price index, and of the regressors of their model using individual ADF tests. They do so only in order to demonstrate the strong cross-sectional correlation remaining in the residuals of the individual ADF regressions, which invalidates the use of the first-generation IPS test, and thus to motivate their resorting to the CADF-based CIPS test. In fact, they do not show the result of an IPS test but only the regression diagnostics.

As every unit root test, the results are sensitive to the order of time series augmentation: the more lags we add, the more confident we are to have effectively filtered out residual serial correlation, but the less degrees of freedom, and hence the less power, we allow to the testing procedure. They consider the first four augmentation orders: following them, below we reproduce the CD statistics and the average pairwise correlation coefficients $\bar{\hat{\rho}}$ for the residuals of the ADF regressions.[3]

Below we explicitly estimate the individual ADF regressions using the `pmg` function: the latter outputs a `pmg` object from which the `pcdtest` function is able to retrieve the residuals as a

3 The results do not correspond exactly to the original paper: for an explanation see Millo (2015).

`pseries`, so it can be directly applied specifying whether one wants the CD statistic (default) or the pairwise correlation coefficients $\bar{\hat{\rho}}$ (then `test` has to be set to `'rho'`).

```
tab5a <- matrix(NA, ncol = 4, nrow = 2)
tab5b <- matrix(NA, ncol = 4, nrow = 2)

for(i in 1:4) {
    mymod <- pmg(diff(log(income)) ~ lag(log(income)) +
                 lag(diff(log(income)), 1:i),
                 data = HousePricesUS,
                 model = "mg", trend = TRUE)
    tab5a[1, i] <- pcdtest(mymod, test = "rho")$statistic
    tab5b[1, i] <- pcdtest(mymod, test =  "cd")$statistic
}

for(i in 1:4) {
    mymod <- pmg(diff(log(price)) ~ lag(log(price)) +
                 lag(diff(log(price)), 1:i),
                 data=HousePricesUS,
                 model="mg", trend = TRUE)
    tab5a[2, i] <- pcdtest(mymod, test = "rho")$statistic
    tab5b[2, i] <- pcdtest(mymod, test =  "cd")$statistic
}

tab5a <- round(tab5a, 3)
tab5b <- round(tab5b, 2)
dimnames(tab5a) <- list(c("income", "price"),
                        paste("ADF(", 1:4, ")", sep=""))
dimnames(tab5b) <- dimnames(tab5a)

tab5a
       ADF(1) ADF(2) ADF(3) ADF(4)
income  0.465  0.443  0.338  0.317
price   0.346  0.326  0.252  0.194
tab5b
       ADF(1) ADF(2) ADF(3) ADF(4)
income  82.84  77.40  57.96  53.21
price   61.73  57.02  43.21  32.52
```

Residual cross-correlation is clearly apparent and motivates employing the CIPS test. In the following we assess the order of integration of prices and income by testing the original series and the differenced ones for unit roots. To do so, the dataset now contained in the `data.frame HousePricesUS` has to be converted into a `pdata.frame` from which the testing function `cipstest` will be able to retrieve the panel indices it needs. The number of `lags` is left at the default value of 2. As for the deterministic component of the CADF regressions, we allow for an intercept (`type='drift'`) in the original series; for the sake of consistency, we then exclude it from the differenced one (`type='none'`).

```
php <- pdata.frame(HousePricesUS)
cipstest(log(php$price), type = "drift")

Pesaran's CIPS test for unit roots

data:  log(php$price)
CIPS test = -2, lag order = 2, p-value = 0.1
alternative hypothesis: Stationarity
cipstest(diff(log(php$price)), type = "none")

Pesaran's CIPS test for unit roots

data:  diff(log(php$price))
CIPS test = -1.8, lag order = 2, p-value = 0.01
alternative hypothesis: Stationarity
```

The CIPS test does not reject a unit root for the original series, while it does for the differenced one[4]. The conclusion is that the price index is integrated of order 1. The same (not reported) happens for income, at which point the crucial issue is whether house prices and income are cointegrated, or otherwise the regression of interest is spurious. A CIPS test of the regression residuals will help shed light on the issue: currently the `cipstest` function only accepts `pseries` objects as arguments; hence, we extract residuals as a `pseries` through the usual `resid.ccep` extractor function prior to feeding them to the unit root test. Given that individual trends have been controlled for at the modeling stage and that by the very nature of regression residuals, the series is not expected to contain a drift (intercept), we eliminate any deterministic component from the CADF regressions by specifying `type='none'`:

```
cipstest(resid(ccemgmod), type="none")

Pesaran's CIPS test for unit roots

data:  resid(ccemgmod)
CIPS test = -2.7, lag order = 2, p-value = 0.01
alternative hypothesis: Stationarity
cipstest(resid(ccepmod), type="none")

Pesaran's CIPS test for unit roots

data:  resid(ccepmod)
CIPS test = -2.2, lag order = 2, p-value = 0.01
alternative hypothesis: Stationarity
```

The unit root hypothesis is rejected for both the residuals of the CCEMG and the CCEP models. The conclusion is that both models represent cointegrating regressions.

---

4 An exact p-value is not available because the test distribution is nonstandard; test results are compared to tabulated critical values from Pesaran (2007), and the nearest significance level is reported, together with a warning.

9

Count Data and Limited Dependent Variables

It is often the case in economics that the dependent variable is not continuous so that OLS estimation is not appropriate. On the one hand, the response may be a count, i.e., it takes only non-negative integer values. In this case, the most commonly used specifications are the Poisson and the NegBin models. On the other hand, the response may exhibit limited dependence. In this case, one can assume that there exists a continuous non-observable variable called y^*. The value of y^* is not observed for some part of the domain or not observed at all. The different cases are depicted in Figure 9.1:

- Figure 9.1a presents the case of a binomial variable ($y = 0, 1$), which indicates the position of y^* relative to a threshold μ,
- Figure 9.1b presents the case of an ordinal variable ($y = 0, 1, 2$), which indicates the position of y^* relative to two thresholds μ_1 and μ_2,
- Figure 9.1c presents the case of a left- truncated variable at μ; on the right of μ, we have $y = y^*$, observations characterized by $y^* < \mu$ are simply not available,
- Figure 9.1d presents the case of a left-censored variable at μ; as for the truncated case, one observes, on the right of μ, $y = y^*$. The sample contains observations for which $y^* < \mu$, but the corresponding values of y^* are unobserved.

Some of these models belong to a broad category called "generalized linear models". More specifically, this concerns:

- the binomial model and especially two particular cases, the logit and the probit models,
- the Poisson model.

The Negbin model is also a generalized linear model if its supplementary parameter is a fixed parameter and is not estimated.

In a cross-section context, both base R and several packages provide the relevant estimators, using the maximum likelihood method:

- probit, logit and Poisson models can be fitted using the `glm` function,
- the NegBin model can be estimated using the `glm.nb` function of the **MASS** package,
- the ordinal model can be fitted using the `polr` function of this same package,
- the censored model can be estimated using the `tobit` function of the **AER** package or the `censReg` function of the **censReg** package,
- the truncated model can be fitted using the `truncreg` function of the **truncreg** package.

The **pglm** package provides similar estimators for panel data. It enables the estimation of binomial and Poisson models and for convenience, also for Negbin and ordinal models, even if strictly speaking these last two are not proper generalized linear models.

Panel Data Econometrics with R, First Edition. Yves Croissant and Giovanni Millo.

Companion website: www.wiley.com/go/croissant/data-econometrics-with-R

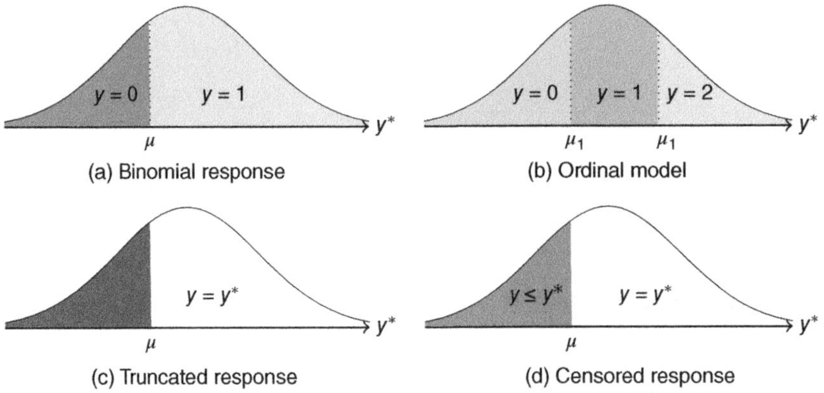

Figure 9.1 Limited dependent variable.

The `pldv` function of the **plm** package provides panel estimators for the case where the response is either truncated or censored.

These models are often estimated using the maximum likelihood method, which requires to make strong hypotheses concerning the distribution of the response. When these hypotheses are not valid, except for very special cases, the estimator is no longer consistent.

This last is a very general drawback of maximum likelihood estimators, but there is also another drawback that is specific to panel data. In linear models, individual effects can be removed using an appropriate transformation (*within* or first differences) or can be directly estimated. This is not the case for most of the models presented in this chapter; the individual effects cannot be removed, and their estimation leads to the *incidental parameter* problem.

When $N \rightarrow +\infty$ for fixed T, for the linear model, the estimation of individual effects is not consistent, as the number of parameters to be estimated grows with N and the variance of the estimators is constant. On the contrary, nevertheless, the estimator of the vector of parameters of interest β is consistent.

Differently from the linear case, for most of the models reviewed in this chapter, when the individual effects are estimated, their inconsistency "contaminates" the estimation of β, which becomes inconsistent as well[1]. This incidental parameter problem leads to abandoning the fixed effects models where the fixed effects are estimated in favor of three alternatives[2]:

- the random effects model, which is always usable: one first writes the individual effects' conditional probabilities and then computes the unconditional probabilities by integrating out the individual effects, making a hypothesis about their distribution,
- a fixed effects model, which uses the notion of *sufficient statistic*: for example, in a logit model, the probability of being unemployed at period t depends on the individual effect, and so does the number of spells of unemployment for every period. By contrast, the ratio of this probability, which is the probability to be unemployed in period t knowing the total number of periods for which the individual is unemployed, does not contain the individual effect. This technique, which is not available for all the models reviewed, enables, like the *within* transformation of the linear models, to get rid of the individual effects,
- for censored or truncated responses, the linear model can be consistently applied if some observations are removed from the sample beforehand (one then speaks of a *trimmed estimator*).

1 For an illustration of this phenomenon in the case of the logit estimator, see Hsiao (2003, pp. 194-195).
2 For a broad view of the estimation non-linear panel models, see Honoré (2002).

In the next sections, we will present the three categories of models previously cited: binomial and ordinal models, truncated and censored models, and count data models. For each of these three sections, we will first briefly describe the estimators used with cross-sectional data. We will then present the estimators appropriate for panel data. We will finally reproduce different empirical examples of these models.

9.1 Binomial and Ordinal Models

9.1.1 Introduction

9.1.1.1 The Binomial Model

We consider a model for which the response is binomial, and we denote without loss of generality the two possible values 0 and 1. We then define a latent variable y^* that is continuous on the real line and is unobserved. The latent variable is linked to the observable binomial variable y by the following rule of observation:

$$
\begin{aligned}
y^* > \mu &\Rightarrow y = 1 \\
y^* \leq \mu &\Rightarrow y = 0
\end{aligned}
$$

The value of the latent variable is the sum of a linear combination of the covariates and an error term. Without loss of generality, if γ includes an intercept, we set $\mu = 0$.

$$
y^* = \gamma^\top z + \epsilon
$$

The variance of ϵ is not identified; it can therefore be set to 1 or to any other arbitrary value. Probabilities for the two possible values of the response are then:

$$
\begin{aligned}
\mathrm{P}(y = 0) &= \mathrm{P}(\epsilon \leq -\gamma^\top z) \\
\mathrm{P}(y = 1) &= \mathrm{P}(\epsilon > -\gamma^\top z)
\end{aligned}
$$

Denoting by F the cumulative density of ϵ, we then have:

$$
\begin{aligned}
\mathrm{P}(y = 0) &= F(-\gamma^\top z) \\
\mathrm{P}(y = 1) &= 1 - F(-\gamma^\top z) \\
&= F(\gamma^\top z)
\end{aligned}
$$

the last expression being valid if the density of ϵ is symmetric. Denoting $q = 2y - 1$, which equals $-1, +1$ for $y = (0, 1)$, the probability of the outcome can be expressed in a compact form:

$$
\mathrm{P}(y) = F(q\gamma^\top z) \tag{9.1}
$$

Two distributions are often used: the normal distribution:

$$
F(\epsilon) = \Phi(\epsilon) = \int_{-\infty}^{\epsilon} \frac{1}{\sqrt{2\pi}} e^{-\frac{1}{2}t^2} dt
$$

which leads to the probit model, and the logistic distribution:

$$
F(\epsilon) = \Lambda(\epsilon) = \frac{e^\epsilon}{1 + e^\epsilon}
$$

which leads to the logit model.

For a sample of size N, the log-likelihood function is obtained by summing the logs of (9.1) for all the observations:

$$
\ln L = \sum_{n=1}^{N} \ln F(q_n \gamma^\top z)
$$

9.1.1.2 Ordered Models

An ordered model is a model for which the response can take J distinct values (with $J > 2$). The construction of the model is very similar to the one of the binomial model. We consider a latent variable, like before equal to the sum of a linear combination of the covariates and an error:

$$y^* = \beta^\top x + \epsilon$$

Denoting $\mu = (\mu_0, \mu_1, \mu_1, \dots, \mu_J)$ a vector of parameters, with $\mu_0 = -\infty$ and $\mu_J = +\infty$, the rule of observation for the different values of y is then:

$$\begin{array}{lllll} y = 1 & \Leftrightarrow & \mu_0 & \leq \beta^\top x + \epsilon \leq & \mu_1 \\ y = 2 & \Leftrightarrow & \mu_1 & \leq \beta^\top x + \epsilon \leq & \mu_2 \\ \vdots & \vdots & \vdots & & \vdots \\ y = J-1 & \Leftrightarrow & \mu_{J-2} & \leq \beta^\top x + \epsilon \leq & \mu_{J-1} \\ y = J & \Leftrightarrow & \mu_{J-1} & \leq \beta^\top x + \epsilon \leq & \mu_J \end{array}$$

Denoting by F the cumulative density of ϵ, the probability for a given value j of y is:

$$\mathrm{P}(y_n = j) = F(\mu_j - \beta^\top x_n) - F(\mu_{j-1} - \beta^\top x_n)$$

The probability of the outcome can be written:

$$\mathrm{P}(y_n) = \sum_{j=1}^{J} 1(y_n = j)[F(\mu_j - \beta^\top x_n) - F(\mu_{j-1} - \beta^\top x_n)] \tag{9.2}$$

For a sample of size N, the log-likelihood function is obtained by summing the logarithms of (9.2) for all the observations:

$$\ln L = \sum_{n=1}^{N} \sum_{j=1}^{J} 1(y_n = j)[F(\mu_j - \beta^\top x_n) - F(\mu_{j-1} - \beta^\top x_n)]$$

As for the binomial model, the most common choices for the distribution of ϵ are the normal and the logistic distributions, which lead respectively to the ordered probit and logit models.

9.1.2 The Random Effects Model

For panel data, we now have repeated observations of y for the same individuals. The latent variable is then defined by:

$$y^*_{nt} = \gamma^\top z_{nt} + \eta_n + \nu_{nt}$$

We assume as usual that the error can be written as the sum of an individual effect η_n and an idiosyncratic term ν_{nt}. Two observations for the same individual are then correlated because of the common term η_n. If the γ vector contains an intercept, we can suppose, without loss of generality, that $\mathrm{E}(\eta) = 0$.

9.1.2.1 The Binomial Model

For a given value of η_n, the probability of the outcome for individual n at period t is defined as before:

$$\mathrm{P}(y_{nt} \mid \eta_n) = F(q_{nt}(\gamma^\top z_{nt} + \eta_n))$$

Denoting $y_n = (y_{n1}, y_{n2}, \dots, y_{nT})$, the joint probability for all the periods for individual n is:

$$\mathrm{P}(y_n \mid \eta_n) = \prod_{t=1}^{T} F(q_{nt}(\gamma^\top z_{nt} + \eta_n))$$

The unconditional probability is obtained by integrating out this expression for η. Assuming that the distribution of η is normal with a standard deviation of σ_η, we obtain:

$$\mathrm{P}(y_n) = \int_{-\infty}^{+\infty} \prod_{t=1}^{T} F(q_{nt}(\gamma^\top z_{nt} + \eta)) \frac{1}{\sqrt{2\pi}\sigma_\eta} e^{-0.5\left(\frac{\eta}{\sigma_\eta}\right)^2} d\eta$$

With the change of variable:

$$v = \frac{\eta}{\sqrt{2}\sigma_\eta} \Rightarrow dv = \frac{d\eta}{\sqrt{2}\sigma_\eta}$$

we obtain

$$\mathrm{P}(y_n) = \frac{1}{\sqrt{\pi}} \int_{-\infty}^{+\infty} \prod_{t=1}^{T} F(q_{nt}(\gamma^\top z_{nt} + \sqrt{2}\sigma_\eta v)) e^{-v^2} dv$$

There is no closed-form for this integrand, but it can be efficiently numerically approximated using Gauss-Hermite quadrature. This method consists in evaluating the function for different values of v (denoted v_r) and computing a linear combination of these evaluations, with weights denoted by w_r. For a fixed number of evaluations R, the values of (v_r, w_r) are tabulated.

$$\mathrm{P}(y_n) \approx \frac{1}{\sqrt{\pi}} \sum_{r=1}^{R} w_r \prod_{t=1}^{T} F(q_{nt}(\gamma^\top z_{nt} + \sqrt{2}\sigma v_r)) \tag{9.3}$$

and the log-likelihood function is obtained by summing over all the individuals the logarithm of (9.3).

Example 9.1 random effects logit model – `Reelection` data set

Brender and Drazen (2008) studied the influence of fiscal policy on the reelection of politicians. It is often suggested that, just before elections, politicians implement more expansionary fiscal policies, i.e., they reduce taxes or increase public spending. A panel of 75 countries is used, with a number of observations varying from 1 to 16. A subsample of these data is also considered when the incumbent is a candidate to the next election (for the other observations, reelection means that the incumbent political party wins the election). This subsample can be selected using the dummy variable `narrow`. The response is `reelect`: it equals 1 in case of reelection and 0 otherwise. The two main covariates are `ddefterm` and `ddefey`. Both variables measure the change in the ratio of government balance (budget surplus) and GDP. The first one is the difference between the two years prior to the elections and the two previous years. For the second, this is the difference between the election year and the previous year. Control variables include the growth rate of GDP `gdppc` and dummies for developing countries `dev`, for new democracies and for majoritarian electoral systems `maj`. The `Reelection` data set is available in the **pder** package.

```
data("Reelection", package = "pder")
```

We first estimate the logit and probit models, with the `glm` function. This function uses the same arguments as `lm`, and a supplementary one called `family`, which indicates the distribution of the response, in our case the binomial distribution. The link between the parameter of the distribution and the linear predictor $\beta^\top x$ is indicated with the `link` argument. The `family` argument can be either a character string (here `'binomial'`), the name of a function (here `binomial`) or a function call (here `binomial()`). The last possibility is the only one that allows to use a link that is not the default one. The logit model is obtained with `link = 'logit'` (the default), the probit model with `link='probit'`. The four following commands all compute the logit model:

```
elect.l <- glm(reelect ~ ddefterm + ddefey + gdppc + dev + nd + maj,
               data = Reelection, family = "binomial", subset = narrow)
l2 <- update(elect.l, family = binomial)
l3 <- update(elect.l, family = binomial())
l4 <- update(elect.l, family = binomial(link = 'logit'))
```

while only the following command allows the estimation of the probit model:

```
elect.p <- update(elect.l, family = binomial(link = 'probit'))
```

The syntax of `pglm` is similar to `glm`. Like for `plm`, there are different ways of describing the structure of the sample:

- by providing a `pdata.frame` to the `data` argument,
- by providing a `data.frame` and using the `index` argument,
- by only providing a `data.frame` if the first two columns of the data contain the individual and the time indexes (which is the case for the `Reelection` data set).

The logit and probit random effects models are estimated below:

```
library("pglm")
elect.pl <- pglm(reelect ~ ddefterm + ddefey + gdppc + dev + nd + maj,
                 Reelection, family = binomial(link = 'logit'),
                 subset = narrow)
elect.pp <- update(elect.pl, family = binomial(link = 'probit'))
```

Estimation results are presented using the `screenreg` function of the **texreg** package:

```
library("texreg")
screenreg(list(logit = elect.l, probit = elect.p,
               plogit = elect.pl, pprobit = elect.pp),
          digits = 3)

==================================================================
             logit         probit         plogit         pprobit
------------------------------------------------------------------
(Intercept)      -1.328 **     -0.822 ***     -1.537 **      -0.942 **
                 (0.410)       (0.248)        (0.489)        (0.294)
ddefterm         14.413         8.381         14.086          8.223
                 (7.746)       (4.685)        (8.211)        (4.853)
```

```
ddefey            14.171 *     8.555 *      13.793 *     8.339
                  (6.660)      (4.039)      (6.998)      (4.257)
gdppc             17.017 *     10.652 *     19.380 *     12.076 **
                  (6.911)      (4.198)      (7.618)      (4.602)
dev                0.822 *      0.504 *      0.893 *      0.541 *
                  (0.358)      (0.218)      (0.430)      (0.258)
nd                 0.683        0.425        0.810        0.495
                  (0.380)      (0.232)      (0.439)      (0.264)
maj                0.768 *      0.472 *      0.847 *      0.515 *
                  (0.314)      (0.192)      (0.381)      (0.230)
sigma                                        0.841 *     -0.518 *
                                            (0.346)      (0.205)
---------------------------------------------------------------------
AIC               343.708      343.851
BIC               368.497      368.640
Log Likelihood   -164.854     -164.926     -163.435     -163.434
Deviance          329.708      329.851
Num. obs.         255          255          255          255
=====================================================================
*** p < 0.001, ** p < 0.01, * p < 0.05
```

The probability of being reelected is larger in developing and newly democratic countries and for majoritarian electoral systems. The growth rate of GDP also has the predicted positive effect on the probability of being reelected. The coefficients of the two fiscal policy covariates are positive, which means that expansionary fiscal policies before elections do not have a systematic positive effect on the probability of the incumbent being reelected. On the contrary, the results indicate that voters tend to sanction such policies.

9.1.2.2 Ordered Models

The line of reasoning is very similar to that of binomial models. The joint probability for an individual n for a given value of the individual effect is:

$$P(y_n \mid \eta_n) = \prod_{t=1}^{T} \sum_{j=1}^{J} 1(y_{nt} = j)[F(\mu_j - \beta^\top x_{nt} - \eta_n) - F(\mu_{j-1} - \beta^\top x_{nt} - \eta_n)]$$

Assuming a normal distribution for the individual effects, the unconditional probability is:

$$P(y_n) = \int_{-\infty}^{+\infty} \prod_{t=1}^{T} \sum_{j=1}^{J} 1(y_{nt} = j)[F(\mu_j - \beta^\top x_{nt} - \eta) - F(\mu_{j-1} - \beta^\top x_{nt} - \eta)] \times \frac{1}{\sqrt{2\pi\sigma_\eta}} e^{-0.5\left(\frac{\eta}{\sigma_\eta}\right)^2} d\eta$$

Using the same change of variable as previously, we obtain:

$$P(y_n) = \frac{1}{\sqrt{\pi\sigma_\eta}} \int_{-\infty}^{+\infty} \prod_{t=1}^{T} \sum_{j=1}^{J} 1(y_{nt} = j) \times [F(\mu_j - \beta^\top x_{nt} - \sqrt{2}\sigma_\eta v) - F(\mu_{j-1} - \beta^\top x_{nt} - \sqrt{2}\sigma_\eta v)] \times e^{-v^2} dv$$

which can be approximated using Gauss-Hermite quadrature:

$$P(y_n) \approx \frac{1}{\sqrt{\pi}\sigma_\eta} \sum_{r=1}^{R} \prod_{t=1}^{T} \sum_{j=1}^{J} 1(y_{nt} = j)$$
$$\times [F(\mu_j - \beta^\top x_{nt} - \sqrt{2}\sigma_\eta v_r) - F(\mu_{j-1} - \beta^\top x_{nt} - \sqrt{2}\sigma_\eta v_r)]$$
$$\times e^{-v_r^2} dv$$

Example 9.2 random effects ordered model – `Fairness` data set
Raux et al. (2009) analyze the perceived fairness of different methods of demand rationing using a survey in which individuals had to indicate their opinion on an ordinal scale concerning different rationing modes for parking places and for fast train seats. The response is `answer` and takes integer values from 0 (very unfair) to 3 (very fair). The main covariate is a factor indicating the rationing mode: peak-load pricing `peak`, administrative rule `admin`, random allocation `lottery`, additive supply `addsupply`, queuing `queuing`, moral rule `moral`, and compensation rule `compensation`. The other covariates are dummies indicating that the rationing is recurring or not `recurring`, that the individual has a diploma `education` and has a driving license `driving`. The `Fairness` dataset is available in the **pglm** package.

```
data("Fairness", package = "pglm")
```

We first use the `polr` function from the **MASS** package to estimate the ordered probit and logit models. We restrict our attention to the rationing of parking places.

```
library("MASS")
parking.ol <- polr(answer ~ recurring + driving + education + rule,
                   data = Fairness, subset = good == "parking",
                   Hess = TRUE, method = "logistic")
parking.op <- update(parking.ol, method = "probit")
```

The "link" is indicated with the `method` argument and we set the `Hess` argument to `TRUE` so that the Hessian, which is necessary to calculate the standard errors of the coefficients, is computed.

We then estimate the random effects ordered models using `pglm`. The following details should be remarked:

- the `family` argument is used, like for `glm`, and an `ordinal` function is added, which allows, setting `link` to either `'probit'` or `'logit'`, the estimation of the probit and the logit ordered models,
- the number of evaluations for the Gauss-Hermite quadrature method is indicated with the argument `R`,
- the `index` is here mandatory, as the second column of `Fairness` is not the time index.

```
parking.opp <- pglm(as.numeric(answer) ~ recurring + driving + education + rule,
                    data = Fairness, subset = good == 'parking',
                    family = ordinal(link = 'probit'), R = 10, index = 'id',
                    model = "random")
parking.olp <- update(parking.opp, family = ordinal(link = 'probit'))
```

Results of the four models are presented using the `screenreg` function:

```
library("texreg")
screenreg(list(ologit = parking.ol, oprobit = parking.op,
          pologit = parking.olp, poprobit = parking.opp),
      digits = 3)

============================================================================
                  ologit        oprobit        pologit        poprobit
----------------------------------------------------------------------------
recurringyes       -0.120         -0.070         -0.077          -0.077
                   (0.075)        (0.044)        (0.059)         (0.059)
drivingno           0.413 ***      0.237 ***      0.255 **        0.255 **
                   (0.101)        (0.060)        (0.080)         (0.080)
educationno        -0.480 ***     -0.280 ***     -0.309 **       -0.309 **
                   (0.138)        (0.079)        (0.105)         (0.105)
ruleadmin          -0.133         -0.061         -0.066          -0.066
                   (0.144)        (0.086)        (0.088)         (0.088)
rulelottery         0.330 *        0.217 *        0.238 **        0.238 **
                   (0.141)        (0.085)        (0.086)         (0.086)
ruleaddsupply       1.892 ***      1.141 ***      1.221 ***       1.221 ***
                   (0.143)        (0.083)        (0.085)         (0.085)
rulequeuing         2.973 ***      1.731 ***      1.848 ***       1.848 ***
                   (0.152)        (0.086)        (0.089)         (0.089)
rulemoral           4.597 ***      2.656 ***      2.837 ***       2.837 ***
                   (0.166)        (0.093)        (0.098)         (0.098)
rulecompensation    4.231 ***      2.458 ***      2.622 ***       2.622 ***
                   (0.162)        (0.091)        (0.096)         (0.096)
(Intercept)                                      -0.269 ***      -0.269 ***
                                                 (0.072)         (0.072)
mu_1                                              1.019 ***       1.019 ***
                                                 (0.038)         (0.038)
mu_2                                              2.515 ***       2.515 ***
                                                 (0.059)         (0.059)
sigma                                             0.529 ***       0.529 ***
                                                 (0.050)         (0.050)
----------------------------------------------------------------------------
AIC               5482.722       5490.689
BIC               5553.360       5561.326
Log Likelihood   -2729.361      -2733.344      -2705.814       -2705.814
Deviance          5458.722       5466.689
Num. obs.         2661           2661           2661            2661
============================================================================
*** p < 0.001, ** p < 0.01, * p < 0.05
```

9.1.3 The Conditional Logit Model

The random effects model is consistent only if the individual effects are uncorrelated with the covariates. If it is not the case, the conditional logit model can be used. It is well known in the statistic literature and has been introduced in panel data econometrics by Chamberlain (1980).

The general presentation of this model is quite complex, but the intuition of it can be perceived using the special case where $T = 2$. We denote $Y_n = y_{n1} + y_{n2}$. Only the individuals for which $Y_n = 1$ can be used to estimate the conditional logit model (more generally, only individuals for which $0 < Y_n < T$ may be used).

For a given period t, the probabilities for the two values of y_{nt} are:

$$\begin{cases} \mathrm{P}(y_{nt} = 0 \mid \eta_n) = \frac{1}{1+e^{\beta^\top x_{nt}+\eta_n}} \\ \mathrm{P}(y_{nt} = 1 \mid \eta_n) = \frac{e^{\beta^\top x_{nt}+\eta_n}}{1+e^{\beta^\top x_{nt}+\eta_n}} \end{cases}$$

or more generally:

$$\mathrm{P}(y_{nt} \mid \eta_n) = \frac{e^{y_{nt}(\beta^\top x_{nt}+\eta_n)}}{1+e^{\beta^\top x_{nt}+\eta_n}}$$

If the idiosyncratic components of the errors are i.i.d., the joint probability for two observations is simply the product of $\mathrm{P}(y_{n1} \mid \eta_n)$ and $\mathrm{P}(y_{n2} \mid \eta_n)$:

$$\mathrm{P}(y_{n1}, y_{n2} \mid \eta_n) = \frac{e^{y_{n1}(\beta^\top x_{n1}+\eta_n)} e^{y_{n2}(\beta^\top x_{n2}+\eta_n)}}{(1+e^{\beta^\top x_{n1}+\eta_n})(1+e^{\beta^\top x_{n2}+\eta_n})}$$

or also, as one and only one of the two y_{nt} equals 1:

$$\begin{aligned} \mathrm{P}(y_{n1}, y_{n2} \mid \eta_n) &= \frac{e^{y_{n1}\beta^\top x_{n1}+y_{n2}\beta^\top x_{n2}+\eta_n}}{(1+e^{\beta^\top x_{n2}+\eta_n})(1+e^{\beta^\top x_{n2}+\eta_n})} \\ &= e^{\eta_n} \frac{e^{y_{n1}\beta^\top x_{n1}+y_{n2}\beta^\top x_{n2}}}{(1+e^{\beta^\top x_{n1}+\eta_n})(1+e^{\beta^\top x_{n2}+\eta_n})} \end{aligned} \quad (9.4)$$

The probability that $Y_n = y_{n1} + y_{n2} = 1$ is equal to the sum of the probabilities of:

- $y_{n1} = 1$ and $y_{n2} = 0$, which is $\frac{e^{\beta^\top x_{n1}+\eta_n}}{1+e^{\beta^\top x_{n1}+\eta_n}} \times \frac{1}{1+e^{\beta^\top x_{n2}+\eta_n}}$,
- $y_{n1} = 0$ and $y_{n2} = 1$, which is $\frac{1}{1+e^{\beta^\top x_{n1}+\eta_n}} \times \frac{e^{\beta^\top x_{n2}+\eta_n}}{1+e^{\beta^\top x_{n2}+\eta_n}}$.

which is therefore:

$$\begin{aligned} \mathrm{P}(Y_n = 1) &= \frac{e^{\beta^\top x_{n1}+\eta_n} + e^{\beta^\top x_{n2}+\eta_n}}{(1+e^{\beta^\top x_{n1}+\eta_n})(1+e^{\beta^\top x_{n2}+\eta_n})} \\ &= e^{\eta_n} \frac{e^{\beta^\top x_{n1}} + e^{\beta^\top x_{n2}}}{(1+e^{\beta^\top x_{n1}+\eta_n})(1+e^{\beta^\top x_{n2}+\eta_n})} \end{aligned} \quad (9.5)$$

Dividing (9.4) by (9.5), one finally obtains the joint probability of y_{n1} and y_{n2} given their sum:

$$\mathrm{P}(y_{n1}, y_{n2} \mid Y_n = 1) = \frac{e^{y_{n1}\beta^\top x_{n1}+y_{n2}\beta^\top x_{n2}}}{e^{\beta^\top x_{n1}} + e^{\beta^\top x_{n2}}} \quad (9.6)$$

This conditional probability is free of the individual effect and the likelihood that uses this expression can therefore be considered as a fixed effects logit model. Note that there is no similar estimator for the probit model.

Example 9.3 conditional logit model – `MagazinePrices` data set

Cecchetti (1986) analyzes price changes, with an application to magazines. His analysis is replicated (and criticized) by Willis (2006). Price changes are costly for two reasons:

- changing prices induce administrative costs,
- in a monopolistic competition context, increasing prices will lead to a loss of customers.

For these two reasons, there is a difference between the optimal price of a good for a given period p^*_{nt} and the actual price p_{nt}. A price change will occur only if the gap between the two becomes greater than a given threshold. More formally, the price will change if:

$$\ln \frac{p^*_{nt}}{p_{nt}} > h^c_{nt}$$

h^c_{nt} is then the minimum relative gap between the optimal and the actual price that would result in a price change. If the price changes, given the infrequency of price changes, the enterprise will set its new price above the optimal price, the relative difference being equal to h^o_{nt}.

Denote $\tilde{t}_n$ the last period when the price of good n has changed. For this period, we have:

$$\ln \frac{p^*_{n\tilde{t}_n}}{p_{n\tilde{t}_n}} = h^o_{n\tilde{t}_n}$$

If the price doesn't change in period t, we have $p_{nt} = p_{n\tilde{t}_n}$. Replacing in the previous equation, we have:

$$\ln \frac{p^*_{nt}}{p^*_{n\tilde{t}_n}} \equiv \Delta \ln P^*_{nt\tilde{t}_n} > h^c_{nt} - h^o_{nt} \tag{9.7}$$

In the context of a simple monopolistic competition model, the demand function for the firm and its cost function are:

$$\begin{cases} Q_{nt} = \left(\frac{p_{nt}}{\bar{p}_t}\right)^a X^b_t \\ C_{nt} = Ae^{\delta t} Q^{\alpha}_{nt} w_t \end{cases}$$

where X_t is the demand faced by the whole industry, w_t the factor price index, and $\bar{p}_t$ the average price in the industry.

Substituting the expression of demand in the cost function, writing the profit function, and setting to zero the first derivative of profit with respect to price, we obtain the following price function:

$$\ln p^*_{nt} = b_0 + b_1 t + b_2 \ln \bar{p}_t + b_3 \ln X_t + b_4 \ln w_t$$

Writing the same price function for the period when the last price change occurred and subtracting both equations, we get:

$$\Delta \ln p^*_{nt\tilde{t}_n} = a_1(t - \tilde{t}_n) + a_2 \ln \frac{\bar{p}_t}{\bar{p}_{\tilde{t}_n}} + a_3 \ln \frac{X_t}{X_{\tilde{t}_n}} + a_4 \ln \frac{w_t}{w_{\tilde{t}_n}}$$

Finally, denoting by $T_{nt} = t - \tilde{t}_{nt}$ the time since the last price change, assuming an identical variation π_{nt} of the average price of the industry and of the inputs and denoting by $\dot{X}_{nt} = \ln \frac{X_t}{X_{\tilde{t}_n}}$ the demand variation for the whole industry since the last price change of enterprise n:

$$\Delta \ln p^*_{nt\tilde{t}_n} = a_1 T_{nt} + a_2 \pi_{nt} + a_3 \dot{X}_{nt}$$

Adding an error term to this expression and inserting it in equation (9.7), we obtain:

$$a_1 T_{nt} + a_2 \pi_{nt} + a_3 \dot{X}_{nt} + \epsilon_{nt} > h^c_{nt} - h^o_{nt}$$

$a_{nt} = h^c_{nt} - h^o_{nt}$ is a specific term for enterprise n at period t, which represents the price change policy. The probability of a price change can then be written:

$$\mathrm{P}(y_{nt} = 1) = \mathrm{P}(a_{nt} + a_1 T_{nt} + a_2 \pi_{nt} + a_3 \dot{X}_{nt} + \epsilon_{nt}) = F(a_{nt} + a_1 T_{nt} + a_2 \pi_{nt} + a_3 \dot{X}_{nt})$$

where F is the cumulative density of ϵ, assumed to be logistic.

Cecchetti (1986) assumes that a_{nt} can be supposed constant for 3 consecutive years. In this case, the period of observation being of 27 years, there are 9 different effects for each magazine. We present below the results of 3 estimations that replicate Table 1 of Willis (2006).

We successively estimate a simple logit, a logit with magazine fixed effects for which the effects are estimated (and therefore suffering from the incidental parameter problem), and a conditional logit model (using the `clogit` function of the **survival**) where three-year magazine fixed effects are removed. The `MagazinePrices` data set is available in the **pder** package.

```
data("MagazinePrices", package = "pder")
logitS <- glm(change ~ length + cuminf + cumsales, data = MagazinePrices,
              subset = included == 1, family = binomial(link = 'logit'))
logitD <- glm(change ~ length + cuminf + cumsales + magazine,
              data = MagazinePrices,
              subset = included == 1, family = binomial(link = 'logit'))
library("survival")
logitC <- clogit(change ~ length + cuminf + cumsales + strata(id),
                 data = MagazinePrices,
                 subset = included == 1)
library("texreg")
screenreg(list(logit = logitS, "FE logit" = logitD,
               "cond. logit" = logitC), omit.coef = "magazine")
```

```
=====================================================
                logit        FE logit     cond. logit
-----------------------------------------------------
(Intercept)        -1.90 ***    -1.18 **
                   (0.14)       (0.42)
length             -0.10 **     -0.07 *      1.02 ***
                   (0.03)       (0.03)      (0.28)
cuminf              6.93 ***     8.83 ***   19.20 *
                   (1.12)       (1.25)      (7.51)
cumsales           -0.36        -1.14        7.60 *
                   (0.98)       (1.06)      (3.46)
-----------------------------------------------------
AIC              1008.90      1028.35      173.44
BIC              1028.63      1230.62
Log Likelihood   -500.45      -473.18
Deviance         1000.90       946.35
Num. obs.        1026         1026        1026
R^2                                          0.20
Max. R^2                                     0.32
Num. events                                213
Missings                                     0
=====================================================
*** p < 0.001, ** p < 0.01, * p < 0.05
```

Note that the coefficient of the length of the period since the last price change has the expected positive sign and is significant only for the conditional logit model.

9.2 Censored or Truncated Dependent Variable

9.2.1 Introduction

It's often the case in economics that the response is only observed on a certain range of values; we then say that the dependent variable is truncated. For example:

- if the response is a proportion, it is necessarily left- truncated on 0 and right-truncated on 1,
- consumption for a good is necessarily positive and therefore left-truncated on 0,
- the demand for a sports event is necessarily lower or equal to the number of seats in the stadium and is therefore right- truncated to this capacity.

From now on, we will consider the most common case, which is a 0 left truncation, but the models we will present easily extend to the case of left or/and right truncations at any value.

As usual, we will assume that the dependent variable can be represented by a latent variable y^* that equals the sum of a linear combination of different covariates and an error term.

$$y^* = \gamma^\top z + \epsilon^*$$

The observed response y equals y^* if it is not in the truncated zone (i.e., here, if it's strictly positive) and equals the truncature (here, 0) otherwise.

$$\begin{array}{l} y^* \leq 0 \Rightarrow y = 0 \\ y^* > 0 \Rightarrow y = y^* \end{array} \tag{9.8}$$

Two kinds of samples can be used to estimate this model:

- a sample is truncated when only observations for which $y > 0$ are available (we therefore don't even know the values of the covariates x for observations for which y is in the truncation zone),
- a sample is censored when it consists of observations for which y^* is either inside or outside the truncation zone.

This latter case is particularly important in econometrics and leads to a model which is called the tobit model (Tobin, 1958). From now, we'll refer to the truncated model when the first kind of sample is used and to the censored model for the second kind of sample.

We'll first analyze why applying a linear regression to a censored or a truncated model leads to inconsistent estimators. We'll then present a non-parametric method that leads, removing some specific observations, to a consistent estimator while making minimal hypotheses on the model errors. We'll conclude this section with the maximum likelihood estimator, which relies on the much stronger hypothesis of homoscedasticity and normal distribution.

9.2.2 The Ordinary Least Squares Estimator

Let f be the density of the distribution of ϵ^* which is supposed, without loss of generality as long as the equation contains an intercept, to be of 0 expected value. We then have:

$$\mathrm{E}(y^* \mid z) = \gamma^\top z + \mathrm{E}(\epsilon^* \mid z) = \gamma^\top z$$

If y^* were observed, OLS would be a consistent estimator for γ. This is not the case when we only observe the truncated variable y. On the truncated sample, we have $y^* > 0$, or $\epsilon^* > -\gamma^\top z$. The distribution of ϵ for the sample is then $f(\epsilon)/\mathrm{P}(\epsilon^* > -\gamma^\top z)$, depicted by the dotted line in Figure 9.2.

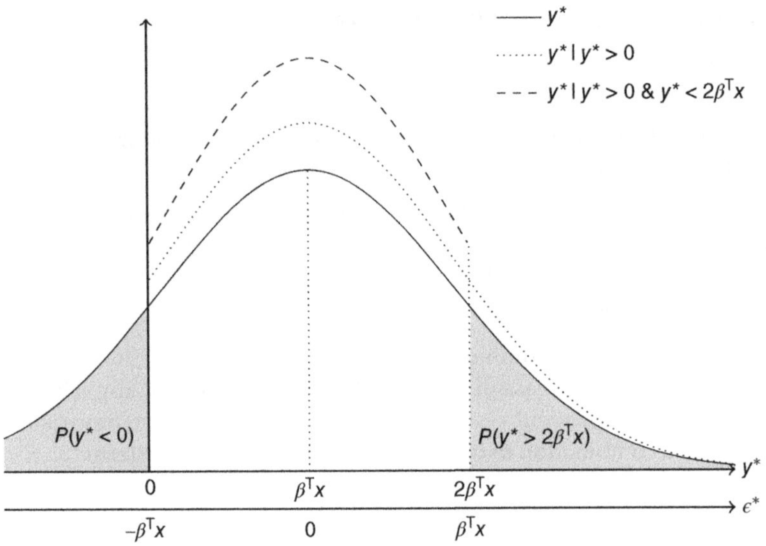

Figure 9.2 Distribution of y^* and ϵ^*.

The distribution of ϵ is not symmetric around 0, and its expected value is positive, because the left side of the distribution, corresponding to values of $\epsilon^* < -\gamma^\top z$, is truncated. We therefore have:

$$
\begin{aligned}
\mathrm{E}(y \mid x, y > 0) &= \mathrm{E}(y^* \mid x, y^* > 0)\\
&= \gamma^\top z + \mathrm{E}(\epsilon^* \mid \epsilon^* > -\gamma^\top z)\\
&= \gamma^\top z + \frac{\int_{-\gamma^\top z}^{+\infty} \epsilon f(\epsilon) d\epsilon}{\mathrm{P}(\epsilon^* > -\gamma^\top z)}
\end{aligned}
$$

which is, for a normal distribution:

$$
\mathrm{E}(y \mid x, y > 0) = \gamma^\top z + \frac{\phi(\gamma^\top z)}{\Phi(\gamma^\top z)}
$$

or, subtracting $\gamma^\top z$:

$$
\mathrm{E}(\epsilon \mid z) = \mathrm{E}(\epsilon^* \mid z, \epsilon^* > -\gamma^\top z) = \frac{\phi(\gamma^\top z)}{\Phi(\gamma^\top z)}
$$

$\mu(z) = \phi(z)/\Phi(z)$ is known as the inverse mills ratio and is a decreasing function of its argument. Computing the derivative with respect to one covariate x_k, we obtain:

$$
\frac{\partial \mathrm{E}(\epsilon \mid z)}{\partial x_k} = -[\gamma^\top z + \mu(\gamma^\top z)](\gamma^\top z)\beta_k
$$

which is negative if $\beta_k > 0$, as $\mu(\gamma^\top z)$ is the average of ϵ for $\epsilon > -\gamma^\top z$ and is therefore greater than $-\gamma^\top z$. The OLS estimator computed on the truncated sample is therefore downward biased.

For the censored sample, we have $y = 0$ for censored observations. We then have:

$$
\begin{aligned}
E(y \mid z) &= \mathrm{P}(\epsilon^* \leq -\gamma^\top z) \times 0 + \mathrm{P}(\epsilon^* > -\gamma^\top z) \times \left(\gamma^\top z + \frac{\int_{-\gamma^\top z}^{+\infty} \epsilon f(\epsilon) d\epsilon}{\mathrm{P}(\epsilon^* > -\gamma^\top z)} \right)\\
&= \mathrm{P}(\epsilon^* > -\gamma^\top z)(\gamma^\top z) + \int_{-\gamma^\top z}^{+\infty} \epsilon f(\epsilon) d\epsilon\\
&= (\gamma^\top z)\Phi(\gamma^\top z) + \phi(\gamma^\top z)
\end{aligned}
$$

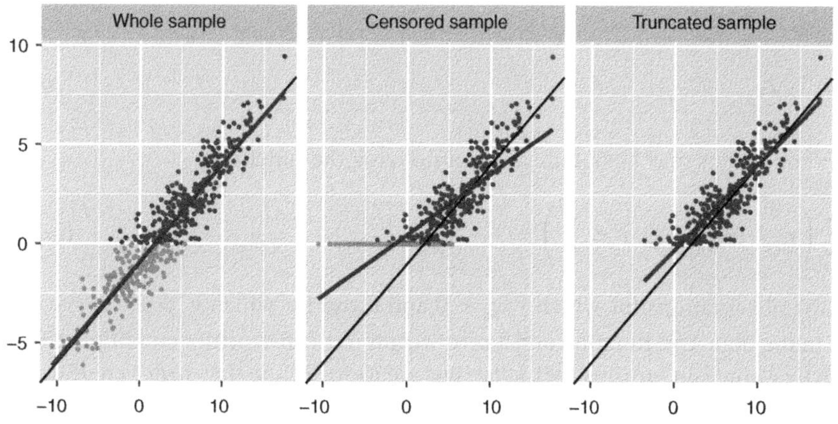

Figure 9.3 OLS bias for the censored and the truncated samples.

where the last expression holds for a normal distribution. Subtracting $\gamma^\top z$, we obtain the expected value of the error of the censored model:

$$E(\epsilon \mid z) = -[1 - \Phi(\gamma^\top z)](\gamma^\top z) + \phi(\gamma^\top z)$$

Computing once again the derivative with respect to a covariate x_k, we have:

$$\frac{\partial \mathrm{E}(\epsilon \mid z)}{\partial x_k} = -[1 - \phi(\gamma^\top z)]\beta_k$$

which, as previously, has the opposite sign of β_k, implying that the **OLS** estimator on the censored sample is downward biased.

The bias of the **OLS** estimator on censored and truncated samples is illustrated on Figure 9.3

9.2.3 The Symmetrical Trimmed Estimator

The **OLS** estimator is inconsistent because the truncation leads to an asymmetric distribution for the errors, for which the expected values depends on x. Powell (1986) proposes to restore the symmetry by removing some observations.

9.2.3.1 Truncated Sample

In the case of the truncated sample, observations for which $y^* \leq 0$, or $\epsilon^* \leq -\gamma^\top z$, are missing. The symmetry may be restored by removing from the right side of the distribution, the observations for which $y^* \geq 2\gamma^\top z$, or $\epsilon^* > \gamma^\top z$. The distributions of y^* and ϵ^* are depicted by the dashed line in Figure 9.2. In this case, we have:

$$\begin{aligned}
\mathrm{E}(y \mid z, y > 0 \;\&\; y < 2\gamma^\top z) &= \mathrm{E}(y^* \mid z, y^* > 0 \;\&\; y^* < 2\gamma^\top z) \\
&= \gamma^\top z + \mathrm{E}(\epsilon^* \mid -\gamma^\top z < \epsilon^* < \gamma^\top z) \\
&= \gamma^\top z + \frac{\int_{-\gamma^\top z}^{+\gamma^\top z} \epsilon f(\epsilon) d\epsilon}{\int_{-\gamma^\top z}^{+\gamma^\top z} f(\epsilon) d\epsilon} \\
&= \gamma^\top z
\end{aligned}$$

A consistent estimator may be obtained using the normal conditions and restricting the sample to observations for which $y < 2\gamma^\top z$. Denoting by $1(v)$ the function that is equal to 1 if v is true and 0 otherwise, we have:

$$\sum_{n=1}^{N} 1(y_n < 2\gamma^\top z)(y_n - \gamma^\top z_n)x_n = 0 \tag{9.9}$$

These first-order conditions may be obtained by minimizing the function:

$$R_N(\gamma) = \sum_{n=1}^{N} \left(y_n - \max\left(\frac{1}{2}y_n, \gamma^\top z_n\right)^2 \right) \tag{9.10}$$

In this case, all the observations for which $\gamma^\top z_n < 0$ and those for which $y_n > 2\gamma^\top z_n$ have a weight equal to $\frac{y_n^2}{4}$ in the objective function and a zero weight in the first-order conditions. The weight in the objective function ensures that fallacious solutions of the first-order conditions like $\gamma = 0$ are excluded.

9.2.3.2 Censored Sample

In the case of the censored sample, symmetry is restored by replacing y_n by $2\gamma^\top z_n$ when $y_n > 2\gamma^\top z_n$ (as y_n^* is replaced by 0 when $y_n^* < -\gamma^\top z_n$). We then have:

$$\sum_{n=1}^{N} 1(\gamma^\top z > 0)(\max(y_n, 2\gamma^\top z_n) - \gamma^\top z_n)x_n = 0 \tag{9.11}$$

These first-order conditions may be obtained by minimizing the following function:

$$S_N(\gamma) = \sum_{n=1}^{N} \left(y_n - \max\left(\frac{1}{2}y_n, \gamma^\top z_n\right)^2 \right) + \sum_{n=1}^{N} 1(y_n > 2\gamma^\top z_n) \left(\left(\frac{1}{2}y_n\right)^2 - \max(0, \gamma^\top z_n)^2 \right) \tag{9.12}$$

Observations for which $\gamma^\top z_n < 0$ now have a weight equal to $\frac{y_n^2}{2}$ in the objective function and a zero weight in the first-order conditions.

9.2.4 The Maximum Likelihood Estimator

If we can assume that the errors are normal and homoscedastic, a more efficient estimator is the maximum likelihood estimator.

9.2.4.1 Truncated Sample

The maximum likelihood estimator for a truncated sample has been proposed by Hansman and Wise (1976). The density of the distribution of y^* is normal, with expected value equal to $\gamma^\top z$ and standard deviation σ. We then have:

$$f(y^*) = \frac{1}{\sigma}\phi\left(\frac{y^* - \gamma^\top z}{\sigma}\right)$$

The probability of y^* being negative is: $\Phi(-\gamma^\top z/\sigma) = 1 - \Phi(\gamma^\top z/\sigma)$.

The density of the distribution of y, denoted f_+, is the zero left-truncated distribution of y^*: We then have:

$$f_+(y) = \frac{f(y)}{P(y^* > 0)} = \frac{1}{\sigma\Phi(\gamma^\top z/\sigma)}\phi\left(\frac{y - \gamma^\top z}{\sigma}\right) \tag{9.13}$$

The log-likelihood function is obtained by summing the logarithms of the density (9.13) for the N observations in the sample:

$$\ln L = -\frac{N}{2}\ln 2\pi - N\ln\sigma - \sum_{n=1}^{N}\left[\ln\Phi(\gamma^\top z_n/\sigma) + \frac{1}{2}\frac{(y_n-\gamma^\top z_n)^2}{\sigma^2}\right] \tag{9.14}$$

9.2.4.2 Censored Sample

When the sample is censored, the distribution of y is a mix of a discrete and a continuous distribution. An observation for which $y_n = 0$ enters the log-likelihood function as:

$$\mathrm{P}(y=0) = \Phi\left(-\frac{\gamma^\top z}{\sigma}\right) = 1-\Phi\left(\frac{\gamma^\top z}{\sigma}\right)$$

while for a positive observation, the contribution to the likelihood is the truncated normal density:

$$f_+(y) = \frac{1}{\sigma\Phi(\gamma^\top z/\sigma)}\phi\left(\frac{y-\gamma^\top z}{\sigma}\right)$$

times the probability that y be positive: $\Phi(\gamma^\top z/\sigma)$. We finally get the log-likelihood function (9.15):

$$\begin{aligned}\ln L =& \sum_{n=1}^{N}\left[1(y_n=0)\ln\left\{1-\Phi\left(\frac{\gamma^\top z_n}{\sigma_\epsilon}\right)\right\}\right]\\ &-\sum_{n=1}^{N}\left[1(y_n>0)\left(\frac{1}{2}\ln(2\pi\sigma^2)+\frac{1}{2}\frac{(y_n-\gamma^\top z_n)^2}{\sigma^2}\right)\right]\end{aligned} \tag{9.15}$$

9.2.5 Fixed Effects Model

Honoré (1992) proposed a symmetrical trimmed estimator that is an extension of Powell (1986)'s estimator to panel data. For now, we consider a panel with only two observations for every individual and one covariate.

$$\begin{cases} y^*_{n1} &= \alpha + \beta^\top x_{n1} + \eta_n + \nu_{n1}\\ y^*_{n2} &= \alpha + \beta^\top x_{n2} + \eta_n + \nu_{n2}\end{cases}$$

The only hypothesis made concerning the errors ν_{n1} and ν_{n2} is that they are identically distributed. The symmetry hypothesis, which was required for the Powell (1986) estimator to be consistent, is not necessary here.

9.2.5.1 Truncated Sample

For the truncated model, only observations for which $y^*_{nt} > 0$ are available. Figure 9.4[3] presents the distribution of y^*_{n1} and y^*_{n2}.

With the hypotheses we've made, these two distributions only differ by their position, y^*_{n1} being centered on $\beta^\top x_{n1} + \eta_n$ and y^*_{n2} on $\beta^\top x_{n2} + \eta_n$. Because of the truncation, the two distributions conditioned to the fact that the observation is in the sample ($y^*_{nt} > 0$), to the values of the covariates (x_n) and to that of the individual effect (η_n) are more substantially different. If $\beta^\top \Delta x_n = \beta^\top(x_{n2} - x_{n1}) > 0$ (Figure 9.4a), the truncated part of the distribution of y^*_{n1} is

3 Inspired by Hsiao (2003, pp. 246–247).

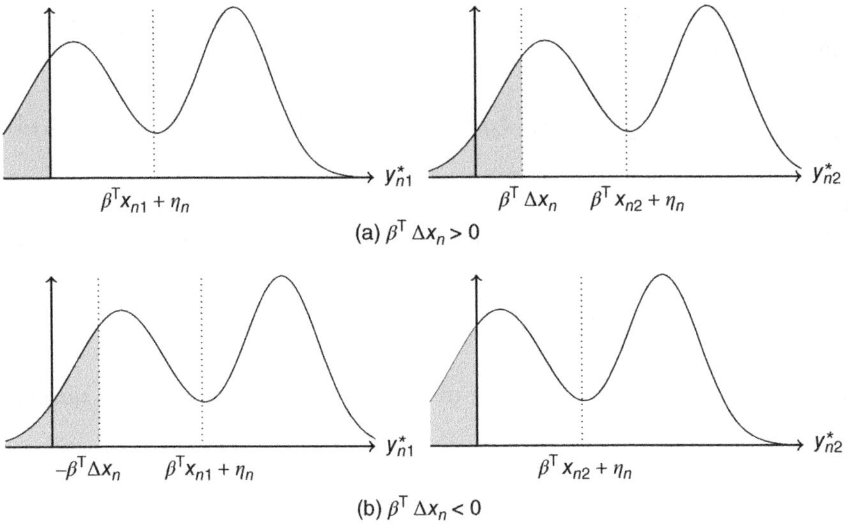

Figure 9.4 Distribution of y^*_{n1} and of y^*_{n2}.

larger than the one of y^*_{n2}. However, identical distributions can be obtained by truncating y^*_{n2} not at 0 (which is the selection rule of the sample) but at $\beta^\top \Delta x_n$. In the case where $\beta^\top \Delta x_n < 0$ (Figure 9.4b), y^*_{n1} is similarly truncated at $-\beta^\top \Delta x_n$.

We then obtain two identical conditional distributions for:

- $y^*_{n1} \mid y^*_{n1} > 0, x_{n1}, \eta_n$ and $y^*_{n2} \mid y^*_{n2} > \beta^\top \Delta x_n, x_{n2}, \eta_n$ in the case when $\beta^\top \Delta x_n > 0$,
- $y^*_{n1} \mid y^*_{n1} > -\beta^\top \Delta x_n, x_{n1}, \eta_n$ and $y^*_{n2} \mid y^*_{n2} > 0, x_{n2}, \eta_n$ in the case when $\beta^\top \Delta x_n < 0$,

More generally, the observations that should be removed to restore symmetry are those for which $y_{n1} > -\beta^\top \Delta x_n$ or $y_{n2} > \beta^\top \Delta x_n$. This situation is depicted in Figure 9.5. When $\beta^\top \Delta x_n > 0$ (9.5a), the joint distribution of (y^*_{n1}, y^*_{n2}) is symmetric around the LL' line which is the 45° line with intercept $\beta^\top \Delta x_n$. Truncating at $(y^*_{n1} > 0)$ and $(y^*_{n2} > \beta^\top \Delta x_n)$, we obtain two symmetric zones A_1 and B_1. The probability of having $y_n = (y_{n1}, y_{n2})$ in zones A_1 or B_1 is the same. This result leads to a first-moment condition:

$$\mathrm{E}[\{1(y_n \in A_1) - 1(y_n \in B_1)\}\Delta x_n] = 0 \tag{9.16}$$

Moreover, by symmetry, in Figure 9.5a:

- the vertical distance between (y_{n1}, y_{n2}) in zone A_1 on the LL' line is $\Delta y_n - \beta^\top \Delta x_n$,
- the horizontal distance between (y_{n1}, y_{n2}) in B_1 and the LL' line is $-(\Delta y_n - \beta^\top \Delta x_n)$,

which can be written as a second-moment condition:

$$\mathrm{E}[\{1(y_n \in A_1 \cup B_1)(\Delta y_n - \beta^\top \Delta x_n)\}\Delta x_n] = 0 \tag{9.17}$$

For a sample of size N, truncated as previously described, the sample analogues of the two moment conditions (9.16) and (9.17) are:

$$\frac{1}{N}\sum_{n=1}^{N} 1(y_{n1} > -\beta^\top \Delta x_n \;\&\; y_{n2} > \beta^\top \Delta x_n)\mathrm{sign}(\Delta y_n - \beta^\top \Delta x_n)\Delta x_n = 0 \tag{9.18}$$

$$\frac{1}{N}\sum_{n=1}^{N} 1(y_{n1} > -\beta^\top \Delta x_n \;\&\; y_{n2} > \beta^\top \Delta x_n)(\Delta y_n - \beta^\top \Delta x_n)\Delta x_n = 0 \tag{9.19}$$

(9.18 and 9.19) are respectively the first-order conditions of the LAD and of the least squares estimator. These first-order conditions may be obtained by maximizing:

$$\begin{aligned} T_n^f(\beta) &= \sum_{n=1}^{N} f(\Delta y_n - \beta^\top \Delta x_n) 1(y_{n1} > -\beta^\top \Delta x_n, y_{n2} > \beta^\top \Delta x_n) \\ &\quad + f(y_{n1}) 1(y_{n1} > -\beta^\top \Delta x_n, y_{n2} < \beta^\top \Delta x_n) \\ &\quad + f(y_{n2}) 1(y_{n1} < -\beta^\top \Delta x_n, y_{n2} > \beta^\top \Delta x_n) \\ &= \sum_{n=1}^{N} f(\psi(y_{n1}, y_{n2}, \beta^\top \Delta x_n)) \end{aligned}$$

with:

$$\psi(z_1, z_2, c) = \begin{cases} z_1 & \text{for } z_2 < c \\ z_2 - z_1 - c & \text{for } -z_1 < c < z_2 \\ z_2 & \text{for } z_1 < -c \end{cases}$$

If $f(x) = x^2$, we obtain the trimmed least squares estimator; if $f(x) = |x|$, we obtain the trimmed least absolute deviations estimator. Only the observations for which $1(y_{n1} > -\beta^\top \Delta x_n \;\&\; y_{n2} > \beta^\top \Delta x_n)$ are included in the first-order conditions, the presence of $f(y_{nt})$ in the objective function excluding trivial solutions.

9.2.5.2 Censored Sample

For the censored sample, observations for which $y_{nt} = 0$ are available, the observation rule for y_{nt} being:

$$\begin{cases} y_{n1} = \max(y_{n1}^*, 0) \\ y_{n2} = \max(y_{n2}^*, 0) \end{cases}$$

From Figure 9.5, we can see that not only A_1 and B_1 are symmetrical but also A_2 defined by $(y_{n1} < -\beta^\top \Delta x_n, y_{n2} > \beta^\top \Delta x_n)$ and B_2 defined by $(y_{n1} > -\beta^\top \Delta x_n, y_{n2} < \beta^\top \Delta x_n)$.

Therefore, to restore symmetry for the censored sample, we have to get rid of the zone for which $y_{n1} < -\beta^\top \Delta x_n$ and $y_{n2} < \beta^\top \Delta x_n$ (the dotted zone on Figure 9.5).

The symmetry between A_2 and B_2 leads to the following moment condition:

$$\mathrm{E}[\{1(y_n \in A_2) - 1(y_n \in B_2)\} \Delta x_n] = 0 \tag{9.20}$$

Moreover, for:

- y_n in A_2, the vertical distance to the limit of the zone is $y_{n2} - \max(0, \beta^\top \Delta x_n)$,
- y_n in A_1, the horizontal distance to the limit of the zone is $y_{n2} - \max(0, \beta^\top \Delta x_n)$

which translates into the following moment condition:

$$\begin{aligned} &\mathrm{E}[\{1(y_n \in A_2)(y_{n2} - \max(0, \beta^\top \Delta x_n) \\ &\quad -1(y_n \in B_2)(y_{n1} - \max(0, -\beta^\top \Delta x_n))\} \Delta x_n] = 0 \end{aligned} \tag{9.21}$$

Using (9.16 and 9.20), we obtain:

$$\mathrm{E}[\{1(y_n \in A_1 \cup B_1) - 1(y_n \in A_2 \cup B_2)\} \Delta x_n] = 0 \tag{9.22}$$

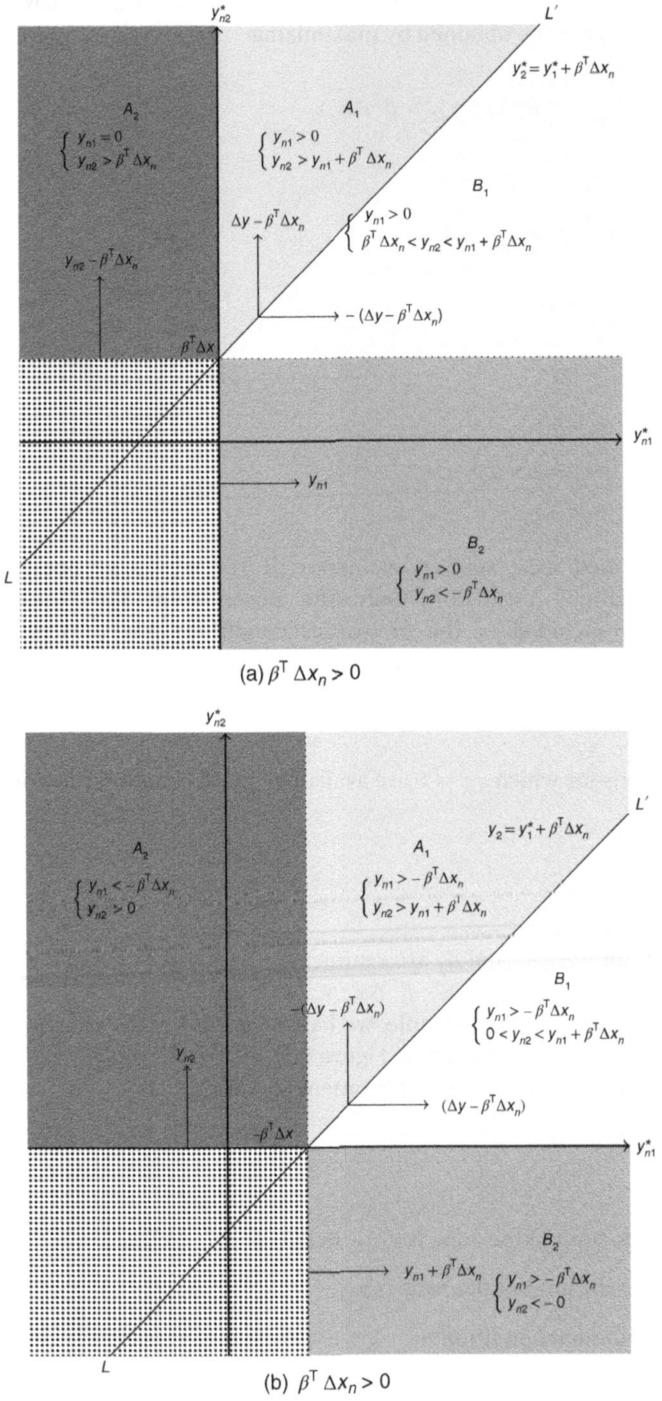

Figure 9.5 Symmetry of the distribution of (y^*_{n1}, y^*_{n2}).

and using (9.17 et 9.21), we obtain:

$$
\begin{aligned}
\mathrm{E}[\{ \quad & 1(y_n \in A_1 \cup B_1)\} \; (\Delta y_n - \beta^\top \Delta x_n) \\
& +1(y_n \in A_2) \qquad (y_{n2} - \max(0, \beta^\top \Delta x_n)) \\
& -1(y_n \in B_2) \qquad (y_{n1} - \max(0, -\beta^\top \Delta x_n))\} \\
& \times \Delta x_n] = 0
\end{aligned}
\tag{9.23}
$$

The sample analogues to (9.22) are the first-order conditions of the following function:

$$
\begin{aligned}
T_n = & \sum_{n=1}^{N} [1 - \{1(y_{n1} < \max(0, -\beta^\top \Delta x_n) \;\&\; 1(y_{n2} < \max(0, \beta^\top \Delta x_n)\}] \\
& \times \mid \Delta y_n - \beta^\top \Delta x_n \mid
\end{aligned}
\tag{9.24}
$$

which is the trimmed **LAD** estimator on the censored sample.

Finally, the sample equivalent of (9.23) are the first- order conditions of the following function:

$$
\begin{aligned}
T_n = & \sum_{n=1}^{N} (\max(y_{n2}, \beta^\top \Delta x_n) - \max(y_{n1}, -\beta^\top \Delta x_n) - \beta^\top \Delta x_n)^2 \\
& - 2 \times 1\{y_{n1} < -\beta^\top \Delta x_n\}(y_{n1} + \beta^\top \Delta x_n) y_{n2} \\
& - 2 \times 1\{y_{n2} < \beta^\top \Delta x_n\}(y_{n2} - \beta^\top \Delta x_n) y_{n1}
\end{aligned}
\tag{9.25}
$$

which is the trimmed least squares estimator for the censored sample. The trimmed **LAD** and least squares estimators have been extended to the case where the dependent variable is two-sided censored or truncated by Alan et al. (2013).

Example 9.4 trimmed tobit model – `LateBudgets` data set

Andersen et al. (2012) study the late adoption of budgets. They use a panel of American states for the 1988-2007 period, for which the date of budget adoption has been collected so that late budget situations can be detected and, in this case, the number of days from the legal limit date can be computed. Among the factors that may explain late budgets, the authors use:

- a shock to the fiscal climate, which is proxied by the annual change of unemployment rate `unempdiff`,
- divided control over the state government: `splitbranch` is a dummy indicating that both chambers are controlled by a different party than the governor's and `splitleg` is a dummy indicating that the two chambers are controlled by different parties.
- variables linked to the cost of a late budget: `elcyear` is a dummy for election years, `deadline` is a factor with levels (`"none"`, `"soft"`, `"hard"`) that indicates if there is a legal date for the end of legislative works,
- `shutdown` indicates whether the state law dictates a shutdown of state government activities in the event of a late budget, `supmaj` that budget adoption requires a super-majority,
- different covariates indicating political and legislative context: the fact that the governor is newly elected `newgov`, the number of years since the incumbent governor took office `govexp`, a dummy for a democrat governor `demgov`, a dummy indicating that the governor is subject to a binding term limit `lameduck`, a 1-to-5 scale for full- vs. part-time legislatures, where 1 corresponds to a part-time "citizen" legislature, and 5 corresponds to a full-time professional legislature `fulltimeleg`, a dummy that indicates that the state law does not allow a budget deficit to be carried over to the next fiscal year `nocarry`,
- several social and demographic covariates: population `pop`, the percentage of African Americans `black`, of college graduates `graduate`, of people older than 65 years `elderly`, of children between 5 and 17 years old `kids`, and the response rate in the 1990 US census `censusrep`, which is used as a proxy for social capital.

In order to investigate whether change of the unemployment rate has an asymmetric effect on budget adoption, two variables are created, indicating positive values of unemployment rise `unemprise` and fall `unempfall`.

```
data("LateBudgets", package = "pder")
LateBudgets$dayslatepos <- pmax(LateBudgets$dayslate, 0)
LateBudgets$divgov <- with(LateBudgets,
                           factor(splitbranch == "yes" |
                                  splitleg == "yes",
                                  labels = c("no", "yes")))
LateBudgets$unemprise <- pmax(LateBudgets$unempdiff, 0)
LateBudgets$unempfall <- - pmin(LateBudgets$unempdiff, 0)
form <- dayslatepos ~ unemprise + unempfall + divgov + elecyear +
    pop + fulltimeleg + shutdown + censusresp + endbalance + kids +
    elderly + demgov + lameduck + newgov + govexp + nocarry +
    supmaj + black + graduate
```

The model is estimated using the `pldv` function, which has a `model` argument with a default value of `'fd'` (for first-difference), which in this context is the fixed effects model of Honoré (1992). Two supplementary arguments can also be specified:

- `objfun` indicates whether one wants to minimize the sum of the least squares of the residuals (`'lsq'`, the default value), or the sum of the absolute values of the residuals (`'lad'`),
- `sample` indicates if the sample is censored (`'censored'`, the default value) or truncated (`'truncated'`).

```
FEtobit <- pldv(form, LateBudgets)
summary(FEtobit)
Oneway (individual) effect First-Difference Model

Call:
pldv(formula = form, data = LateBudgets)

Unbalanced Panel: n = 48, T = 2-20, N = 730
Observations used in estimation: 682

Residuals:
   Min. 1st Qu.  Median    Mean 3rd Qu.    Max.
 -107.8   -15.1     5.2     7.6    26.4   168.5

Coefficients:
            Estimate Std. Error t-value Pr(>|t|)
unemprise      9.042     10.944    0.83    0.409
unempfall    -31.641      6.887   -4.59  5.2e-06 ***
divgovyes     19.793      8.767    2.26    0.024 *
elecyear     -24.505     10.190   -2.40    0.016 *
pop           -0.683      2.512   -0.27    0.786
endbalance    -3.856     62.829   -0.06    0.951
kids           0.774      4.547    0.17    0.865
elderly       60.880      2.669   22.81  < 2e-16 ***
demgovyes     -6.371      6.770   -0.94    0.347
```

```
lameduckyes  -22.032      4.043   -5.45  7.1e-08 ***
newgovyes      5.606     10.532    0.53    0.595
govexp         3.395     38.894    0.09    0.930
---
Signif. codes:
0 '***' 0.001 '**' 0.01 '*' 0.05 '.' 0.1 ' ' 1

Total Sum of Squares:    328000
Residual Sum of Squares: 996000
R-Squared:         0.0255
Adj. R-Squared: 0.00953
F-statistic: -40.8339 on 11 and 670 DF, p-value: 1
```

As can be seen from the results, the economic situation influences the timing of budget adoption. The effect is asymmetric, an increase of the unemployment rate having stronger impact than a drop in the unemployment rate. Divided control over the government (measured by `divgov`) has a significantly positive effect on late budget adoptions.

9.2.6 The Random Effects Model

The trimmed estimator has two useful features: it is robust to non-normality and heteroscedasticity, on the one hand, and to correlation between the individual effects and the covariates, on the other hand, the individual effects being wiped out by the first- difference transformation. However, if the errors are normal and homoscedastic and if the individual effects are also normal and uncorrelated with the covariates, the maximum likelihood estimator is consistent and more efficient.

For panel data with individual effects, the latent variable writes:

$$y^*_{nt} = \gamma^\top z_{nt} + \eta_n + \nu_{nt}$$

9.2.6.1 Truncated Sample

The density of $y_{nt} \mid z_{nt}, \eta_n$ is:

$$f_+(y_{nt} \mid z_{nt}, \eta_n) = \frac{\phi\left(\frac{y_{nt}-\gamma^\top z_{nt}-\eta_n}{\sigma_\nu}\right)}{\sigma_\nu \Phi\left(\frac{\beta^\top x_{nt}+\eta_n}{\sigma_\nu}\right)}$$

The joint density of $y_n = y_{n1} \dots y_{nt}$ is, assuming the independence of the errors:

$$f_+(y_n \mid z_n, \eta_n) = \prod_{t=1}^{T} \frac{\phi\left(\frac{y_{nt}-\gamma^\top z_{nt}-\eta_n}{\sigma_\nu}\right)}{\sigma_\nu \Phi\left(\frac{\gamma^\top z_{nt}+\eta_n}{\sigma_\nu}\right)} \tag{9.26}$$

Assuming that the distribution of individual effects is normal with a standard deviation equal to σ_η, the unconditional joint density is obtained by integrating out (9.26) for the individual effects:

$$f_+(y_n \mid z_n) = \frac{1}{\sqrt{2\pi}\sigma_\eta} \int_{-\infty}^{+\infty} \prod_{t=1}^{T} \frac{\phi\left(\frac{y_{nt}-\gamma^\top z_{nt}-\eta}{\sigma_\nu}\right)}{\sigma_\nu \Phi\left(\frac{\gamma^\top z_{nt}+\eta}{\sigma_\nu}\right)} e^{-\frac{1}{2}\frac{\eta^2}{\sigma_\eta^2}} d_\eta \tag{9.27}$$

Using the change of variable $v = \frac{\eta}{\sqrt{2}\sigma_\eta}$, we obtain:

$$f_+(y_n \mid x_n) = \frac{1}{\sqrt{\pi}} \int_{-\infty}^{+\infty} \prod_{t=1}^{T} \frac{\phi\left(\frac{y_{nt}-\gamma^\top z_{nt}-\sigma_\eta\sqrt{2}v}{\sigma_\nu}\right)}{\sigma_\nu \Phi\left(\frac{\gamma^\top z_{nt}+\sigma_\eta\sqrt{2}v}{\sigma_\nu}\right)} e^{-v^2} d_v \tag{9.28}$$

which can be approximated by the Gauss-Hermite quadrature method:

$$f_+(y_n \mid x_n) \approx \frac{1}{\sqrt{\pi}} \sum_{r=1}^{R} w_r \prod_{t=1}^{T} \frac{\phi\left(\frac{y_{nt}-\gamma^\top z_{nt}-\sigma_\eta\sqrt{2}v_r}{\sigma_\nu}\right)}{\sigma_\nu \Phi\left(\frac{\gamma^\top z_{nt}+\sigma_\eta\sqrt{2}v_r}{\sigma_\nu}\right)} \tag{9.29}$$

The log-likelihood function for the truncated model is then simply obtained by summing the logarithms of (9.29) for all individuals:

$$\ln L = \sum_{n=1}^{N} \ln f_+(y_n \mid x_n) \tag{9.30}$$

9.2.6.2 Censored Sample

In this case, the conditional distribution of y_{nt} is either given by a probability or by a density:

$$g(y_{nt} \mid z_{nt}, \eta_n) = \frac{1}{\sigma_\nu}\phi\left(\frac{y_{nt} - \gamma^\top z_{nt} - \eta_n}{\sigma_\nu}\right) 1(y_{nt} > 0) + \Phi\left(-\frac{\gamma^\top z_{nt} + \eta_n}{\sigma_\nu}\right) 1(y_{nt} = 0)$$

Using a similar reasoning as for the truncated model, individual n contributes to the likelihood with a product of probabilities and/or densities:

$$g(y_n \mid x_n) \approx \frac{1}{\sqrt{\pi}} \sum_{r=1}^{R} w_r \prod_{t=1}^{T} \left\{ \frac{1}{\sigma_\nu}\phi\left(\frac{y_{nt} - \gamma^\top z_{nt} - \sigma_\eta\sqrt{2}v_r}{\sigma_\nu}\right) 1(y_{nt} > 0) + \Phi\left(-\frac{\gamma^\top z_{nt} + \sigma_\eta\sqrt{2}v_r}{\sigma_\nu}\right) 1(y_{nt} = 0) \right\} \tag{9.31}$$

The log-likelihood function for the censored sample is obtained by summing over all the individuals the logarithm of (9.31):

$$\ln L = \sum_{n=1}^{N} \ln g(y_n \mid z_n) \tag{9.32}$$

Example 9.5 random effects censored model – `Donor` data set
Landry et al. (2012) study the dynamic of behaviors of donors to public utility organizations and more specifically to the "Center for Natural Hazards Research at East Carolina University" (ECU). A first door-to-door campaign was realized in 2004. During this campaign, two kinds of treatment were used: a standard "simply ask for money" treatment, called VCM, and a treatment with a lottery with which potential donors can receive a gift. The second campaign took place in 2006. Some of the donors of the first campaign had been solicited, and three treatments were used, described in the factor variable `treatment` with three levels: `"vcm"` for a "simply ask

for money" treatment and `"sgift"` if a small gift (a bookmark) or `"lgift"` if a large gift (a book) were given to the potential donors. The main objective of the article is to study whether people who initially give to charities are more willing to give again than others. The response is the amount of the gift; it is therefore left-censored at 0. In the article, the authors present results of linear regressions with solicitors' fixed effects. In the online appendix, the same equations are estimated using a random effects tobit model. Two equations are estimated, both employing `treatment` and a dummy for previous donors `prcontr` as explanatory variables, the second adding an interaction term between the two.

```
data("Donor", package = "pder")
library("plm")
library("texreg")
T3.1 <- plm(donation ~ treatment +  prcontr, Donor, index = "id")
T3.2 <- plm(donation ~ treatment * prcontr - prcontr, Donor, index = "id")
T5.A <- pldv(donation ~ treatment +  prcontr, Donor, index = "id",
             model = "random", method = "bfgs")
T5.B <- pldv(donation ~ treatment * prcontr - prcontr, Donor, index = "id",
             model = "random", method = "bfgs")
screenreg(list(OLS = T3.1, Tobit = T5.A, OLS = T3.2, Tobit = T5.B),
          reorder.coef = c(1:3, 7:9, 4:6))
```

```
=============================================================================
                            OLS          Tobit          OLS          Tobit
-----------------------------------------------------------------------------
treatmentsgift             -0.41         2.36          0.06          3.53
                           (0.61)       (1.86)        (0.66)        (2.04)
treatmentlgift              1.79 **      6.36 ***      2.07 **       7.66 ***
                           (0.64)       (1.93)        (0.68)        (2.08)
prcontryes                  1.29 *       5.74 **
                           (0.59)       (1.79)
treatmentvcm:prcontryes                                3.14 **      10.78 **
                                                      (1.13)        (3.41)
treatmentsgift:prcontryes                              0.20          4.47
                                                      (0.95)        (2.88)
treatmentlgift:prcontryes                              1.05          3.23
                                                      (1.00)        (3.01)
(Intercept)                            -15.16 ***                  -16.05 ***
                                        (1.89)                      (1.97)
sd.nu                                   16.40 ***                   16.36 ***
                                        (0.80)                      (0.80)
sd.eta                                   4.05 ***                    3.92 ***
                                        (1.11)                      (1.10)
-----------------------------------------------------------------------------
R^2                         0.02                       0.02
Adj. R^2                   -0.02                      -0.01
Num. obs.                1039         1039          1039         1039
Log Likelihood                       -1498.38                    -1496.84
=============================================================================
*** p < 0.001, ** p < 0.01, * p < 0.05
```

The average gift (including censored observations) is 2.5\$. The first column indicates that previous donors give on average 1.3\$ more. A large gift increases the donation by 1.8%, while a small gift has no effect on donation. The third column distinguishes the treatment effect for previous donors and the others. For the `vcm` treatment, the gift of previous donors is much larger (about 3\$). On the contrary, there is no difference between previous donors and other people when a gift is proposed by solicitors. The random effects tobit models are presented in columns 2 and 4. The results are very similar but more difficult to interpret, as the expected value of the response for the tobit model is:

$$E(y \mid z) = \gamma^\top z \Phi\left(\frac{\gamma^\top z}{\sigma}\right) + \sigma\phi\left(\frac{\gamma^\top z}{\sigma}\right)$$

For example, for someone who didn't give previously and who received the `vcm` treatment, $\gamma^\top z = -15.13$. With $\sigma = 16.40$, we obtain an expected donation of 1.57. For someone who made a donation previously and who also received the `vcm` treatment, we have $\gamma^\top z = -15.16 + 5.74 = -9.42$, and the expected donation is 2.88. The effect for previous donors is therefore equal to $2.88 - 1.57 = 1.31$, which is very close to the linear regression coefficient.

9.3 Count Data

We now consider the case where the response is a count. We will first briefly review the estimation of count data models in a cross-sectional context, and then we will describe specific estimators for panel data.

9.3.1 Introduction

The two most widely used models when the response is a count are the Poisson and the NegBin models.

9.3.1.1 The Poisson Model

We first suppose that the response follows a Poisson distribution of parameter θ_n (which is the mean and the variance of the variable). Under this distributional assumption, the probability of observing a value y_n is:

$$\mathrm{P}(y_n) = \frac{e^{-\theta_n}\theta_n^{y_n}}{y_n!}$$

Using the logarithmic link, the Poisson parameter is the exponential of the linear predictor:

$$\theta_n = e^{\gamma^\top z_n}$$

which leads to the following probability for observation n:

$$\mathrm{P}(y_n \mid x_n) = \frac{e^{-e^{\gamma^\top z_n}} e^{\gamma^\top z_n y_n}}{y_n!}$$

Taking the logarithm of this probability and summing over all individuals, we obtain the following log-likelihood function:

$$\ln L = -\sum_{n=1}^{N} e^{\gamma^\top z} + \sum_{n=1}^{N} \gamma^\top z y_n - \sum_{n=1}^{N} \ln y_n!$$

9.3.1.2 The NegBin Model

Count data often exhibit excess dispersion, i.e., the variance is greater than the mean. In this case, the NegBin model is more appropriate than the Poisson model.

Suppose that y_n is a random variable that follows a Poisson distribution of parameter $\theta_n = \alpha_n \lambda_n$ (with $\lambda_n = e^{\gamma^\top z_n}$ in the case of a logarithmic link), α_n being a random variable.

The conditional probability of y_n is:

$$P(y_n \mid x_n, \alpha_n, \beta) = \frac{e^{-\theta_n}\theta_n^{y_n}}{y_n!} = \frac{e^{-\alpha_n\lambda_n}(\alpha_n\lambda_n)^{y_n}}{y_n!}$$

Let now suppose that α_n follows a gamma distribution. If β contains an intercept, the mean of α is not identified and therefore a one-parameter distribution, which imposes a unit mean, is chosen.

$$f(\alpha) = \frac{\delta^\delta}{\Gamma(\delta)} e^{-\delta\alpha}\alpha^{\delta-1}$$

Integrating out this conditional probability using the density of α, we obtain:

$$P(y_n \mid x_n) = \int_0^{+\infty} \frac{e^{-\alpha\lambda_i}(\alpha\lambda_i)^{y_i}}{y_i!} \frac{\delta^\delta}{\Gamma(\delta)} e^{-\delta\alpha}\alpha^{\delta-1} d\alpha$$

$$P(y_n \mid x_n) = \left(\frac{\delta_n}{\lambda_n + \delta_n}\right)^{\delta_n} \left(\frac{\lambda_n}{\lambda_n + \delta_n}\right)^{y_n} \frac{\Gamma(y_n + \delta_n)}{\Gamma(y_n + 1)\Gamma(\delta_n)}$$

To understand the meaning of δ_n, the first two moments of y_n are computed. For a given value of α_n, we have, as for the Poisson model: $E(y_n \mid \alpha_n) = V(y_n \mid \alpha_n) = \theta_n = \alpha_n\lambda_n$. The unconditional mean is $E_\alpha(\alpha\lambda_n) = \lambda_n$, because the expected value of α equals 1.

To compute the unconditional variance, the variance decomposition formula is applied:

$$V(y_n) = E_\alpha(\alpha\lambda_n) + V_\alpha(\alpha\lambda_n) = \lambda_n + \frac{1}{\delta_n}\lambda_n^2$$

A general formula for δ_n is:

$$\delta_n = \frac{\lambda_n^{2-k}}{\nu}$$

For $k = 1$, we get the Negbin1 model, with $\delta_n = \lambda_n/\nu$ and $V(y_n) = \lambda_n(1 + \nu)$. In this case, the variance is proportional to the mean.

For $k = 2$, we obtain the Negbin2 model, with $\delta_n = 1/\nu$ and $V(y_n) = \lambda_n + \nu\lambda_n^2$; here, the variance is a quadratic function of the mean.

9.3.2 Fixed Effects Model

Fixed effects Poisson and NegBin models are proposed by Hausman et al. (1984).

9.3.2.1 The Poisson Model

The fixed effects Poisson model is very specific, as it doesn't suffer from the incidental parameter problem and can therefore be obtained either by estimating the individual effects or by using a sufficient statistic[4].

In a panel context, the Poisson parameter for individual n in period t is written:

$$\theta_{nt} = \eta_n \lambda_{nt} = \eta_n e^{\beta^\top x_{nt}}$$

---

4 See Cameron and Trivedi (1998, chap. 9).

which means that the individual effect is multiplicative. For a given value of the individual effect, the probability of observing y_{nt} is:

$$\mathrm{P}(y_{nt} \mid x_{nt}, \eta_n, \beta) = \frac{e^{-\theta_{nt}}\theta_{nt}^{y_{nt}}}{y_{nt}!} = \frac{e^{-\eta_n\lambda_{nt}}(\eta_n\lambda_{nt})^{y_{nt}}}{y_{nt}!}$$

Let $Y_n = \sum_{t=1}^T y_{nt}$ be the sum of all the values of the response for individual n and $\Lambda_n = \sum_{t=1}^T \lambda_{nt}$ the sum of the Poisson parameters. A sum of Poisson variables follows a Poisson distribution with parameter equal to the sum of the parameters of the summed variables. We therefore have:

$$\mathrm{P}(Y_n \mid x_n, \eta_n, \beta) = \frac{e^{-\eta_n\Lambda_n}(\eta_n\Lambda_n)^{Y_n}}{Y_n!} \tag{9.33}$$

Let $y_n = (y_{i1}, y_{i2}, \dots, y_{nt})$ be the vector of values of y for individual n. We then have:

$$\mathrm{P}(y_n \mid x_n, \eta_n, \beta) = \frac{e^{-\eta_n\sum_{t=1}^T\lambda_{nt}}\prod_{t=1}^T(\eta_n\lambda_{nt})^{y_{nt}}}{\prod_{t=1}^T y_{nt}!} = \frac{e^{-\eta_n\Lambda_i}\eta_n^{Y_n}\prod_{t=1}^T\lambda_{nt}^{y_{nt}}}{\prod_{t=1}^T y_{nt}!} \tag{9.34}$$

Applying Bayes' theorem, we obtain:

$$\mathrm{P}(y_n \mid x_n, \eta_n, \beta) = \mathrm{P}(y_n \mid x_n, \eta_n, \beta, Y_n)\mathrm{P}(Y_n \mid x_n, \eta_n, \beta)$$

i.e., the joint probability of the components of y_n is the product of the conditional probability of y_n given Y_n and the marginal distribution of Y_n. This conditional probability is:

$$\mathrm{P}(y_n \mid x_n, \eta_n, \beta, Y_n) = \frac{\mathrm{P}(y_n \mid x_n, \eta_n, \beta)}{\mathrm{P}(Y_n \mid x_n, \eta_n, \beta)}$$

which implies:

$$\mathrm{P}(y_n \mid x_n, \beta, Y_n) = \frac{Y_n!}{\Lambda_n^{Y_n}}\prod_{t=1}^T\frac{\lambda_{nt}^{y_{nt}}}{y_{nt}!} \tag{9.35}$$

As for the logit model, Y_n is a sufficient statistic, which means that it allows to get rid of the individual effects. Taking the logarithm of this expression and summing over all individuals, we obtain the *within* Poisson model:

$$\ln L(y \mid x, \beta, Y) = \sum_{n=1}^N\left(\ln Y_n! - Y_n\ln\sum_{t=1}^T\lambda_{nt} + \sum_{t=1}^T(y_{nt}\ln\lambda_{nt} - \ln y_{nt}!)\right) \tag{9.36}$$

or:

$$\begin{aligned}\ln L(y \mid x, \beta, Y) &= \sum_{n=1}^N\left(\ln Y_n! - \sum_{t=1}^T\ln y_{nt}! + \sum_{t=1}^T y_{nt}\ln\frac{\lambda_{nt}}{\sum_{t=1}^T\lambda_{nt}}\right)\\ &\propto \sum_{n=1}^N\left(\sum_{t=1}^T y_{nt}\ln\frac{\lambda_{nt}}{\sum_{t=1}^T\lambda_{nt}}\right)\end{aligned} \tag{9.37}$$

As stated previously, the Poisson model is not affected by the incidental parameter problem, as the same estimator may be obtained by estimating the individual effects. To show this result, we take the logarithm of the joint probability for the T observations of y for individual n (equation 9.34), in order to obtain the log-likelihood function:

$$\ln\mathrm{P}(y_n \mid x_n, \eta_n, \beta) = -\eta_n\sum_t\lambda_{nt} + \sum_t y_{nt}\ln(\eta_n\lambda_{nt}) - \sum_t\ln y_{nt}! \tag{9.38}$$

The first-order condition for η_n to maximize the log- likelihood function is:

$$\frac{\partial \ln \mathrm{P}_n}{\partial \eta_n} = -\sum_t \lambda_{nt} + \frac{1}{\eta_n}\sum_t y_{nt} = 0$$

which implies that: $\eta_n = \frac{\sum_t y_{nt}}{\sum_t \lambda_{nt}}$.

Introducing this expression in (9.38) and summing over all n, we obtain the concentrated log-likelihood function:

$$\ln L_{\text{conc}}(y \mid x, \beta) = \sum_n \left(-Y_n + Y_n \ln Y_n + \sum_t y_{nt} \frac{\lambda_{nt}}{\sum_t \lambda_{nt}} - \sum_t \ln y_{nt}! \right)$$

$$\propto \sum_{n=1}^{N} \left(\sum_{t=1}^{T} y_{nt} \ln \frac{\lambda_{nt}}{\sum_{t=1}^{T} \lambda_{nt}} \right) \tag{9.39}$$

The two log-likelihood functions (9.37) and (9.39) are proportional, they therefore lead to the same estimators of β. Moreover, if a logarithmic link is chosen, we have: $\lambda_{nt} = e^{\beta^\top x}$. The likelihood is in this case proportional to:

$$\Pi_{n=1}^{N} \Pi_{t=1}^{T} \left(\frac{e^{\beta^\top x_{nt}}}{\sum_t e^{\beta^\top x_{nt}}} \right)^{y_{nt}}$$

which is similar to the likelihood of a multinomial logit model for which N individuals must choose one among L mutually exclusive alternatives. The difference is that in this latter model y_{nt} is either equal to 0 or to 1, and $\sum_t y_{nt} = 1$, as in our context each y_{nt} is a natural integer.

9.3.2.2 Negbin Model

Hausman et al. (1984) also propose a fixed effects NegBin model. We just present below without demonstration the joint probability for individual n:

$$\mathrm{P}(y_n \mid x_n, \beta, Y_n) = \left(\prod_{t=1}^{T} \frac{\Gamma(\lambda_{nt} + y_{nt})}{\Gamma(\lambda_{nt})\Gamma(y_{nt} + 1)} \right) \frac{\Gamma(\Lambda_n)\Gamma(Y_n + 1)}{\Gamma(\Lambda_n + Y_n)} \tag{9.40}$$

9.3.3 Random Effects Models

9.3.3.1 The Poisson Model

Hausman et al. (1984) also proposed a *between* and a random effects Poisson model, integrating out the relevant probabilities (9.33 et 9.34 respectively). A gamma distribution hypothesis is made for the individual effects, with the following density:

$$f(x, a, b) = \frac{a^b}{\Gamma(b)} e^{-ax} x^{b-1}$$

with

$$\Gamma(z) = \int_0^{+\infty} t^{z-1} e^{-t} dt$$

the gamma function. The expected value and the variance of x are respectively:

$$\mathrm{E}(x) = \frac{b}{a} \text{ and } \mathrm{V}(x) = \frac{b}{a^2}$$

If the model contains an intercept, the expected value is not identified and we can then suppose, without restriction, that it is equal to 1, which implies $a = b$. We then obtain a gamma distribution with one parameter (denoted δ):

$$f(\alpha) = \frac{\delta^\delta}{\Gamma(\delta)} e^{-\delta\alpha} \alpha^{\delta-1}$$

Integrating out the conditional probabilities (9.33 and 9.34), we obtain the unconditional probabilities for the between and the random effects models:

$$\mathrm{P}(Y_n \mid z_n, \beta) = \int_0^{+\infty} \mathrm{P}(Y_n, z_n, \alpha, \gamma) f(\alpha) d\alpha = \frac{\Lambda_n^{Y_n}}{Y_n!} \frac{\delta^\delta}{\Gamma(\delta)} \frac{\Gamma(Y_n + \delta)}{(\Lambda_n + \delta)^{Y_n+\delta}}$$

$$\mathrm{P}(y_n, x_n, \beta) = \int_0^{+\infty} \mathrm{P}(y_n, x_n, \alpha, \beta) f(\alpha) d\alpha = \prod_{t=1}^{T} \frac{\lambda_{nt}^{y_{nt}}}{y_{nt}!} \frac{\delta^\delta}{\Gamma(\delta)} \frac{\Gamma(Y_n + \delta)}{(\Lambda_n + \delta)^{Y_n+\delta}}$$

which leads to the following log-likelihood functions:

$$\begin{aligned} \ln L(Y \mid z, \gamma) = \sum_{n=1}^{N} \Bigg[& Y_n \ln \sum_t \lambda_{nt} - \ln Y_n! + \delta \ln \delta \\ & - \ln \Gamma(\delta) + \ln \Gamma(Y_n + \delta) \\ & - (Y_n + \delta) \ln \left(\sum_{t=1}^{T} \lambda_{nt} + \delta \right) \Bigg] \end{aligned} \tag{9.41}$$

$$\begin{aligned} \ln L(y \mid z, \gamma) = \sum_{n=1}^{N} \Bigg[& \sum_t (y_{nt} \ln \lambda_{nt} - \ln y_{nt}!) + \delta \ln \delta \\ & - \ln \Gamma(\delta) + \ln \Gamma(Y_n + \delta) \\ & - (Y_n + \delta) \ln \left(\sum_{t=1}^{T} \lambda_{nt} + \delta \right) \Bigg] \end{aligned} \tag{9.42}$$

9.3.3.2 The NegBin Model

In addition to the Poisson model, Hausman et al. (1984) also proposed *between* and random effects NegBin models. We just present below without demonstration the joint probability for individual n.

$$\mathrm{P}(Y_n \mid x_n, \gamma) = \frac{\Gamma(\Lambda_n + Y_n)}{\Gamma(\Lambda_n)\Gamma(Y_n + 1)} \frac{\Gamma(a + b)\Gamma(a + \Lambda_n)\Gamma(b + Y_n)}{\Gamma(a)\Gamma(b)\Gamma(a + b + \Lambda_n + Y_n)} \tag{9.43}$$

$$\mathrm{P}(y_n, x_n, \gamma) = \frac{\Gamma(a + b)\Gamma(a + \Lambda_n)\Gamma(b + Y_n)}{\Gamma(a)\Gamma(b)\Gamma(a + b + \Lambda_n + Y_n)} \left(\prod_{t=1}^{T} \frac{\Gamma(\lambda_{nt} + y_{nt})}{\Gamma(\lambda_{nt}) + \Gamma(y_{nt} + 1)} \right) \tag{9.44}$$

Example 9.6 fixed effects NegBin model – `GiantsShoulders` data set

Furman and Stern (2011) assess the impact of a scientific institution, a biological resource center, whose objective is to certify and disseminate knowledge, on knowledge accumulation. More specifically, they are interested in the ACTT (American Type Culture Collection), which collects, certifies, and distributes biological organisms. The authors are interested in the citations of publications for which the results are hosted by the ACTT, and they try to estimate the causal effect of ACTT hosting. There is an obvious selection problem, because it is natural to think that some of the best pieces of research will end up to be hosted by ACTT and that the same would be heavily cited because of their quality even if they were not hosted by the ACTT.

In order to identify the causal effect of ACTT hosting on knowledge dissemination, the authors use two strategies:

- the first is that there is often a long lag between publication and hosting, and this lag is mostly exogenous,
- the second consists in matching every hosted article to a similar (same journal, date, and subject) non-hosted article.

The `GiantsShoulders` data set is available in the **pder** package.

```
data("GiantsShoulders", package = "pder")
head(GiantsShoulders)
  pair article brc pubyear brcyear year citations
1  184    1184 yes    1983    1994 1983         0
2  184    1184 yes    1983    1994 1984        31
3  184    1184 yes    1983    1994 1985        89
4  184    1184 yes    1983    1994 1986       105
5  184    1184 yes    1983    1994 1987        84
6  184    1184 yes    1983    1994 1988        75
```

The response is `citations`, the annual number of citations of the article. Each article is identified by the variable `article` and by the pair of articles it belongs to `pair`. For each pair, an article is hosted by the ATCC and the other is not, which is indicated by the variable `brc`. Years of observation, publication, and hosting are indicated by the variables `year`, `puyear`, and `brcyear`.

Figure 1 in Furman and Stern (2011), reproduced here in Figure 9.6, presents the average number of citations for hosted and non-hosted articles as a function of publication age. It is computed using the **dplyr** and the **ggplot2** packages.

```
library("dplyr")
library("ggplot2")
GiantsShoulders <- mutate(GiantsShoulders, age = year - pubyear)
cityear <- summarise(group_by(GiantsShoulders, brc, age),
                     cit = mean(citations, na.rm = TRUE))
ggplot(cityear, aes(age, cit)) + geom_line(aes(lty = brc)) +
    geom_point(aes(shape = brc)) + scale_x_continuous(limits = c(0, 20))
```

As can be seen, the number of citations increases the first year, to reach a maximum at about the third or fourth year and then decreases. Figure 9.6 also shows that hosted articles are much more cited that non-hosted articles.

To estimate the marginal causal effect of the hosting institution, two covariates are constructed for hosted articles:

- `window` is 1 around the hosting date, more precisely for a three-year period centered on the hosting year,
- `post_brc` is 1 for articles hosted for more than a year.

To reproduce the results exactly, we use annual fixed effects for years after 1979 and 5-year effects for the 1970-74 and 1975-79 periods. We also introduce fixed effects for the age of the articles (omitting the 31 years age dummy).

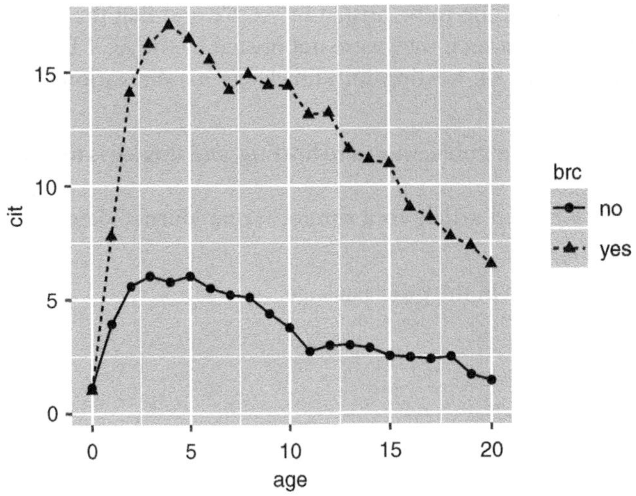

Figure 9.6 Average annual citations by age, BRC versus control articles.

```
GiantsShoulders <- mutate(GiantsShoulders,
                          window = as.numeric( (brc == "yes") &
                                               abs(brcyear - year) <= 1),
                          post_brc = as.numeric( (brc == "yes") &
                                               year - brcyear > 1),
                          age = year - pubyear)
GiantsShoulders$age[GiantsShoulders$age == 31] <- 0
GiantsShoulders$year[GiantsShoulders$year %in% 1970:1974] <- 1970
GiantsShoulders$year[GiantsShoulders$year %in% 1975:1979] <- 1975
```

In the two first columns, a linear model is estimated. The first model contains only age fixed effects, and the second one add pairs and years fixed effects. The results are similar; the selection effect of hosting is about 50% of more citations, and the marginal effect is 35% for the hosting period and about 50% for latter years.

The other two columns present the results of the fixed effects NegBin model. Pair (column 3) and article (column 4) fixed effects are alternatively used.

```
library("pglm")
t3c1 <- lm(log(1 + citations) ~ brc + window + post_brc + factor(age),
           data = GiantsShoulders)
t3c2 <- update(t3c1, . ~ .+  factor(pair) + factor(year))
t3c3 <- pglm(citations ~ brc + window + post_brc + factor(age) + factor(year),
           data = GiantsShoulders, index = "pair",
           effect = "individual", model = "within", family = negbin)
t3c4 <- pglm(citations ~ window + post_brc + factor(age) + factor(year),
             data = GiantsShoulders, index = "article",
             effect = "individual", model = "within", family = negbin)
screenreg(list(t3c2, t3c3, t3c4),
          custom.model.names = c("ols: age/year/pair-FE",
                                 "NB:age/year/pair-FE",
                                 "NB: age/year/article-FE"),
          omit.coef="(factor)|(Intercept)", digits = 3)
```

	ols: age/year/pair-FE	NB:age/year/pair-FE	NB: age/year/article-FE
brcyes	0.501 ***	0.752 ***	
	(0.057)	(0.073)	
window	0.385 ***	0.352 ***	0.565 ***
	(0.074)	(0.082)	(0.065)
post_brc	0.535 ***	0.538 ***	0.810 ***
	(0.063)	(0.079)	(0.056)
R^2	0.538		
Adj. R^2	0.522		
Num. obs.	4857	4857	4857
RMSE	0.829		
Log Likelihood		-10759.180	-9632.404

*** p < 0.001, ** p < 0.01, * p < 0.05

9.4 More Empirical Examples

Charness and Villeval (2009) investigate the difference in behavior between senior and junior workers in terms of risk aversion, competition, and cooperation. They conduct an experiment during which every participant can invest his or her initial endowment in a public good game, which is the explanatory variable of their econometric analysis. This variable is left- (null contribution) and right- (contribution of the full endowment) censored. As the participants are observed during 16 periods, they use a random effects tobit model. The `Seniors` data set is available in package **pder**.

Michalopoulos and Papaioannou (2016) explore the consequences of ethnic partitioning, which is one aspect of the "scramble for Africa" during which European countries partitioned Africa without caring much about the boundaries of ethnic groups. Their pseudo-panel consists of 825 ethnic groups belonging to 49 countries. The authors estimate a Negbin model, where the response is the number of conflicts in an ethnicity-country homeland, the major covariate being a dummy for partitioned ethnic areas. They introduce country fixed effects and estimate a specification where the fixed effects are estimated and not wiped out using a sufficient statistic. Partitioned ethnicities experience an increase of 57% in political violence compared to other areas. The data are available in package **pder** as `ScrambleAfrica`.

Bardhan and Mookherjee (2010) analyze the political determinants of land reform in West Bengal, India. More specifically, they use yearly data on 89 villages for the 1978-1998 period. The response is the percentage of land or of households affected by land reform; it is highly censored, as it is 0 for more than 80% of the sample. The main covariate is the presence of a left-wing coalition at the head of the local government. The authors use the trimmed least absolute deviation estimator of Honoré (1992) and don't find any significant effect of the left-wing government variable on the strength of land reform. The `LandReform` data are available in package **pder**

Brandts and Cooper (2006) analyze how financial incentives can be used to overcome a history of coordination failure. For this purpose, they conduce an experiment where "firms," composed of four "employees" have an output that is related to the lowest level of effort implemented by the employees. The individual or the lowest firm level effort is the response and, as the same employees/firms are observed during 30 different rounds, panel data techniques are used. The level of effort being ordinal, the authors use ordered probit models, with firm random effects and nested random effects respectively when the analysis is at the firm or at the employee level. Their dataset is available in the **pder** package as `CoordFailure`.

Farber et al. (2016) conducted an audit study to analyze the determinants of callbacks to job applications. They sent four fake resumes for 1,118 job openings and the response is a dummy indicating a callback, the covariates being the unemployment spell duration, the age, and the fact that the worker has held a low level interim job. They estimate a random effects and a conditional logit model with job opening effects. The `Callbacks` data are to be found in the **pder** package.

Bazzi (2017) investigates the influence of income on migration. At the household and at the village level, the response being in the first case a dummy that indicates whether a person in the household migrated during the given year and in the second case the percentage of the population of the village that has migrated. The main covariates are rainfall, rice price shock, and wealth at the household level and at the village level, indicators of the shape of wealth distribution. The author uses a conditional logit at the household level and the two-sided trimmed least absolute deviations estimator (see Alan et al., 2013) at the village level. The `IncomeMigrationV` (village level) and `IncomeMigrationH` (household level) datasets are also included in the **pder** package.

Vella and Verbeek (1998) estimate the union premium for young men. In a first step, they estimate a dynamic random effect probit model for union membership. The `UnionWage` dataset is available in the **pglm** package.

Hausman et al. (1986) and Cincer (1997) study the dynamic relationship between patents and R&D using yearly panels of firms. They fit different count data models, including conditional Poisson and Negbin models. The data sets they used are available as `PatentsRDUS` and `PatentsRD` in the **pglm** packages.

10

Spatial Panels

10.1 Spatial Correlation

If the cross-sectional dimension of a dataset has any form of ordering, or if a distance is defined over each pair of observations (here: *spatial units*), one can use spatial methods to account for the possibility that correlation be stronger between "nearby" ones. The most commonly used definitions of proximity are either distance- or neighborhood-related. Neighborhood depends on the spatial units being arranged in a topological space on a regular or irregular grid, an example of the latter being state or regional borders in geography.[1] On the subject, see Anselin (1988, Ch. 3).

This subject is most relevant in nonrandom samples such as countries within a geographical region, or regions within one country; but spatial methods can also be employed wherever some kind of distance between observations is defined, be it in a geographic space or perhaps in an economic, demographic, or psychological one. Hence spatial methods, although more common in the former context, can be relevant in random samples too, such as, e.g., in household surveys.

10.1.1 Visual Assessment

Correlation in bidimensional space can be multifaceted, and in some ways more complicated to assess than correlation in time, which has a single dimension and often an obvious direction. Therefore, preliminary data analysis based on visual assessments, while always important and perhaps underutilized in econometric practice (Kleiber and Zeileis, 2008), is all the more useful in a spatial context. In the first part of this section we present an example of visual assessment of spatial correlation drawing on R's map plotting facilities; next, we proceed to formal statistical tests.

Example 10.1 Visual assessment of spatial correlation – `HousePricesUS` data set
Visualizing data on a choropleth map is often the first step toward assessing the correlation of data in a geographical space. Plotting statistical maps is a complex subject that is out of the scope of the present book and is made easier by a number of dedicated packages: below we provide an example of plotting maps with **ggplot2**, adapting the example in package **fiftystater**

1 Contiguity/neighborhood is straightforward in irregular grids, while on regular ones (as in the literature on lattice processes) different definitions can apply, e.g., in the now-standard chess-related terminology, "queen contiguity" when units sharing either a border or a vertex are defined as neighbors; "rook" contiguity if considering only pairs sharing a border; and so on.

Panel Data Econometrics with R, First Edition. Yves Croissant and Giovanni Millo.

Companion website: www.wiley.com/go/croissant/data-econometrics-with-R

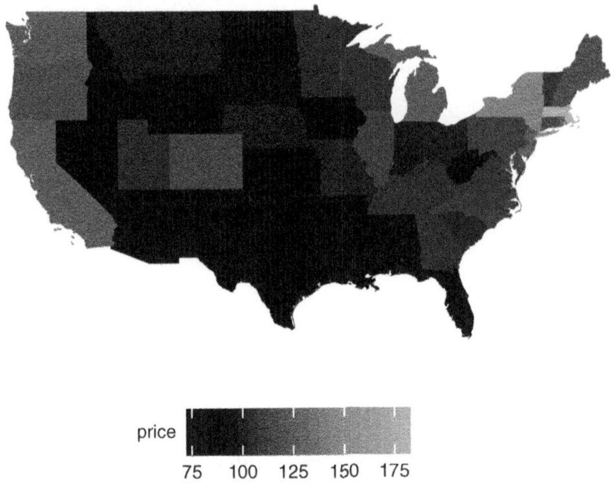

Figure 10.1 Growth of house prices indexes in the USA between 1980 and 2000.

(Murphy, 2016) for displaying the growth of house prices indices in the USA between 1980 (=100) and 2000 (darker is lower):

```
data("HousePricesUS", package="pder")
library("ggplot2")
data("fifty_states", package = "fiftystater")
houses00 <- subset(HousePricesUS, year == 2000)
houses00$name <- tolower(houses00$name)
p <- ggplot(houses00, aes(map_id = name)) +
    geom_map(aes(fill = price), map = fifty_states) +
    expand_limits(x = fifty_states$long, y = fifty_states$lat) +
    coord_map() +
    scale_x_continuous(breaks = NULL) +
    scale_y_continuous(breaks = NULL) +
    labs(x = "", y = "") +
    theme(legend.position = "bottom",
          panel.background = element_blank()) +
    theme(legend.text = element_text(size = 6),
          legend.title= element_text(size = 8),
          axis.title = element_text(size = 8))
p <- p + scale_fill_gradient2(low = "grey30", high = "grey5")
p
```

Clusters of low-growth regions are evident by their darker color and the opposite (Figure 10.1): in general, shades tend to distribute nonrandomly, nearby states tending to behave similarly. Formal testing for spatial correlation is likely to corroborate this first impression.

10.1.2 Testing for Spatial Dependence

One first issue when confronted with spatially referenced data is to determine whether spatial dependence exists, i.e., whether "nearby" units (according to the chosen metric) are more correlated than distant ones. The raw data are tested for spatial dependence in order to inform and justify the use of spatial estimation methods; then, after estimation, the residuals are tested again to determine whether the model has been able to effectively account for the spatial features of the process at hand.

10.1.2.1 CD p Tests for Local Cross-sectional Dependence

A very flexible way of assessing whether dependence in the cross-section of a panel dataset is spatially related goes through a particularization of the CD test for general cross-sectional dependence described in Chapter 4. The latter is in principle completely a-spatial, being based on a scaled average of the pairwise correlation coefficients $\hat{\rho}_{nm}$ between observations (or residuals). Still, the CD can be restricted to those pairs of observations satisfying one given criterion: most frequently, a contiguity-based neighborhood one but also that distance be under a given cutoff level.

The *local* variant of the CD test, called CD(p) test (Pesaran, 2004), takes into account an appropriate subset of *neighboring* cross-sectional units to check the null of no cross-sectional dependence against the alternative of *local* cross-sectional dependence, i.e., dependence between neighbors only. To do so, the pairs of neighboring units are selected by means of a binary proximity matrix, in which zeros correspond to pairs of observations that are not neighbors. The latter is used for discarding the correlation coefficients relative to pairs of observations that are not neighbors in computing the CD statistic. The test is then defined as:

$$\text{CD} = \sqrt{\frac{1}{\sum_{n=1}^{N-1}\sum_{m=n+1}^{N} w(p)_{nm}}}\left(\sum_{n=1}^{N-1}\sum_{m=n+1}^{N}[w(p)]_{nm}\sqrt{T_{nm}}\hat{\rho}_{nm}\right)$$

where $[w(p)]_{nm}$ is the (n, m)-th element of the p-th order proximity matrix, so that if any pair n, m are not neighbors, $[w(p)]_{nm} = 0$ and $\hat{\rho}_{nm}$ is eliminated from the summation; and T_{nm} is the number of time series observations in common between individuals n and m (T if the panel is balanced).[2]

The same procedure can be applied to the LM and SCLM tests described in section 4.3.1. The *local* version of either test can be computed supplying an $n \times n$ matrix (of any type coercible to `logical`), providing information on whether any pair of observations are neighbors or not, to the `w` argument of `pcdtest`. If `w` is supplied, only neighboring pairs will be used in computing the test; else, `w` will default to `NULL`, and all observations will be used. The matrix needs not really be binary, so commonly used "row-standardized" matrices can be employed as well: it is enough that neighboring pairs correspond to nonzero elements in `w`[3].

10.1.2.2 The Randomized W Test

The CD(p) test is flexible and well behaved in small samples; moreover it does not suffer the biggest drawback of its global sibling, which does not have any power under zero-mean dependence and therefore cannot be employed, for example, on cross-sectionally demeaned data – or equivalently on the residuals of a model containing time fixed effects. Nevertheless, it does not tolerate serial correlation and can be sensitive to non-spatial types of dependence. In fact, if cross-sectional dependence of the non-spatial type is present and a CD(p) test is

2 This more general formula is easily seen to reduce to formula (14) in Pesaran (Pesaran, 2004) for the special case considered in that paper, where a regular ordering of observations is assumed so that the n-th cross-sectional observation is a neighbor to the $(n-1)$-th and to the $(n+1)$-th.

3 A row-standardized proximity matrix is one transformed so that all rows sum to 1. The very comprehensive package **spdep** for spatial dependence analysis (see Bivand, 2008) contains features for creating, lagging, and manipulating *neighbor list* objects of class `nb`, that can be readily converted to and from proximity matrices by means of the `nb2mat` function. Higher orders of the CD(p) test can be obtained lagging the corresponding `nb`s through `nblag`.

performed, it will be based on a subset of spatially related pairs from a population of correlated ones; it is therefore likely to yield a false positive result (a type I error) favoring spatial dependence.

The idea underlying the CD(p) test, that not all pairs of neighbors are correlated but only those in a specific spatial relationship are and that the latter are identified through the W matrix, gives rise to another testing procedure that is remarkably robust to all the above confounding features. The RW test of Millo (2017a) employs a permutation procedure to produce a large number of randomized neighborhood matrices and then compares the CD(p) statistic under the true spatial ordering with the population of those under the randomized ones. If spatial dependence is absent, the observations must be *exchangeable* in the cross-section: then, the true CD(p) will not take an extreme value with respect to the randomization-based ones, and the null hypothesis of no spatial dependence will hold. As usual, the share of randomized statistics more extreme than the true one will be the pseudo-p-value of the test. In the majority of situations, the alternative hypothesis is of positive spatial dependence. In this case a one-tailed test will be appropriate. Given a panel-indexed vector x, call $\tau_h^* = \text{CD}_p(W_h^*|x)$ the randomized statistic from the h-th draw, with $h = 1 \dots B$; and $\hat{\tau} = \text{CD}_p(W|x)$ the one under the true W. If the alternative is positive spatial dependence, the pseudo-p-value of the one-tailed RW_p^+ test is then

$$\text{p-value}^*(\text{RW}_p^+) = \frac{\sum_{h=1}^{B} 1[\tau_h^* \geq \hat{\tau}]}{B+1} \tag{10.1}$$

where 1[.] is the indicator function. The null of no spatial dependence in x would be rejected at, say, 5% significance if p-value*(RW) < 0.05, meaning that the actual CD_p value is more extreme than the 95th quantile of the distribution of randomized values.

Negative spatial autocorrelation is less common in empirical practice but can be relevant, e.g., in the description of competitive processes (see Griffith and Arbia, 2010; Elhorst and Zigova, 2014). In this case it may happen that the distribution of randomized statistics be shifted in the opposite direction by positive global dependence so that the value of the true test statistic be *less* extreme, and the one-tailed procedure would not work. A two-tailed test is then needed, which is easily accomplished by taking absolute values *and* cross-sectionally demeaning the data so that the average of the factors, and hence the average global correlation, is re-centered on zero:

$$\text{p-value}^*(\text{RW}_p^{symm}) = \frac{\sum_{h=1}^{B} 1[|\tau_h^*| \geq |\hat{\tau}|]}{B+1}. \tag{10.2}$$

To take heed of possible asymmetries in the (re-centered) distribution of randomized statistics, one can go the safest way employing the asymmetric version of the test:

$$\text{p-value}^*(\text{RW}_p) = 2 \times \min\left(\frac{\sum_{h=1}^{B} 1[\tau_h^* \leq \hat{\tau}]}{B+1}, \frac{\sum_{h=1}^{B} 1[\tau_h^* > \hat{\tau}]}{B+1}\right). \tag{10.3}$$

Example 10.2 Spatial dependence – `HousePricesUS` data set

In their analysis of the income elasticity of house prices across continental US states, Holly et al. (2010) employ CD tests to assess cross-sectional dependence in the raw data and in the residuals from the various regression models they estimate. Below we present an assessment of spatial dependence in the house prices index from their dataset employing a binary neighborhood matrix. As was the case for the a-spatial CD test, if analyzing raw data, then the `data.frame` must be pre-transformed into a `pdata.frame`, so the testing function can find the appropriate indices:

```
data("usaw49", package="pder")
library("plm")
php <- pdata.frame(HousePricesUS)
pcdtest(php$price, w = usaw49)

Pesaran CD test for local cross-sectional dependence
in panels

data:  php$price
z = 37, p-value <2e-16
alternative hypothesis: cross-sectional dependence
```

The local CD(p) test finds a strong, statistically very significant average correlation between neighboring pairs. There is little doubt that the original data are correlated in the cross section; it remains to be ascertained whether said correlation is truly spatial or due to common factor influence in the process originating the data. An RW test will determine whether there is any spatial correlation proper left after controlling for cross-sectional correlation:

```
library("splm")
rwtest(php$price, w = usaw49, replications = 999)

Randomized W test for spatial correlation of order 1

data:  formula
p-value = 0.002
alternative hypothesis: twosided
```

The spatial correlation according to the "true" neighborhood matrix is the most extreme in the distribution of statistics obtained from drawing 999 more random orderings next to the original one, leaving little doubt about the presence of a spatial component in the process generating the data. The same is found when analyzing the explanatory variable, income. The question becomes then, after estimating the model, whether there is any spatial correlation remaining in the residuals after explaining house prices through income, or whether the spatial structure of income effectively explained away that in the dependent variable, house prices. Holly et al. (2010) estimate a common correlated effects (CCE) model of house prices vs income in order to control for unobservable common factors perturbating the relationship of interest. CCE will effectively "defactor" the model residuals, so that any purely spatial process will now be detectable without the confounding effect of the former so that a CD(p) test of the model residuals will reveal it. The same goes for the RW test, but the latter will control for factor structures so that it can be applied without defactoring as well. Residuals from a `pmg` object are a regular `pseries` so that `pcdtest` and `rwtest` can be directly applied:

```
mgmod <- pmg(log(price) ~ log(income), data = HousePricesUS)
ccemgmod <- pmg(log(price) ~ log(income), data = HousePricesUS, model = "cmg")
pcdtest(resid(ccemgmod), w = usaw49)

Pesaran CD test for local cross-sectional dependence
in panels
```

```
data:  resid(ccemgmod)
z = 28, p-value <2e-16
alternative hypothesis: cross-sectional dependence
rwtest(resid(mgmod), w = usaw49, replications = 999)

Randomized W test for spatial correlation of order 1

data:  formula
p-value = 0.002
alternative hypothesis: twosided
```

Any way we look at it, substantial spatial dependence is still present in the model residuals even after controlling for cross-sectional common factors.

10.2 Spatial Lags

The basic tool of spatial econometrics is the definition of a *spatial lag*. Given an observation and a distance metric, the spatial lag of that observation is usually defined as some kind of weighted average of the observations that are considered "near" to it according to the given metric: $\sum_{m=1}^{N} w_{nm} z_{mt}$. Either a distance or a neighborhood matrix is commonly employed to provide the weights. In the neighborhood case, for each pair of observations n, m, the matrix will have an element $w_{n,m} = 1$ if the two are neighbors, i.e., if they share a common border like Germany and Austria (*first-order neighborhood*) or if there are at most p other observations separating them (*p-th order neighborhood*), so that Italy and Germany are second-order neighbors. In the distance-based case, the generic element will be dependent on some inverse function of the distance $d(n, m)$ between them, usually the reciprocal: $w_{n,m} = 1/d(n, m)$. It is customary to set a cutoff point at some distance $\bar{d}$ beyond which one does not expect any influence to be present so that $w_{n,m} = 0$ if $d(n, m) \geq \bar{d}$[4]. In both cases, it is customary to *standardize* W so that the rows sum to one: $\sum_{n=1}^{N} w_{n\ m} = 1\ \forall m$. Then, for each z_n, Wz_n will contain, respectively, the simple average of values in neighboring locations or a distance-weighted average of all z_m for which $w_{nm} \neq 0$.

In all of the following, we will refer to the simpler neighborhood-based definition of proximity. All techniques illustrated in this chapter are nevertheless applicable as well in the case of distance-based weights. The spatial weights matrix can be based on definitions of *distance* not based on geographical position but defined instead in some other kind of space, like e.g., one where dimensions are corresponding to some set of economic or demographic or psychological characteristics. The technical aspects of estimation do not vary with respect to the case of geographical distance, or neighborhood, as long as the fundamental hypothesis of exogeneity of W holds. One desirable feature of geographic space is that it is exogenous, unlike, e.g., bilateral (contemporaneous) trade-based weights in a model of international commerce, which would be generated inside the same economic system to be modeled.

It is important to recall that the hypothesis of exogenous and time-invariant W will be maintained throughout this chapter. Spatial lags in a panel setting can be written compactly in

4 The choice of the contiguity matrix is one of the most controversial subjects in spatial econometrics (see Anselin, 1988, p. 19). Binary contiguity has the advantage of simplicity, of imposing a minimum of a priori structure and of making the interpretation of a spatial lag straightforward as the average value of neighbors; therefore (Anselin, 1988, p. 21) it is often preferred for spatial error structures and in general where the focus is on testing for spatial effects rather than on precisely estimating a theoretically well-defined spatial process.

vector form stacking observations by time first, in the now-standard notation, as $(W \otimes I_N)z$. The concept of spatial lag has some analogies with the familiar time lag but also important differences, the most important one being that while time is directed, space is generally not; hence the idea of predeterminedness and the fact that usually (although not always) the past is expected to influence the future but not vice versa do not apply. Dependence in space is usually circular, and the influence from "nearby" observations gives rise to feedback effects that importantly affect estimation. In particular, as will be clear in the following, a spatial lag of the dependent variable is endogenous by construction, and a model including it will require more sophisticated techniques than (ordinary or generalized) least squares in order to be consistently estimated.

10.2.1 Spatially Lagged Regressors

Suppose that the need to account for space in the specification has been established either a priori, in the economic model, or because spatial dependence has been detected in the data or in the residuals of an estimated model.

One first way to consider the influence of neighboring spatial units is to take into account spatial lags of the explanatory variables. The economic meaning of spatially lagged regressors is to account for explicit spatial influences from relevant explanatory variables in nearby spatial units. Spatial lags WX can easily be added to the specification and, provided X was exogenous to begin with, pose no additional problem in estimation of this model.

As a first example of augmenting a model with a spatial lag, let us consider the case of a spatially lagged regressor representing (if W is row-standardized) *the average of X at neighboring locations.*

Example 10.3 Spatially lagged explanatory variables – `Cigarette` data set
Baltagi and Griffin (2001) consider demand for cigarettes across 46 US states over the years 1963-1992 in the framework of the rational addiction model. Next to the original dynamic model, static versions have become an ubiquitous example in papers and textbooks. The demand for cigarettes (`sales`) is estimated as a function of real per capita income (`ndi/cpi`), cigarette price in the given state (`price`), and minimum price in neighboring states (`pimin`), this last term accounting for cross-border smuggling. The model is then estimated by fixed effects through `plm` ; we use `coeftest` for a compact representation of the estimation output.

```
library("plm")
library("splm")
data("Cigar", package = "plm")
fm <- log(sales) ~ log(price) + log(pimin) + log(ndi / cpi)
femod <- plm(fm, Cigar)
library("lmtest")
coeftest(femod)

t test of coefficients:

             Estimate Std. Error t value Pr(>|t|)
log(price)    -0.7513     0.0462   -16.3   <2e-16 ***
log(pimin)     0.4946     0.0456    10.8   <2e-16 ***
log(ndi/cpi)   0.6801     0.0368    18.5   <2e-16 ***
```

```
---
Signif. codes:
0 '***' 0.001 '**' 0.01 '*' 0.05 '.' 0.1 ' ' 1
```

A natural application of the spatial lag operator in this context is to substitute `pimin` with an average of the prices in neighboring states, to account for the smuggling effect across all borders. The spatial lag operator, when applied to `price` using a binary contiguity, row-standardized matrix, produces exactly this average price. We read in the relevant W matrix, standardize it, and check that the row sums are actually all 1:

```
data("usaw46", package = "pder")
wcig <- usaw46 / apply(usaw46, 1, sum)
summary(apply(wcig, 1, sum))
   Min. 1st Qu.  Median    Mean 3rd Qu.    Max.
      1       1       1       1       1       1
```

In a cross-sectional setting, spatial lags are very easy to construct as WX. In a panel setting, every cross-section has to be premultiplied by W; or, equivalently, a larger block-diagonal neighborhood matrix $W_{NT} = \mathrm{I}_T \otimes W$ has to be employed. Let us construct a spatial (panel) lag of the price variable. Remembering that panel data are usually ordered by `state`, `year` with the first being the "slow" index, we can proceed by making a reordered copy of `Cigar`, extracting the variable `price` and lagging it through premultiplication by $\mathrm{I}_T \otimes W$; then adding it to the dataset:

```
cig <- Cigar[order(Cigar$year, Cigar$state),]
wp <- kronecker(diag(1, 30), wcig) %*% cig$price
Cigar$wp <- wp[order(cig$state, cig$year)]
```

or, much faster although less intuitive, by reversing the Kronecker product:

```
Cigar$wp <- kronecker(wcig, diag(1,30)) %*% Cigar$price
```

Now wp is a regular regressor, which we can add to the specification in lieu of `pimin`, redoing all the previous steps, estimating the alternative model to appreciate the difference:

```
fm2 <- update(fm, . ~. - log(pimin) + log(wp))
femod2 <- plm(fm2, Cigar)
coeftest(femod2)

t test of coefficients:

             Estimate Std. Error t value Pr(>|t|)
log(price)    -0.8292     0.0528   -15.7   <2e-16 ***
log(ndi/cpi)   0.6294     0.0371    17.0   <2e-16 ***
log(wp)        0.5874     0.0537    10.9   <2e-16 ***
---
Signif. codes:
0 '***' 0.001 '**' 0.01 '*' 0.05 '.' 0.1 ' ' 1
```

To automate the tedious construction of spatial panel lags, a function `slag` is provided, needing a `pseries` and either a proximity matrix or an equivalent `listw` object to represent the spatial ordering of observations. The `slag` operator can be employed directly in formulae:

```
lwcig <- mat2listw(wcig)
fm3 <- update(fm, . ~. - log(pimin) +
                    log(slag(price, listw=lwcig)))
```

The somewhat cumbersome syntax deriving from the need to specify W can be avoided, e.g., defining a small convenience function where the given W matrix is hardwired, as follows (the output is the same):

```
wx <- function(x) slag(x, listw = lwcig)
fm3.alt <- update(fm, . ~. - log(pimin) + log(wx(price)))
```

As it turns out, substituting the minimum price in neighboring states with the average price of neighbors has little effect on the model results.

10.2.2 Spatially Lagged Dependent Variables

A more direct, although much more problematic, way of incorporating spatial structure in an econometric model is through inclusion of spatial lags of the dependent variable. The model is then:

$$y = \lambda(I_T \otimes W_N)y + Z\gamma + \epsilon$$

where W_N is the $N \times N$ spatial weights matrix of known constants whose diagonal elements are set to zero, and λ is the corresponding spatial parameter.

This is called the *spatial lag* model proper. From a theoretical viewpoint, it is appropriate whenever one expects the *outcome* of one observation to influence the *outcomes* of neighboring ones, such as, e.g., for the spreading of a disease, where one unit being positive has a direct effect on the likelihood of neighboring units to be so too.

Another example is if (within-period) strategic interaction is expected to happen, e.g., each country takes the tax rates of neighbors into account in setting its own and may react within the same time period, as in Franzese and Hays (2006). In this case, one might expect positive spatial correlation. In a microeconomic setting, the effect of a spatial lag term could be expected to turn out positive is in copycatting behavior, when e.g., buying a product sparks imitation hereby raising the propensity of neighbors to follow suit. A negative spatial lag can instead be consistent with the idea of free riding: if one can reap advantage from the actions of neighbors through some kind of externality, then this will lower his or her own effort: an example is labor market training in the European Union, where trained labor can easily commute across borders (Franzese and Hays, 2008).

Spatial-lag-type dependence has been evocatively termed “substantial” (Franzese and Hays, 2007) as opposed to spatial error dependence, which in the same context is described as “nuisance,” to be controlled for the sake of precision in estimation but devoid of theoretical meaning. This is not necessarily true, as spatial error dependence can have substantial meaning too, for example in the context of economic shock diffusion (see e.g. Holly et al., 2010), and can be a subject of the analysis in its own right.

The spatial lag process, and by extension the model with a spatial lag plus regressors, is universally known by the acronym **SAR**, for "spatially autoregressive." The Wy term is inherently endogenous; in a reduced form, the model becomes nonlinear:

$$y = (I_T \otimes \lambda W)^{-1}[Z\gamma + \epsilon]$$

so that maximum likelihood estimation (**ML**) is called for. Only as a very first approximation, it can be of interest to estimate the so-called "spatial **OLS**".

10.2.2.1 Spatial OLS

Ordinary least squares estimation is consistent, under the usual exogeneity conditions on X, for models with spatially lagged regressors, in which case it is also efficient provided that the standard hypotheses of homoscedasticity and incorrelation hold; in fact, adding WX may eliminate the spatial correlation in error terms and effectively make **OLS** the efficient estimator. Even in the case of the spatial error model, **OLS** remain consistent, although inefficient, for γ.

As a first approximation, and in cases where **ML** and **GM** are problematic (one for all, dynamic panels), the so-called *spatial-***OLS** method has been advocated: adding the spatial lag of the dependent variable Wy to the right-hand side regressors. This solution is in general not advisable because Wy is endogenous by construction, and therefore the estimator is hopelessly biased; yet simulation studies have shown how the magnitude of the bias can be limited in real-world cases, to the point of making this computationally simple solution relatively viable in some applied settings (see Franzese and Hays, 2007).

10.2.2.2 ML Estimation of the SAR Model

An appropriate way to estimate a **SAR** model, provided the errors ϵ are i.i.d. normal, is by **ML**. Let us start from the cross-sectional case where Z is $N \times (K+1)$ and y is a vector of length N. Denoting $A = I_N - \lambda W$, the model becomes $Ay = Z\gamma + \epsilon$ so that $\epsilon = Ay - Z\gamma$. Expressing the usual likelihood function of the linear model in terms of the transformed y requires adding the Jacobian of the transformation, i.e., the determinant of A, therefore the log-likelihood becomes:

$$\ln L = -\frac{N}{2}\ln(2\pi) + ln|A| - \frac{1}{2}\epsilon^\top \epsilon$$

and this likelihood is to be optimized with respect to γ and λ, efficient optimization strategies having been outlined in the seminal book of Anselin (1988). The pure-**SAR** panel case, pooling the data without any individual feature, just substitutes NT for N, $I_T \otimes W$ for W and $A = I_{NT} - I_T \otimes \lambda W$ so that it could be estimated with the `lagsarlm` function from package **spdep**. Nevertheless, it is always preferable for computational reasons to resort to specific methods for spatial panels when available.

Example 10.4 Spatial lag model – `HousePricesUS` data set

In the house prices application of Holly et al. (2010), the authors estimate a **SAR** model of defactored residuals to assess the presence and the degree of spatial correlation net of the influence of common factors. We replicate their analysis by estimating a pure-**SAR** model (no regressors but an intercept) of the residuals from the **CCEMG** model.[5] With respect to common-factor robust testing (see previous example), this has the additional advantage of explicitly estimating a spatially autoregressive coefficient measuring the intensity of the spatial effect.

5 The authors follow a slightly different procedure than we do here, explicitly estimating out common factors through a principal components procedure: see the R replication in Millo (2015). The end result is nevertheless remarkably similar.

The CCE residuals from pmg (or, equivalently, pcce) are a regular `pseries`; it is easy to make a dataframe in **plm**-compliant format by stacking the individual and time indexes next to the residuals themselves, stripped of their panel attributes through the `as.numeric.pseries` converter function. The function `spreml` can then be used for estimation, specifying `lag=TRUE` and `errors='ols'` for a pure SAR model without any panel features:

```
e <- resid(ccemgmod)
edat <- data.frame(ind = attr(e, "index")[[1]],
                   tind = attr(e, "index")[[2]],  e = as.numeric(e))
sarmod.e <- spreml(e ~ 1, data = edat, w = usaw49, lag = TRUE, errors = "ols")
summary(sarmod.e)$ARCoefTable
         Estimate Std. Error t-value   Pr(>|t|)
lambda    0.6498    0.02037    31.9 2.685e-223
```

The spatial correlation in the (defactored) residuals is estimated at 65% and is statistically significant at any confidence level. The simpler spatial-OLS model can be easily estimated with the help of the `slag` function:

```
library("lmtest")
coeftest(plm(e ~ slag(e, listw = usaw49) - 1, data = edat, model = "p"))

t test of coefficients:

                        Estimate Std. Error t value
slag(e, listw = usaw49)   0.8831     0.0251    35.2
                        Pr(>|t|)
slag(e, listw = usaw49)   <2e-16 ***
---
Signif. codes:
0 '***' 0.001 '**' 0.01 '*' 0.05 '.' 0.1 ' ' 1
```

The bias in the SAR coefficient is evident, yet this simple procedure can be enough for detecting a problem or as a very first assessment.

10.2.3 Spatially Correlated Errors

The other main specification in the literature, the *spatial error*, is instead appropriate when one expects the *innovation* relative to one observation to influence the *outcomes* of neighboring ones, as would be the case for an economic shock of some kind to a given region (fully) influencing the relevant dependent variable in that region and also propagating – with distance-decaying intensity – toward nearby ones; or for a location-related measurement error, by its nature affecting nearby observations in a similar way. Another reason for spatially correlated errors is misspecification resulting from the omission of a spatially correlated variable. This specification is called SEM, for "spatial error model".

The model is then the familiar linear model with regressors:

$$y = Z\gamma + \epsilon$$

where ϵ is a vector of spatially autocorrelated idiosyncratic errors that follows a spatial autoregressive process of the form

$$\epsilon = \rho W\epsilon + \nu$$

with ρ as the spatial autoregressive parameter, W_N the spatial weights matrix and $v \sim \text{IID}(0, \sigma_v^2)$. As can be seen, the **SEM** model is nothing but a linear model with a **SAR** process in the errors instead of in the response. The likelihood for the cross-sectional **SEM** model is:

$$\ln L = -\frac{N}{2}\ln(2\pi) + \ln \mid B \mid -\frac{1}{2}\epsilon^\top B^\top B\epsilon$$

where $B = \text{I}_N - \rho W$. As for the **SAR** case, pooling the data is accomplished by substituting NT to N, the extended proximity matrix $\text{I}_T \otimes W$ for W and $B = \text{I}_{NT} - \text{I}_T \otimes \rho W$.

It is typical in the literature to estimate either of the two specifications, **SAR** or **SEM**, although in principle they can be combined. The subject of choosing between the spatial lag and the spatial error models by means of diagnostic testing will be treated in the following; it should nevertheless be borne in mind that the specification of one or the other spatial model should always be informed by the a priori beliefs of the researcher and the economic model she postulates for the phenomenon at hand. In fact, while some empirical cases happen to be sufficiently clear-cut for an exclusively data-driven decision to be taken, most of the time model uncertainty – regarding the specification of regressors, of the neighborhood structure (the W matrix), or that of the spatial process in either response or error – is so pervasive that one can hardly rely on statistical procedures alone in order to conduct a specification search.

Nevertheless, from a diagnostic rather than modeling viewpoint, a general result is that the omission of a spatially correlated relevant regressor would show up as spatially correlated errors, and the same would happen for the omission of a spatially lagged dependent variable; much as would happen in time series data with omitted dynamics showing up in residual autocorrelation. Generality stops here, though, because while the symptoms of either neglected spatial lag or error processes are similar, the consequences on the properties of estimators are different already. In fact, an omitted spatial lag renders the estimator inconsistent, while an omitted spatial process in the error merely results in inefficiency and invalid inference.

Example 10.5 Spatial error – `RiceFarms` data set

The `RiceFarms` dataset contains observations from 171 rice farms in Indonesia, observed over six growing seasons, three wet and three dry, between 1975 and 1983. The farms are located in six different villages of the Chimanuk River basin in West Java. According to Druska and Horrace (2004), two villages are in flatlands on the north coast of the island, three in the highlands (600-1100 m) in the central part of West Java, and the last is in the center of the island with an average altitude of 375 meters. Roads and more in general proximity to big cities are extremely heterogeneous.

In this geographical setting, one can expect both village-level heterogeneity and spatial correlation between farms belonging to the same village. Spatial dependence is easier to justify for the error terms, due to spillovers across neighboring farms in idiosyncratic factors and climate conditions; more difficult to find reasons for the inclusion of a spatial lag of the dependent variable, as it seems less realistic for the outcome in one farm to influence those of neighbors.[6]

With respect to the original analysis, our production frontier equation will relate rice output to three inputs only: `seed`, labor hours `totlabor`, and land `size`, all in logs.

A contiguity matrix `riceww` is provided, where for each farm all other farms from the same village are defined as neighbors. The **SEM** panel model is then explicitly augmented with village fixed effects and time fixed effects to account for the influence of the different growing seasons. It is estimated through the `spreml` function, setting the `lag` to `FALSE` and the `errors` to `'sem'`:

6 A complete specification analysis of the original model is to be found in Millo (2014).

```
data("RiceFarms", package = "splm")
data("riceww", package = "splm")
library("spdep")
ricelw <- mat2listw(riceww)
Rice <- pdata.frame(RiceFarms, index = "id")
```

```
riceprod <- log(goutput) ~ log(seed) + log(totlabor) +
    log(size) + region + time

rice.sem <- spreml(riceprod, data = Rice, w = riceww,
                   lag = FALSE, errors = "sem")

summary(rice.sem)
ML panel with, spatial error correlation

Call:
spreml(formula = riceprod, data = Rice, w = riceww, lag = FALSE,
    errors = "sem")

Residuals:
    Min.   1st Qu.    Median   3rd Qu.      Max.
-1.06858 -0.23300  0.00581  0.23481  1.48962

Error variance parameters:
    Estimate Std. Error t-value Pr(>|t|)
rho   0.5627     0.0518    10.9   <2e-16 ***

Coefficients:
                  Estimate Std. Error t-value Pr(>|t|)
(Intercept)        5.85413    0.19469   30.07  < 2e-16 ***
log(seed)          0.16626    0.02475    6.72  1.8e-11 ***
log(totlabor)      0.24822    0.02758    9.00  < 2e-16 ***
log(size)          0.59776    0.02800   21.35  < 2e-16 ***
regionlangan      -0.09779    0.09137   -1.07    0.285
regiongunungwangi -0.14048    0.08422   -1.67    0.095.
regionmalausma    -0.11865    0.08650   -1.37    0.170
regionsukaambit    0.00723    0.09372    0.08    0.938
regionciwangi     -0.01381    0.08465   -0.16    0.870
time2             -0.04745    0.07885   -0.60    0.547
time3             -0.18551    0.07886   -2.35    0.019 *
time4             -0.34722    0.07883   -4.40  1.1e-05 ***
time5              0.15818    0.07886    2.01    0.045 *
time6              0.13805    0.07881    1.75    0.080.
---
Signif. codes:
0 '***' 0.001 '**' 0.01 '*' 0.05 '.' 0.1 ' ' 1
```

Somewhat surprisingly, the village fixed effects show up as all but unimportant. On the converse, spatial error correlation between farms belonging to the same village (estimated by the ρ coefficient) is substantial and highly significant.

10.3 Individual Heterogeneity in Spatial Panels

Cross-sectional spatial specifications are readily extended to the case of a pooled panel dataset, as above, but in the case of spatial panels, just as in the general case, it becomes of primary interest to model heterogeneity and persistence at the individual level. Again, the most popular device is the inclusion of individual, time-invariant effects in the model, and again the crucial distinction is whether said effects can be assumed independent from the model regressors or not. From a statistical viewpoint, the approach detailed in the previous chapters when speaking of non-spatial panels is still valid, but there are also specific considerations to be made for spatial applications. For example, as the random effects hypothesis is considered consistent with sampling individuals from a potentially infinite population, some (Elhorst and Fréret (2009) for example) have dismissed its plausibility in spatial econometric contexts, where sampling most typically takes place over a fixed set of countries or regions.

Spatial methods are nevertheless of interest also in contexts much akin to random sampling. For one, applications on survey data can be devised where individual units are located into some non-geographic space, defined by their attributes and a distance function. Among the geographically referenced data proper, the same random samples of firms or households can be located and recorded as points in the landscape (Bell and Bockstael, 2000). In this sense, the `RiceFarms` dataset is a good candidate for random effects: many locations with similar characteristics, plausibly drawn from the same distribution, although lacking latitude and longitude information, are grouped in a way that naturally defines a neighborhood. Another case, where this time data are located as points in geographical space, are the ever more popular spatial applications from experimental contexts in life sciences, of which we will see an example later in the chapter.

Moreover, from a computational viewpoint random effects turn out to be a more general case with respect to fixed effects.

10.3.1 Random versus Fixed Effects

As detailed in the previous chapters and recalled above, unobserved individual heterogeneity is dealt with in different ways depending on the statistical properties of the individual effects, the crucial distinction becoming whether one can assume them to be uncorrelated with the regressors or not. If uncorrelated, then individual effects can be considered as a component of the error term. If not, then the latter strategy leads to inconsistency; the individual effects will have to be estimated or, more frequently, eliminated by first differencing or time-demeaning the data. In the spatial setting, the standard solution to the fixed effects case has long been time-demeaning: in the framework of Elhorst (2003), fixed effects estimation of spatial panel models is accomplished as pooled ML estimation on time-demeaned data. Nevertheless, Elhorst's procedure has been questioned by Anselin et al. (2008) because time-demeaning alters the properties of the joint distribution of errors, introducing serial dependence. As it turns out, despite the misspecification of the likelihood, the only parameter affected is the variance of the error term, the other estimators remaining consistent.[7]

To solve the problem, Lee and Yu (2010a, 3.2) suggest either a different orthonormal transformation of the data, or an ex-post correction of the estimated variance (see also Lee and Yu, 2012). For all this, ML estimation of spatial panel models with individual fixed effects is encompassed by the ML estimator for the pooled case, after a suitable transformation of the data and,

7 See Lee and Yu (2010b, p. 257) for a discussion of the issue, and Millo and Piras (2012, p. 33) for an evaluation if its practical significance through Monte Carlo simulation.

in the case one uses the simpler within transformation, an appropriate ex-post correction of the error variance estimate.

Example 10.6 Spatial fixed effects – `RiceFarms` data set
Spatial fixed effects panels can be estimated through the general wrapper function for maximum likelihood estimation, `spml` for "spatial panel by maximum likelihood", leaving the `model` argument at the default value of `'within'` (in the case `model` is either of `'random'` or `'pooling'`, the lower level function `spreml` seen in previous examples is called; while here a special infrastructure is used). The spatial structure in the error can be `'none'` or either of `'b'` or `'kkp'`, which makes a difference only in the random effects case. A spatial lag can be included setting `lag` to `TRUE`. Village fixed effects must be omitted here because of collinearity, while time ones can be implicitly added to estimation by specifying `effect='twoways'`, again consistently with the syntax of `plm`.

```
riceprod0 <- update(riceprod, . ~. - region - time)
semfemod <- spml(riceprod0, Rice, listw = ricelw,
                 lag = FALSE, spatial.error = "b")
summary(semfemod)
Spatial panel fixed effects error model


Call:
spml(formula = riceprod0, data = Rice, listw = ricelw, lag = FALSE,
    spatial.error = "b")

Residuals:
   Min. 1st Qu.  Median 3rd Qu.    Max.
-1.0195 -0.2105  0.0222  0.2127  1.3298

Spatial error parameter:
    Estimate Std. Error t-value Pr(>|t|)
rho   0.7913     0.0249    31.8   <2e-16 ***

Coefficients:
              Estimate Std. Error t-value Pr(>|t|)
log(seed)       0.1342     0.0226    5.94  2.8e-09 ***
log(totlabor)   0.2505     0.0267    9.38  < 2e-16 ***
log(size)       0.5419     0.0273   19.84  < 2e-16 ***
---
Signif. codes:
0 '***' 0.001 '**' 0.01 '*' 0.05 '.' 0.1 ' ' 1
```

A Hausman-type test will determine whether the individual effects are to be treated as fixed or can be assumed incorrelated with the regressors, employing a more efficient random effects specification:

```
Rice <- pdata.frame(RiceFarms, index = "id")
sphtest(riceprod0, Rice, listw = ricelw)

Hausman test for spatial models
```

```
data:  x
chisq = 2.6, df = 3, p-value = 0.4
alternative hypothesis: one model is inconsistent
```

The random effects hypothesis being not rejected, random effects methods are in order.

10.3.2 Spatial Panel Models with Error Components

While fixed effects estimation of spatial panels can be performed in the framework of the pooled spatial models, after transforming out the individual effects by a *within* transformation, treating the individual effects as random introduces substantial complications in the specification of the likelihood.

We consider a general static panel model that includes a spatial lag of the dependent variable and spatial autoregressive disturbances:

$$y = \lambda(I_T \otimes W_N)y + Z\gamma + \epsilon$$

The disturbance vector is the sum of two terms:

$$\epsilon = (j_T \otimes I_N)\eta + \nu$$

η being the individual effect and ν a vector of spatially autocorrelated idiosyncratic errors that follow a spatial autoregressive process of the form

$$\nu = \rho(I_T \otimes W_N)\nu + \zeta$$

with ρ as the spatial autoregressive parameter, W_N the spatial weights matrix and $\zeta \sim \text{IID}(0, \sigma^2_\zeta)$. The spatial weights matrices in the lag and the error term can differ (see the following). $I_N - \rho W_N$ is assumed non-singular.

10.3.2.1 Spatial Panels with Independent Random Effects

In a random effects specification, the unobserved individual effects are assumed uncorrelated with the other explanatory variables in the model and can therefore be safely treated as components of the error term.[8] In this case, $\eta \sim \text{IID}(0, \sigma^2_\eta)$, and the error term can be rewritten as:

$$\nu = (I_T \otimes B_N^{-1})\zeta$$

where $B_N = (I_N - \rho W_N)$. As a consequence, the composite error term becomes

$$\epsilon = (j_T \otimes I_N)\eta + (I_T \otimes B_N^{-1})\zeta$$

and its variance-covariance matrix is:

$$\Omega_\epsilon = \sigma^2_\eta (J_T \otimes I_N) + \sigma^2_\zeta [I_T \otimes (B_N^\top B_N)^{-1}]. \tag{10.4}$$

In deriving several Lagrange multiplier (LM) tests, Baltagi et al. (2003b) consider a panel data regression model that is a special case of the model presented above in that it does not include a spatial lag of the dependent variable. Elhorst (2003), Elhorst and Fréret (2009) define a taxonomy for spatial panel data models both under the fixed and the random effects assumptions. Following the typical distinction made in cross-sectional models, they define the fixed as well as the random effects panel data versions of the spatial error and spatial lag models. However, unlike Case (1991), they do not consider a model including both the spatial lag of the dependent variable and a spatially autocorrelated error term. Therefore, the models reviewed

8 See, e.g., Assumption RE.1.b in Wooldridge (2010, 10.4.1).

in Elhorst (2003), Elhorst and Fréret (2009) can also be seen as special cases of this more general specification.

Following the treatment in Millo (2014), on which this part of the chapter is based, we label the combined model containing both a spatial lag and a spatial error process **SAREM**. (This is also often called **SARAR**, because of the two spatial autoregressive processes, one in the response and one in the errors.) If a random individual effect is also part of the composite error term, then we will add the suffix **RE**. Although **SAR** and **SEM**, combined with either **FE** or **RE**, are by far the most popular specifications, the literature has also dealt with different types of spatial diffusion processes in the errors other than the autoregressive one, most notably the spatial moving average.[9] We do not consider them here.

10.3.2.2 Spatially Correlated Random Effects

A different specification for the disturbances was considered in Kapoor et al. (2007). They assume that spatial correlation applies to both the individual effects and the remainder error components. Although the two data-generating processes look similar, they do imply different spatial spillover mechanisms governed by a different structure of the implied variance-covariance matrix. In this case, commonly referred to as **KKP**, the composite disturbance term

$$\epsilon = (j_T \otimes \mathrm{I}_N)\eta + \nu$$

follows a first-order spatial autoregressive process of the form:

$$\epsilon = \rho(\mathrm{I}_T \otimes W_N)\epsilon + \zeta$$

It follows that the variance-covariance matrix of ϵ is:

$$\Omega_\epsilon = (\mathrm{I}_T \otimes B_N^{-1})\Omega_\nu(\mathrm{I}_T \otimes ({B_N}^\top)^{-1}) \tag{10.5}$$

where $\Omega_\nu = [\sigma_\zeta^2 \mathrm{I}_T + \sigma_\eta^2 J_T] \otimes \mathrm{I}_N$ is the typical variance-covariance matrix of a one-way error component model. The variance matrix in (10.5) is simpler than the one in (10.4), and therefore its inverse is easier to calculate, as will be discussed below. As Baltagi et al. (2013) observe, the economic meaning of the two models is also different: in the first model only the time-varying components diffuse spatially; in the second, spatial spillovers too have a permanent component. Lee and Yu (2012, 2.4) illustrate the difference between this latter specification and **SEMRE** through the likelihood of the *between* model. We label this latter alternative specification **SEM2RE**, and its extension to including a spatial lag (see Mutl and Pfaffermayr, 2011) **SAREM2RE**.

10.3.3 Estimation

To review the theory of maximum likelihood estimation of spatial panel models with random effects, we will start from models with a spatially lagged dependent variable, spatial error correlation, and a general covariance structure for the error, as described by Anselin (1988), without any panel structure (although it must be noted that in his book Anselin (1988) already considered a **SEM** panel with random effects, deriving the model likelihood, as a special case). Following Millo (2014), we will introduce random effects as just one particular type of error covariance structure, thus comprising spatial panels in Anselin's general framework.[10]

---

9 The spatial moving average process is $\nu = \zeta + \rho(\mathrm{I}_T \otimes W_N)\zeta$, see e.g.Fingleton (2008)

10 This approach, with respect to the more common Elhorst (2003) one, based on the combination of partial demeaning with a spectral decomposition of the error covariance matrix, lends itself more easily to generalization and in particular facilitates considering random effects and serial correlation together with the spatial effects.

10.3.3.1 Spatial Models with a General Error Covariance

Maximum Likelihood estimation with a general error covariance matrix has been outlined in Magnus (1978) (see also Anselin et al., 2008). If the error ϵ is distributed as $N(0,\Omega)$ then the log-likelihood is

$$\ln L = -\frac{N}{2}\ln 2\pi - \frac{1}{2}\ln|\Omega| - \frac{1}{2}\epsilon^{\top}\Omega^{-1}\epsilon.$$

Particularizing this likelihood w.r.t. the case at hand, and adding a *spatial filter* if needed, provides a general framework for ML estimation of the models of interest. Anselin (1988), the classic reference on spatial econometric model estimation by ML, outlines the general procedure for a model with spatial lag, spatial errors, and possibly nonspherical residuals as follows. Let us restrict the analysis, for the moment, to one cross-section and let our model be:

$$\begin{aligned} y &= \lambda W_1 y + Z\gamma + \epsilon \\ \epsilon &= \rho W_2 \epsilon + \zeta \end{aligned} \tag{10.6}$$

with $\zeta \sim N(0,\Omega)$ and, in general, $\Omega \neq \sigma^2 \mathrm{I}$. Two special cases of this general model are often found in applied literature: if $\rho = 0$ one has the spatial autoregressive (SAR) model, while if $\lambda = 0$, the spatial (autoregressive) error (SEM) model. Both usually include the hypothesis of spherical remainder errors: $\Omega = \sigma^2\mathrm{I}$. Introducing the now-standard simplifying notation $A = \mathrm{I} - \lambda W_1$, $B = \mathrm{I} - \rho W_2$ the model becomes:

$$\begin{aligned} Ay &= Z\gamma + \epsilon \\ B\epsilon &= \zeta \end{aligned}$$

where W_1, W_2 are potentially different spatial weights matrices.[11] If there exists Ω such that $\xi = \Omega^{-\frac{1}{2}}\zeta$ and $\xi \sim N(0,\sigma_\xi^2\mathrm{I})$, and B is invertible, then $\epsilon = B^{-1}\Omega^{\frac{1}{2}}\xi$ and the model (10.6) can be written as

$$Ay = Z\gamma + B^{-1}\Omega^{\frac{1}{2}}\xi$$

or, equivalently,

$$\Omega^{-\frac{1}{2}}B(Ay - Z\gamma) = \xi$$

with ξ a "well-behaved" error.

Still following Anselin, making the estimator operational requires the transformation from the unobservable ξ to observables. Expressing the likelihood function in terms of y requires calculating the Jacobian of the transformation $J = \det\left(\frac{\partial\xi}{\partial y}\right) = |\Omega^{-\frac{1}{2}}BA| = |\Omega^{-\frac{1}{2}}||B||A|$. These determinants are to be added to the log-likelihood, which becomes

$$\ln L = -\frac{N}{2}\ln(2\pi) - \frac{1}{2}\ln|\Omega| + \ln|B| + \ln|A| - \frac{1}{2}\xi^{\top}\xi$$

where the difference w.r.t. the usual likelihood of the classic linear model is given by the terms of the Jacobian.[12] The likelihood is thus a function of γ, λ, ρ, and parameters in Ω.

It will be convenient for our purposes, and without loss of generality, to scale the overall errors' covariance writing it as $B^{\top}\Omega B = \sigma_\zeta^2\Sigma$ (the latter expression is in fact more general, as it does not constrain the heteroscedastic error term ζ to be spatially lagged, through premultiplication by B, in its entirety. In our case, only the error covariance of the SEM2

11 The above notation expressing a spatial lag model as $Ay = Z\gamma + \epsilon$ or, equivalently provided A is invertible, $y = A^{-1}(Z\gamma + \epsilon)$ is well known in the literature as "spatial filtering" representation.

12 The Jacobian is simply $J = 1$ in the classical case, see Greene (2003), B.41).

specification can be separated into a heteroscedastic error term and a spatial filter and therefore straightforwardly written as $B^{\top}\Omega B$, while the more common SEM specification cannot). This likelihood can be concentrated w.r.t.γ and the error variance σ_{ζ}^2, by substituting $\zeta = (\hat{\sigma}_{\zeta}^2\Sigma)^{-\frac{1}{2}}(Ay - Z\hat{\gamma})$

$$\ln L = -\frac{N}{2}\ln(2\pi\hat{\sigma_{\zeta}^2}) - \frac{1}{2}\ln|\Sigma| + \ln|A| - \frac{1}{2\hat{\sigma_{\zeta}^2}}(Ay - Z\hat{\gamma})'\Sigma^{-1}(Ay - Z\hat{\gamma}) \tag{10.7}$$

and a closed-form GLS solution for $\hat{\gamma}$ and $\hat{\sigma}_{\zeta}^2$ is available for any given set of spatial and other covariance parameters

$$\begin{cases} \hat{\gamma} = (Z^{\top}\Sigma^{-1}Z)^{-1}Z^{\top}\Sigma^{-1}Ay \\ \hat{\sigma}_{\zeta}^2 = \dfrac{(Ay - Z\hat{\gamma})'\Sigma^{-1}(Ay - Z\hat{\gamma})}{N} \end{cases} \tag{10.8}$$

so that a two-step procedure is possible that alternates optimization of the concentrated likelihood and GLS estimation. From here on, we explicitly consider the (balanced) panel structure of the data: N individuals observed over T time periods.

10.3.3.2 General Maximum Likelihood Framework

Building on the framework from Anselin (1988) outlined above, explicitly particularizing and operationalizing it with respect to a number of possible error covariance structures, all specifications outlined above can be estimated without the need to pre-transform the data as has been customary in the literature since Elhorst (2003). Random effects will instead be considered as one feature of the errors' covariance, just like spatial (or, later on in the chapter, serial) correlation (see Millo, 2014). Considering the spatial dependence features together with all the other sources of heteroscedasticity and correlation instead of separating it clearly, as done in the original Anselin framework, has the advantage of keeping some components of the error term (most notably, the random effects) out of the spatial dependence, which can remain a feature of the idiosyncratic error only, in accordance with most applications in the literature; but also some clear computational disadvantages, as will be discussed below. We will also consider the alternative specification where the individual effects are lagged together with the idiosyncratic errors, as in Kapoor et al. (2007), which one can straightforwardly express in terms of Anselin's original expression $\mathrm{E}(\epsilon\epsilon^{\top}) = B^{\top}\Omega B$, also extending the structure of Ω to include serial correlation. This latter will turn out to be easier to compute, especially on large examples.

First we will discuss the combination of a spatial lag with any error covariance structure; then we will review the most significant among the latter; lastly we will give an example of operationalization through the use of analytical expressions for the inverse and determinant of the error covariance matrix Σ.

Optimization will generally be subject to box constraints according to the following rules: the spatial lag and spatial errors coefficients λ and ρ will be bounded between $1/\omega_{\min}$ and 1, where $\omega_{\min}$ is the smallest characteristic root of W;[13] the serial correlation coefficient will be constrained to the usual stationarity condition $|\psi| < 1$ and the variance ratio of the random effects ϕ to be non-negative.

Spatial Lag Although both the SAR and the SEM specifications are popular in the literature, estimation generally focuses on one effect only, and there are few applications allowing for both of them to be present in the estimated model, one notable exception being the pioneering work

---

13 The standard conditions are reviewed in Elhorst (2008, Footnote 1).

of Case (1991). It is nevertheless straightforward, at least as far as expressing the likelihood is concerned, to combine a spatial lag with any error structure, including spatial dependence ones.

The general likelihood for the spatial lag panel model combined with any error covariance structure Σ is a panel version of (10.7):

$$\ln L = -\frac{NT}{2}\ln(2\pi\sigma_\zeta^2) - \frac{1}{2}\ln|\Sigma| + T\ln|A| - \frac{1}{2\sigma_\zeta^2}[(\mathrm{I}_T \otimes A)y - Z\gamma]'\Sigma^{-1}[(\mathrm{I}_T \otimes A)y - Z\gamma] \tag{10.9}$$

The usual iterative procedure a la Oberhofer and Kmenta (1974) can be employed to obtain the maximum likelihood estimates. Starting from an initial value for the spatial lag parameter λ and the error covariance parameters, we obtain estimates for γ and σ_ζ^2 from the first-order conditions:

$$\begin{cases} \hat{\gamma} = (Z^\top\Sigma^{-1}Z)^{-1}Z^\top\Sigma^{-1}(\mathrm{I}_T \otimes A)y \\ \hat{\sigma_\zeta^2} = [(\mathrm{I}_T \otimes A)y - Z\gamma]^\top\Sigma^{-1}[(\mathrm{I}_T \otimes A)y - Z\gamma]/NT \end{cases} \tag{10.10}$$

The likelihood can be concentrated and maximized with respect to the parameters in A and Σ. The estimated values thereof are then used to update the expression for Σ^{-1}. These steps are then repeated until convergence. In other words, for a specific Σ the estimation can be operationalized by a two-step iterative procedure that alternates between GLS (for γ and σ_ζ^2) and concentrated likelihood (for the remaining parameters) until convergence.

This general scheme can be applied to the random effects case, where it provides a simple and effective equivalent to the usual partial time-demeaning procedure, as well as to all the more complicated error covariance specifications discussed in the following.

For example, the spatial autoregressive model with random effects SARRE can be written as a combination of spatial filtering on the regressand and a random effects structure in the errors:

$$(\mathrm{I}_T \otimes A)y = Z\gamma + \epsilon$$
$$\epsilon = (j_T \otimes \eta) + \nu$$

hence it can be estimated by "plugging into" the general likelihood (10.9) the particular scaled error covariance $\Sigma_{\mathrm{RE}} = \phi(J_T \otimes \mathrm{I}_N) + \mathrm{I}_{NT}$ characterized by one parameter: $\phi = \sigma_\eta^2/\sigma_\nu^2$, the ratio of the variance of the individual effect over that of the idiosyncratic error.

Example 10.7 Spatial lag and RE – `RiceFarms` data set

In the following, a SARRE version of the `RiceFarms` model is estimated, for the sake of illustration and comparison. This specification has little economic underpinning, there not being many reasons why the output of one farm should depend on the output of neighboring ones (here, firms from the same village).

```
sarremod.ml <- spml(riceprod0, Rice, listw = ricelw,
                    model = "random", lag = TRUE, spatial.error = "none")
summary(sarremod.ml)
ML panel with spatial lag, random effects

Call:
spreml(formula = formula, data = data, index = index, w = listw2mat(listw),
    w2 = listw2mat(listw2), lag = lag, errors = errors, cl = cl)
```

```
Residuals:
   Min. 1st Qu.  Median    Mean 3rd Qu.    Max.
   1.59    2.52    2.81    2.78    3.05    4.02

Error variance parameters:
    Estimate Std. Error t-value Pr(>|t|)
phi   0.3690     0.0701    5.27  1.4e-07 ***

Spatial autoregressive coefficient:
       Estimate Std. Error t-value Pr(>|t|)
lambda   0.4132     0.0268    15.4   <2e-16 ***

Coefficients:
              Estimate Std. Error t-value Pr(>|t|)
(Intercept)     2.7731     0.1834   15.12  < 2e-16 ***
log(seed)       0.1415     0.0253    5.58  2.4e-08 ***
log(totlabor)   0.2740     0.0280    9.78  < 2e-16 ***
log(size)       0.5231     0.0295   17.72  < 2e-16 ***
---
Signif. codes:
0 '***' 0.001 '**' 0.01 '*' 0.05 '.' 0.1 ' ' 1
```

Nevertheless, the **SAR** parameter λ turns out significant and relatively large in magnitude. This will prove to be a feature of model specification, more precisely of neglecting the "true" source of spatial dependence: the **SEM** term. More on this in the next examples. Individual effects are in turn detected, witness the significant variance ratio parameter ϕ, although $\hat{\sigma}^2_\eta$ is estimated at little over one third of $\hat{\sigma}^2_\nu$.

Error Structures As already discussed, the spatial error, random effects model gives rise to two possible specifications, depending on the interaction between the spatial autoregressive effect and the individual error components: the **SEMRE** specification first analyzed by Anselin (1988) where only the idiosyncratic error is spatially correlated:

$$
\begin{aligned}
y &= Z\gamma + \epsilon \\
\epsilon &= (j_T \otimes \eta) + \nu \\
\nu &= \rho(\mathrm{I}_T \otimes W)\nu + \zeta
\end{aligned}
$$

with the scaled errors' covariance (denoting $\bar{J}_T = J_T/T$ and $\bar{\mathrm{I}}_T = \mathrm{I}_T - \bar{J}_T$):

$$\Sigma_{\mathbf{SEMRE}} = \bar{J}_T \otimes (T\phi \mathrm{I}_N + (B^\top B)^{-1}) + \bar{\mathrm{I}}_T \otimes (B^\top B)^{-1}$$

and that of Kapoor et al. (2007) where the same spatial process applies both to the individual and the idiosyncratic error component:

$$
\begin{aligned}
y &= Z\gamma + \epsilon \\
\epsilon &= (j_T \otimes \eta) + \nu \\
\epsilon &= \rho(\mathrm{I}_T \otimes W)\epsilon + \zeta
\end{aligned}
$$

where the scaled errors' covariance is:

$$\Sigma_{\mathbf{SEM2RE}} = (\phi J_T + \mathrm{I}_T) \otimes (B^\top B)^{-1}.$$

Example 10.8 Baltagi or KKP random effects SEM – `RiceFarms` data set
In the following, both the SEMRE and SEM2RE models are estimated, again, on the `RiceFarms` dataset.

```
semremod.ml <- spml(riceprod0, Rice, listw = ricelw,
                    model = "random", lag = FALSE, spatial.error = "b")
summary(semremod.ml)
ML panel with, random effects, spatial error correlation

Call:
spreml(formula = formula, data = data, index = index, w = listw2mat(listw),
    w2 = listw2mat(listw2), lag = lag, errors = errors, cl = cl)

Residuals:
   Min. 1st Qu.  Median 3rd Qu.    Max.
-1.1858 -0.2563  0.0119  0.2476  1.3683

Error variance parameters:
    Estimate Std. Error t-value Pr(>|t|)
phi   0.2955     0.0565    5.23  1.7e-07 ***
rho   0.7748     0.0271   28.57  < 2e-16 ***

Coefficients:
              Estimate Std. Error t-value Pr(>|t|)
(Intercept)     5.6983     0.1797   31.70  < 2e-16 ***
log(seed)       0.1520     0.0235    6.47  9.8e-11 ***
log(totlabor)   0.2562     0.0272    9.42  < 2e-16 ***
log(size)       0.5757     0.0275   20.96  < 2e-16 ***
---
Signif. codes:
0 '***' 0.001 '**' 0.01 '*' 0.05 '.' 0.1 ' ' 1
sem2remod.ml <- spml(riceprod0, Rice, listw = ricelw,
                    model = "random", lag = FALSE, spatial.error = "kkp")
summary(sem2remod.ml)
ML panel with, spatial RE (KKP), spatial error correlation

Call:
spreml(formula = formula, data = data, index = index, w = listw2mat(listw),
    w2 = listw2mat(listw2), lag = lag, errors = errors, cl = cl)

Residuals:
   Min. 1st Qu.  Median 3rd Qu.    Max.
-1.1855 -0.2563  0.0119  0.2478  1.3703

Error variance parameters:
    Estimate Std. Error t-value Pr(>|t|)
phi   0.2959     0.0569     5.2    2e-07 ***
rho   0.7686     0.0277    27.8   <2e-16 ***
```

```
Coefficients:
                 Estimate Std. Error t-value Pr(>|t|)
(Intercept)       5.6986     0.1864   30.57  < 2e-16 ***
log(seed)         0.1518     0.0236    6.44  1.2e-10 ***
log(totlabor)     0.2564     0.0273    9.41  < 2e-16 ***
log(size)         0.5763     0.0275   20.94  < 2e-16 ***
---
Signif. codes:
0 '***' 0.001 '**' 0.01 '*' 0.05 '.' 0.1 ' ' 1
```

The differences are minimal. Random effects are significant, albeit weak in magnitude; while in accordance with the original work of Druska and Horrace (2004), very strong spatial error correlation is detected. The limited importance of the RE component makes the distinction between the two specifications scarcely relevant.

10.3.3.3 Generalized Moments Estimation

The computational intensity of ML estimation, which in the simpler models is related mostly to the need to recompute the determinants at each optimization step, has long been a limiting factor in practical applications. Samples of cross-sectional size in the hundreds were the practical maximum for the simple SAR or SEM models at the end of the 20th century, both because of the difficulty in obtaining a result at all and of the numerical unreliability of the latter if any because of precision problems (Kelejian and Prucha, 1999, Bell and Bockstael, 2000). Today, much more powerful computers have extended the scope of ML methods, but on the other hand the increasing availability of GIS data has brought forward a new generation of estimation problems of ever increasing size (an early survey and examples in Bell and Bockstael, 2000).

This has prompted researchers to explore alternative estimation strategies. Kelejian and Prucha (1999) proposed the generalized moments (GM) method, which, despite being asymptotically equivalent to ML under normality of the errors, is consistent irrespective of the latter; computationally, moreover, it does not require the numerically cumbersome calculation of the determinants.

The GM estimator for the cross-sectional SEM model (see also Bell and Bockstael, 2000) is based on the following three moments of the error term:

$$E\begin{bmatrix} \frac{1}{N}\epsilon^\top \epsilon \\ \frac{1}{N}\epsilon^\top W^\top W\epsilon \\ \frac{1}{N}\epsilon^\top W\epsilon \end{bmatrix} = \begin{bmatrix} \sigma^2 \\ \sigma^2 \frac{1}{N}\mathrm{tr}(W^\top W) \\ 0 \end{bmatrix} \tag{10.11}$$

The estimation strategy is based on the idea of estimating the spatial autoregressive coefficient ρ based on the residuals from a consistent estimator (here, OLS) and then using it in a feasible GLS analysis. With respect to maximum likelihood, the GM estimator has the additional advantage of not relying on a normality assumption for the errors. One drawback is that standard errors are not available for the ρ parameter.

The Kelejian and Prucha (1999) GM estimator has first been extended to the panel case by Druska and Horrace (2004), then by Kapoor et al. (2007) who estimated the above described SEM2 model with RE, a specification which, after them, is known as KKP. In order to perform feasible GLS, one does now need consistent estimates of the spatial autoregressive parameter ρ

and the two variance components of the composite error, σ^2_η and σ^2_ν. The **SEM2REGM** estimator a la **KKP** estimates them based on six moment conditions, using the **OLS** residuals $\hat{\epsilon}$, which are still consistent in this setting:

$$E\begin{bmatrix} \frac{1}{N(T-1)}\nu^\top Q_0 \nu \\ \frac{1}{N(T-1)}\bar{\nu}^\top Q_0 \bar{\nu} \\ \frac{1}{N(T-1)}\bar{\nu}^\top Q_0 \nu \\ \frac{1}{N}\nu^\top Q_1 \nu \\ \frac{1}{N}\bar{\nu}^\top Q_1 \bar{\nu} \\ \frac{1}{N}\bar{\nu}^\top Q_1 \nu \end{bmatrix} = \begin{bmatrix} \sigma^2_\nu \\ \sigma^2_\nu \frac{1}{N}\mathrm{tr}(W^\top W) \\ 0 \\ \sigma^2_\eta \\ \sigma^2_\eta \frac{1}{N}\mathrm{tr}(W^\top W) \\ 0 \end{bmatrix} \qquad (10.12)$$

where $\nu = \epsilon - \rho\bar{\epsilon}$, $\bar{\nu} = \bar{\epsilon} - \rho\bar{\bar{\epsilon}}$, $\bar{\epsilon} = (\mathrm{I}_T \otimes W_N)\epsilon$, and $\bar{\bar{\epsilon}} = (\mathrm{I}_T \otimes W_N)\bar{\epsilon}$; and $Q_0 = \mathrm{I}_T - \frac{J_T}{T} \otimes \mathrm{I}_N$ and $Q_1 = \frac{J_T}{T} \otimes \mathrm{I}_N$ are, respectively, a time-demeaning and a time-averaging matrix.

The moment conditions are now redundant and can be employed in different ways. The simplest is to consider only the first three moment conditions. The second way is to employ all six moments in estimating the three unknown parameters, weighing them through a covariance matrix calculated under the assumption of normally distributed errors. The third and last proceeds like the second, using all available moments but employs a simplified weighting matrix.

GM methods have been extended to the other relevant specifications in spatial econometrics. Spatial fixed effects models can also be estimated in this framework, through a modification of the **KKP** procedure suggested by Mutl and Pfaffermayr (2011) and consisting in replacing the **OLS** residuals, inconsistent under the fixed effects assumption, with spatial **2SLS** *within* residuals; the spatial parameter ρ is estimated by an adaptation of the simplified KKP procedure (first three moment conditions only) and used in a spatial Cochrane-Orcutt transformation of the *within*-transformed variables. The **GM** method has also been extended to the **SAR** and **SAREM** models, so that now any combination of spatial lag and error, with individual effects of either random or fixed type, can be estimated through this numerically very efficient method (see Millo and Piras, 2012).

Example 10.9 Spatial GM – `RiceFarms` data set

The function spgm (for "spatial panel by **GM**") is the general wrapper for **GM** estimation in **splm**, and the counterpart to `spml`. The `model` is specified as either `'within'` or `'random'`, as usual; analogously, a **SAR** term is added by setting `lag` to `TRUE`; differently from `spml`, whether to include a **SEM** term is a binary choice because only **KKP**-type random effects are available: hence the `spatial.error` argument can only be `TRUE` or `FALSE`.

```
semremod.gm <- spgm(riceprod0, Rice, listw = ricelw,
                    lag = FALSE, spatial.error = TRUE)
summary(semremod.gm)
Spatial panel fixed effects GM model


Call:
spgm(formula = riceprod0, data = Rice, listw = ricelw, lag = FALSE,
    spatial.error = TRUE)

Residuals:
     Min.   1st Qu.    Median   3rd Qu.      Max.
-0.841575 -0.163147  0.000527  0.167523  1.355049
```

```
Estimated spatial coefficient, variance components and theta:
           Estimate
rho          0.7807
sigma^2_v    0.0801

Coefficients:
                Estimate Std. Error t-value Pr(>|t|)
log(seed)         0.1346     0.0227    5.94  2.9e-09 ***
log(totlabor)     0.2508     0.0268    9.36  < 2e-16 ***
log(size)         0.5418     0.0274   19.77  < 2e-16 ***
---
Signif. codes:
0 '***' 0.001 '**' 0.01 '*' 0.05 '.' 0.1 ' ' 1
```

Comparing the result of the **SEMFE** model by **GM** below with the previous **ML** example one can see that there is no substantial difference between the estimated coefficients, despite the moderate size of the sample; lastly, as observed, the **GM** method does not provide an estimate of dispersion for ρ; hence no significance testing is possible.

10.3.4 Testing

10.3.4.1 LM Tests for Random Effects and Spatial Errors

Requiring only the estimation of the restricted specification, Lagrange multiplier (**LM**) tests in the tradition of Breusch and Pagan (1980) are particularly appealing in a spatial random effects setting because of the computational difficulties related to **ML** estimation of encompassing models.

Baltagi et al. (2003b) derived joint, marginal and conditional tests for all combinations of random effects and spatial correlation. Starting from the random effects model with **SEM** errors (**SEMRE**), the error term can be written as:

$$\epsilon = (j_T \otimes \mathrm{I}_N)\eta + (\mathrm{I}_T \otimes B^{-1})\nu \tag{10.13}$$

and the (unscaled) variance covariance matrix of the errors as:

$$\Omega = \sigma_\eta^2 (J_T \otimes \mathrm{I}_N) + \sigma_\nu^2 (\mathrm{I}_T \otimes (B'B)^{-1}). \tag{10.14}$$

The hypotheses under consideration are:

1) $\mathrm{H}_0^a : \lambda = \sigma_\eta^2 = 0$ under the alternative that at least one component is not zero
2) $\mathrm{H}_0^b : \sigma_\eta^2 = 0$ assuming no spatial correlation, under the one-sided alternative that the variance component is greater than zero
3) $\mathrm{H}_0^c : \lambda = 0$ assuming no random effects, under the two-sided alternative that the spatial autocorrelation coefficients is different from zero
4) $\mathrm{H}_0^d : \lambda = 0$ assuming the possible existence of random effects, under the two-sided alternative that the spatial autocorrelation coefficient is different from zero
5) $\mathrm{H}_0^e : \sigma_\eta^2 = 0$ assuming the possible existence of spatial autocorrelation and the one-sided alternative that the variance component is greater than zero

The joint **LM** test for the first hypothesis of no random effects and no spatial autocorrelation (H_0^a) is given by:

$$\mathrm{LM}_{\lambda,\eta} = \frac{NT}{2(T-1)} G^2 + \frac{N^2 T}{b} H^2 \tag{10.15}$$

where $G = \hat{\epsilon}^\top (J_T \otimes \mathrm{I}_N)\hat{\epsilon}/\hat{\epsilon}^\top\hat{\epsilon} - 1$, $H = \hat{\epsilon}^\top(\mathrm{I}_T \otimes (W + W^\top)/2)\hat{\epsilon}/\hat{\epsilon}^\top\hat{\epsilon}$, $b = \mathrm{tr}(W + W^\top)^2/2$ and $\hat{\epsilon}$ denotes OLS residuals. The marginal LM test for random effects assuming no spatial correlation is given by:

$$\mathrm{LM}_1 = \sqrt{\frac{NT}{2(T-1)}}G. \tag{10.16}$$

An alternative standardized version with better finite sample properties can be obtained by centering and scaling the one-sided LM statistic:

$$SLM_1 = \frac{\mathrm{LM}_1 - \mathrm{E}(LM1)}{\sqrt{\mathrm{V}(\mathrm{LM}_1)}} \tag{10.17}$$

Analogously, the marginal LM test of no spatial autocorrelation assuming no random effects is given by:

$$\mathrm{LM}_2 = \sqrt{\frac{N^2T}{b}}H. \tag{10.18}$$

which also admits a standardized form with better properties:

$$SLM_2 = \frac{\mathrm{LM}_2 - \mathrm{E}(\mathrm{LM}_2)}{\sqrt{\mathrm{V}(\mathrm{LM}_2)}}. \tag{10.19}$$

SLM_1 and SLM_2 are asymptotically normally distributed as $N \to \infty$ for fixed T, under H_0^b and H_0^c respectively. Based on the latter, a one-sided joint test statistic for $\mathrm{H}_0^a : \lambda = \sigma_\eta^2 = 0$ can be derived as:

$$\mathrm{LM}_H = (\mathrm{LM}_1 + \mathrm{LM}_2)/\sqrt{2} \tag{10.20}$$

which is asymptotically distributed as a standard normal. In practical applications LM_1 can turn out negative, especially when the random effects variance is small, and the same applies to LM_2 when the spatial autocorrelation coefficient is small. A test for the joint null hypothesis can therefore be based on the following decision rule:

$$\chi_m^2 = \begin{cases} \mathrm{LM}_1^2 + \mathrm{LM}_2^2 & \text{if } \mathrm{LM}_1 > 0, \mathrm{LM}_2 > 0 \\ \mathrm{LM}_1^2 & \text{if } \mathrm{LM}_1 > 0, \mathrm{LM}_2 \le 0 \\ \mathrm{LM}_2^2 & \text{if } \mathrm{LM}_1 \le 0, \mathrm{LM}_2 > 0 \\ 0 & \text{if } \mathrm{LM}_1 \le 0, \mathrm{LM}_2 \le 0 \end{cases}$$

Under the null the test statistic χ_m^2 has a mixed χ^2-distribution given by:

$$\chi_m^2 = (1/4)\chi^2(0) + (1/2)\chi^2(1) + (1/4)\chi^2(2) \tag{10.21}$$

When using LM_2, one is assuming that random regional effects do not exist. However, especially when the random effect variance is actually large, this may lead to incorrect inference. For this reason Baltagi et al. (2003b) derived a conditional LM test for spatial autocorrelation allowing for the random effects variance to be non-zero. The expression for the test assumes the following form:

$$\mathrm{LM}_{\lambda|\rho} = \frac{\hat{D}(\lambda)^2}{[(T-1) + \hat{\sigma}\nu^4/\hat{\sigma}_\iota^4]b} \tag{10.22}$$

where $\hat{D}(\lambda)^2 = \frac{1}{2}\hat{\epsilon}^\top \left[\frac{\hat{\sigma}_\nu^4}{\hat{\sigma}_\iota^4}(\bar{J}_T \otimes (W^\top + W)) + \frac{1}{\hat{\sigma}^4\nu}(E_T \otimes (W^\top + W))\right]\hat{\epsilon}$. Also, $\hat{\sigma}_\iota^4 = \hat{\epsilon}^\top(\bar{J}_T \otimes \mathrm{I}_N)\hat{\epsilon}/N$, $\hat{\sigma}^4\nu = \hat{\epsilon}^\top(E_T \otimes \mathrm{I}_N)\hat{\epsilon}/N(T-1)$ and $E_T = \mathrm{I}_T - \bar{J}_T$.

Contrarily to previous tests that use OLS residuals, the residuals $\hat{\epsilon}$ come from the ML estimation of a one-way error component model. This last point, on the converse, makes the implementation slightly more complicated. A one-sided test is simply obtained by taking the square root of 10.22. The resulting test statistics are asymptotically distributed as a standard normal. Similarly, when using LM_1, one is assuming no spatial error correlation. This assumption may lead to incorrect inference particularly when it is not the case that λ is close to zero. A conditional LM test allowing for spatial error correlation can be derived as:

$$\begin{aligned}\text{LM}_{\eta|\rho} &= (\hat{D}_\eta)^2 \\ &\times \left(\frac{2\hat{\sigma}v^4}{T}\right)(TN\hat{\sigma}v^4ec - N\hat{\sigma}v^4d^2 - T\hat{\sigma}v^4g^2e + 2\hat{\sigma}v^4ghd - \hat{\sigma}v^4h^2c)^{-1} \\ &\times (N\hat{\sigma}v^4c - \hat{\sigma}v^4g^2) \end{aligned} \quad (10.23)$$

where $g = \text{tr}[(W^\top\hat{B} + \hat{B}^\top W)(\hat{B}^\top\hat{B})^{-1}]$, $h = \text{tr}[\hat{B}^\top\hat{B}]$, $d = \text{tr}[(W^\top\hat{B} + \hat{B}^\top W)]$, $c = \text{tr}[((W^\top\hat{B} + \hat{B}^\top W)(\hat{B}^\top\hat{B})^{-1})^2]$ and $e = \text{tr}[(\hat{B}^\top\hat{B})^2]$. A one-sided test can be defined by taking the square root of 10.23 based on ML residuals. The test statistic is again asymptotically normally distributed.

Example 10.10 BSK tests – `RiceFarms` data set

In the `RiceFarms` case, it is easy to assume the presence of farm individual effects, perhaps representing parcel quality, farmer's ability, or other time-invaryiing idiosyncrasies. In the following we test for either random farm effects or spatial correlation in the remainder errors, drawing on the specification from the previous examples (i.e., controlling for village and time fixed effects).

The main function to perform the joint, marginal, and conditional tests for random effects and spatial error correlation is `bsktest`. It will take a pair of `formula`, `data` arguments, plus a `listw` object representing the spatial ordering and the `test` to be performed.

The joint test (`test = 'LMH'`) is of little use, because it will reject in the presence of either effect, giving no further directions:

```
bsktest(riceprod, data = Rice, listw = ricelw, test = "LMH")

Baltagi, Song and Koh LM-H one-sided joint test

data:  log(goutput) ~ log(seed) + log(totlabor) + log(size) + region +     time
LM-H = 310, p-value <2e-16
alternative hypothesis: Random Regional Effects and Spatial autocorrelation
```

More interestingly, the conditional test for random farm effects allowing for spatial error correlation (`test = 'CLMmu'`) does in turn reject:

```
bsktest(riceprod, data = Rice, listw = ricelw, test = "CLMmu")

Baltagi, Song and Koh LM*- mu conditional LM test
(assuming lambda may or may not be = 0)

data:  log(goutput) ~ log(seed) + log(totlabor) + log(size) + region +     time
LM*-mu = 11, p-value <2e-16
alternative hypothesis: Random regional effects
```

as does the conditional spatial test, allowing for random effects:

```
bsktest(riceprod, data = Rice, listw = ricelw, test = "CLMlambda")

Baltagi, Song and Koh LM*-lambda conditional LM test
(assuming sigma^2_mu >= 0)

data:  log(goutput) ~ log(seed) + log(totlabor) + log(size) + region +     time
LM*-lambda = 21, p-value <2e-16
alternative hypothesis: Spatial autocorrelation
```

A comprehensive **SEMRE** specification is appropriate.

10.3.4.2 Testing for Spatial Lag vs Error

If a researcher has a strong reason to expect a spatial-data- generating process to be of the **SAR** (or, respectively, **SEM**) kind, then her only problem is to determine whether said spatial effect is present. Then she can either proceed general to specific, estimating the **SAR** (**SEM**) model and assessing the significance of the spatial coefficient, or specific to general, testing from the non-spatial model toward the spatial alternative. In a **ML** framework, the optimal **LM** tests for one effect assuming the other out are called *marginal.* They are dependent on the above hypothesis and will be inconsistent if it is violated; in case only the "other" effect is actually present, they will usually yield a type I error.

As outlined above, although empirical practice has mostly concentrated on either the **SAR** or the **SEM** model, estimation of **SAREM** models containing both a spatial lag and a spatial error is possible. Therefore, if the researcher does not have a strong prior in favor of either, an empirical strategy can be to start from the most general **SAREM** specification, together with the appropriate kind of individual heterogeneity, and let the data tell us which of the two spatial processes – if any and if not both – did actually generate the observed sample, by looking at the significance diagnostic for either spatial coefficient.

One drawback of this strategy is its computational demands and lesser stability than estimating the simpler models; another is that it does not allow the inclusion of a full set of spatially lagged regressors, a specification approach that has become increasingly popular in recent years.

Lagrange multiplier tests for **SAR** (**SEM**) can be either of the conditional type, allowing for the presence of **SEM** (**SAR**) tout court, or of the locally robust type, allowing for a limited deviation from zero of the **SEM** (**SAR**) coefficient. The former are optimal under the standard assumptions of the **ML** framework detailed above, and provided the general **SAREM** model holds; and they require residuals from the restricted **SEM** (**SAR**) model. The second kind have suboptimal statistical properties with respect to the optimal conditional tests, and under the above hypotheses on the data-generating process, they are not guaranteed to hold if misspecification is "too far away," i.e., if the **SAR** (**SEM**) coefficient is of sizable magnitude (and how far is far, i.e., whether 0.1 or 0.4 is tolerable, is an empirical question); moreover the currently available robust **LM** tests have been developed in a cross-sectional framework and do not explicitly incorporate panel features. On the other hand, they are computationally simpler being based on the residuals of the non-spatial model, and they allow including spatially lagged regressors; hence their remarkable success in applied practice.

Marginal vs Locally Robust* LM *Tests The original **LM** tests for either spatial lag or error (Burridge, 1980; Anselin, 1988) were derived in a cross-sectional context, as tests for, respectively, H_0 :

$\rho = 0$ vs $H_A : \rho \neq 0$ assuming $\lambda = 0$ (henceforth LM_ρ); and $\mathrm{H}_0 : \lambda = 0$ vs $H_A : \lambda \neq 0$ assuming $\rho = 0$ (henceforth LM_λ). i.e., both can only be employed assuming that the "other" effect is not present. Otherwise, each test has power against the "wrong" alternative as well; therefore, these procedures are of limited value in the model selection process.

Based on the general local robustness framework of Bera and Yoon (1993), in a cross sectional context, Anselin et al. (1996) derived robust LM statistics for $\mathrm{H}_0 : \rho = 0$ allowing for $\lambda \neq 0$ (henceforth $R\mathrm{LM}_\rho$) and, respectively, for $\mathrm{H}_0 : \lambda = 0$ allowing for $\rho \neq 0$ (henceforth $R\mathrm{LM}_\lambda$). These procedures have since been successfully employed in specification searches to discriminate between SAR and SEM models, as formalized in Florax et al. (2003).

Marginal Spatial* LM *Tests In the context of pooled cross sections, without allowing for any correlation feature across either time or cross section (i.e., setting $\eta = 0$ and $\psi = 0$ in equation 10.4.1.1), any cross- sectional test can be straightforwardly applied to the pooled dataset. The LM tests of Anselin et al. (1996) (LM) are simply rewritten for the pooled dataset, stacked by cross section, and drawing on an enlarged version of the weights matrix obtained by replicating the cross-sectional W_N over the main diagonal so that $W_{NT} = \mathrm{I}_T \otimes W_N$ (see Anselin et al., 2008). The pooled LM_ρ test becomes:

$$\mathrm{LM}_\rho = \frac{(\hat{\epsilon}^\top(\mathrm{I}_T \otimes W)y/\hat{\sigma}_\epsilon{}^2)^2}{J} \tag{10.24}$$

where $\hat{\epsilon}$ are the OLS residuals and

$$J = \frac{1}{\hat{\sigma}_\epsilon^2}(((\mathrm{I}_T \otimes W)Z\hat{\gamma})^\top(\mathrm{I}_{NT} - X(X^\top X)^{-1}X(\mathrm{I}_T \otimes W)Z\hat{\gamma})^\top + TT_W\hat{\sigma}_\epsilon^2)$$

and

$$T_W = \mathrm{tr}(WW + W^\top W)$$

(Elhorst, 2010, Formulae 11 to 13). In turn, the pooled LM_λ test is:

$$\mathrm{LM}_\lambda = \frac{(\hat{\epsilon}^\top(\mathrm{I}_T \otimes W)\hat{\epsilon}/\hat{\sigma}_\epsilon^2)^2}{TT_W} \tag{10.25}$$

Locally Robust Spatial* LM *Tests The robust LM tests of Anselin et al. (1996) can in turn be straightforwardly adapted to the (pooled) panel case, as per Elhorst (2014, Ch. 2.3):

$$R\mathrm{LM}_{\rho|\lambda} = \frac{(\hat{\epsilon}^\top(\mathrm{I}_T \otimes W)y/\hat{\sigma}_\epsilon^2)^2 - \hat{\epsilon}^\top(\mathrm{I}_T \otimes W)\hat{\epsilon}/\hat{\sigma}_\epsilon^2)^2}{J - TT_W}$$

$$R\mathrm{LM}_{\lambda|\rho} = \frac{(\hat{\epsilon}^\top(\mathrm{I}_T \otimes W)\hat{\epsilon}/\hat{\sigma}_\epsilon^2)^2 - TT_W/J\hat{\epsilon}^\top(\mathrm{I}_T \otimes W)y/\hat{\sigma}_\epsilon^2)^2}{TT_W(1 - TT_W/J)}$$

using, again, the OLS residuals $\hat{\epsilon}$ (Elhorst, 2010, Formulae 14-15).

Moreover, according to Bera et al. (2009), the LM test for the joint null hypothesis $\mathrm{H}_0 : \rho = \lambda = 0$ versus $H_A : \rho \neq 0$ or $\lambda \neq 0$ is equal to the sum of the marginal test for one effect and the locally robust test for the other:

$$\mathrm{LM}_{\rho\lambda} = \mathrm{LM}_\rho + R\mathrm{LM}_{\lambda|\rho} = \mathrm{LM}_\lambda + R\mathrm{LM}_{\rho|\lambda}$$

so that the RLM tests can also be obtained indirectly by subtracting the marginal test for the "other" effect from the joint test.

The `slmtest` function, specifying `test = 'lml'` (`'lme'`) will perform either the marginal test for SAR (SEM) assuming no SEM (SAR) component in the data-generating process or the locally robust version if specifying `test = 'rlml'` (`'rlme'`).

Example 10.11 Robust LM tests for SAR or SEM – `RiceFarms` data set
As we have seen, in the rice farms example there are reasons for assuming out a spatial lag model from the beginning. Nevertheless, it is sensible to check this assumption. The standard versions of the (pooled) LM test for SAR (SEM), as observed, is not robust to the presence of a SEM (SAR) term, i.e., of the "other" effect. Robust LM tests instead allow for "local" deviations from zero of the "other" parameter. Of course, the extent of the tolerated deviation is uncertain; still, from this example the difference between the false positive given by the non-robust SAR test (`test = 'lml'`) and the locally robust counterpart (`test = 'rlml'`) clearly stands out:

```
local.rob.LM <- matrix(ncol = 4, nrow = 2)
tests <- c("lml", "lme", "rlml", "rlme")
dimnames(local.rob.LM) <- list(c("LM test", "p-value"),
                              tests)
for(i in tests) {
    local.rob.LM[1, i] <- slmtest(riceprod, data = Rice,
                              listw=ricelw, test = i)$statistic
    local.rob.LM[2, i] <- slmtest(riceprod, data = Rice,
                              listw=ricelw, test = i)$p. value
    }
round(local.rob.LM, 4)
           lml   lme   rlml  rlme
LM test 39.28 244.8 0.1654 205.7
p-value  0.00   0.0 0.6842   0.0
```

The robust test favors the SEM model over the SAR.

It shall be kept in mind, moreover, that none of the above procedures allow for individual effects; one approximate solution is to demean the data. The `Within` function – which can be used directly in the model formula – will subtract time means, thus eliminating any individual effect, of either random or fixed type:

```
local.rob.LMw <- matrix(ncol = 4, nrow = 2)
wriceprod <- Within(log(goutput)) ~ Within(log(seed)) +
    Within(log(totlabor)) + Within(log(size)) +
    region + time
dimnames(local.rob.LMw) <- list(c("LM test", "p-value"),
                             c("lml", "lme", "rlml", "rlme"))
for(i in c("lml", "lme", "rlml", "rlme")) {
    local.rob.LMw[1, i] <- slmtest(wriceprod, data = Rice,
                              listw=ricelw, test = i)$statistic
    local.rob.LMw[2, i] <- slmtest(wriceprod, data = Rice,
                              listw=ricelw, test = i)$p. value
    }
round(local.rob.LMw, 4)
          lml   lme  rlml  rlme
LM test 125.2 604.3 1.538 480.6
p-value   0.0   0.0 0.215   0.0
```

The result is unchanged, but now we are more confident in it because we have controlled, although in an *ad hoc* way, for individual effects.

Likelihood-Based Tests Given that estimation of the full SAREM model is possible (see the extensive discussion in Millo, 2014), one could directly employ the encompassing model as a specification device, relying on the Wald restriction tests from the general model as an alternative specification strategy instead of looking at the RLM tests. This strategy has the drawback of being computationally more intensive but also some important advantages: the Wald z-tests for significance of ρ and λ are optimal; there is no need for robustification, as the "other" spatial effect is explicitly accounted for in the model, as can be random individual effects; lastly, estimation of the encompassing model also provides the magnitudes of the spatial coefficients together with the significance level of the zero-restriction tests so that their substantial importance can be assessed.

As usual, two kinds of tests are possible from the estimated encompassing model: Wald-type tests, requiring only an estimate of the latter, and likelihood ratio tests, requiring both the encompassing and the restricted.

Wald Tests Wald-type tests are z-tests for significance of the relevant parameter in the encompassing model. Thus, from ML estimates of the general SAREM-RE model,

$$\text{Wald}_{\rho|\lambda} = \frac{\hat{\rho}}{\sqrt{\hat{V}(\hat{\rho})}} \sim N(0,1)$$

and symmetrically for $\text{Wald}_{\lambda|\rho}$. Importantly, the test can be made conditional to (i.e., valid in the presence of) individual random effects by including them in the specification. As observed, fixed individual effects can be eliminated through data transformation in two ways, both familiar from the spatial panel literature: either through time-demeaning (*within* transformation) (Elhorst, 2003) or by forward orthogonal deviations (Lee and Yu, 2010a). The former induces residual serial correlation, which can nevertheless be considered (i.e., estimated out) in the encompassing model; while the latter preserves the features of the original errors covariance matrix (Debarsy and Ertur, 2010, p. 7).

Example 10.12 Wald tests for SEM vs SAR – `RiceFarms` data set

In the following we estimate the full SAREMRE model in order to test whether it is possible to simplify it, in a general-to-specific fashion. The `spml` function is the highest-level wrapper for spatial panel estimation by maximum likelihood, allowing for either fixed, random, or no effects (in the random or none cases, it calls `spreml` internally). Its syntax is mostly consistent with that of `plm`. We select `model='random'` and `spatial.error='b'` for "Baltagi," which selects the SEMRE specification (`'kkp'` would estimate the SEM2RE).

```
saremremod <- spml(riceprod, data = Rice, listw = ricelw, lag = TRUE,
                   model = "random", spatial.error = "b")
summary(saremremod)
ML panel with spatial lag, random effects, spatial error correlation

Call:
spreml(formula = formula, data = data, index = index, w = listw2mat(listw),
    w2 = listw2mat(listw2), lag = lag, errors = errors, cl = cl)

Residuals:
   Min. 1st Qu.  Median    Mean 3rd Qu.    Max.
 -1.154  -0.321  -0.076  -0.090   0.149   1.351
```

```
Error variance parameters:
    Estimate Std. Error t-value Pr(>|t|)
phi   0.2967     0.0568     5.23  1.7e-07 ***
rho   0.6281     0.0790     7.96  1.8e-15 ***

Spatial autoregressive coefficient:
        Estimate Std. Error t-value Pr(>|t|)
lambda   -0.0134     0.1755   -0.08     0.94

Coefficients:
                   Estimate Std. Error t-value Pr(>|t|)
(Intercept)          5.9684     0.1957   30.50  < 2e-16 ***
log(seed)            0.1531     0.0235    6.51  7.6e-11 ***
log(totlabor)        0.2492     0.0271    9.19  < 2e-16 ***
log(size)            0.5784     0.0274   21.08  < 2e-16 ***
regionlangan        -0.0926     0.1051   -0.88    0.378
regiongunungwangi   -0.1567     0.0969   -1.62    0.106
regionmalausma      -0.1572     0.0995   -1.58    0.114
regionsukaambit     -0.0243     0.1078   -0.23    0.822
regionciwangi       -0.0267     0.0973   -0.27    0.784
time2               -0.0612     0.0813   -0.75    0.452
time3               -0.1911     0.0813   -2.35    0.019 *
time4               -0.3650     0.0813   -4.49  7.1e-06 ***
time5                0.1626     0.0813    2.00    0.045 *
time6                0.1325     0.0813    1.63    0.103
---
Signif. codes:
0 '***' 0.001 '**' 0.01 '*' 0.05 '.' 0.1 ' ' 1
```

From the estimation results, we gather that the SEMRE is the better specification: the estimated SAR term $\hat{\lambda}$ is not significant, while the SEM coefficient $\hat{\rho}$ is of considerable magnitude and highly significant.

LR Tests Likelihood ratio tests are based on the likelihoods from the general and the restricted model. The test statistic is a simple transform of the difference in likelihoods:

$$2[\ln L(\hat{\theta}) - \ln L(\tilde{\theta})] \sim \chi^2_m$$

where $\hat{\theta}$ is the full vector of ML parameter estimates from the unrestricted model and $\tilde{\theta}$ from the restricted one, and m the number of restrictions. Thus,

$$\mathrm{LR}_{\rho|\lambda} = 2[\ln L(\hat{\rho}, \hat{\lambda}, \hat{\beta}) - \ln L(\tilde{\lambda}, \tilde{\beta})] \sim \chi^2_1$$

and symmetrically for $LR_{\lambda|\rho}$. Again, including random effects in the estimated models makes the test conditional to these effects, while fixed effects can be transformed out as detailed in the previous paragraph but always keeping in mind the effects of the transformation on the error properties.

Example 10.13 LR tests for SEM vs SAR – `RiceFarms` data set
The restriction test for the SAR term is performed as:

```
ll1 <- saremremod$logLik
ll0 <- spml(riceprod, data = Rice, listw = ricelw, lag = FALSE,
                model = "random", spatial.error = "b")$logLik
LR <- 2 * (ll1 - ll0)
pLR <- pchisq(LR, df = 1, lower.tail = FALSE)
pLR
[1] 0.9121
```

The p-value from the LR spatial lag test is very high, and not unlike the (asymptotically equivalent) result from the Wald restriction test in the previous example.

10.4 Serial and Spatial Correlation

It is possible to generalize the structure of the errors further by introducing serial correlation in the remainder of the error term, together with spatial correlation and random effects. Baltagi et al. (2007) do so in the context of the Anselin SEMRE, specifying the model errors as the sum of an individual, time-invariant component and an idiosyncratic one that is spatially autocorrelated, as above, but also has serial correlation in the remainder:

$$\begin{aligned} \nu &= \rho(\mathrm{I}_T \otimes W_N)\nu + \zeta \\ \zeta_t &= \psi\zeta_{t-1} + \xi_t \end{aligned} \tag{10.26}$$

where ξ is i.i.d.. The combination of this more general error structure, termed SEMSRRE because of the addition of Serially autoRegressive errors, with a spatially lagged dependent variable and the estimation of the most general model SAREMSRRE can still be dealt with in the general ML framework outlined above.

10.4.1 Maximum Likelihood Estimation

The model combining spatial and serial correlation with individual effects can be estimated by maximum likelihood, through an extension of the framework outlined in the previous sections of this chapter.

10.4.1.1 Serial and Spatial Correlation in the Random Effects Model

Generalizing the structure of the errors further by introducing serial correlation in the remainder of the error term, together with spatial correlation and random effects, Baltagi et al. (2007) derived a number of conditional and marginal LM tests for the different effects, possibly allowing for the presence of the other ones. Based on their work, Millo (2014) extended the model to include a SAR term. The errors of the SAREM model are specified as in the previous paragraph, so that the full model is:

$$\begin{aligned} y &= \lambda(\mathrm{I}_T \otimes W)y + Z\gamma + \epsilon \\ \epsilon &= (j_T \otimes \eta) + \nu \\ \nu &= \rho(\mathrm{I}_T \otimes W)\nu + \zeta \\ \zeta_t &= \psi\zeta_{t-1} + \xi_t \end{aligned}$$

To derive the likelihood, Baltagi et al. (2007) suggest a Prais-Winsten transformation of the model with random effects and spatial autocorrelation. Following their simplifying notation, define: $V_\psi = \frac{1}{1-\psi^2} V_1$ with:

$$V_1 = \begin{bmatrix} 1 & \psi & \psi^2 & \dots & \psi^{T-1} \\ \psi & 1 & \psi & \dots & \psi^{T-2} \\ \vdots & \vdots & \vdots & \ddots & \vdots \\ \psi^{T-1} & \psi^{T-2} & \psi^{T-3} & \dots & 1 \end{bmatrix}; \tag{10.27}$$

then the expression for the scaled error covariance matrix Σ can be written as

$$\Sigma_{\mathbf{SEMSRRE}} = \phi(J_T \otimes \mathrm{I}_N) + V_\psi \otimes (B^\top B)^{-1}.$$

While in principle the inverse and determinant of Σ can be calculated by brute force, in practice it is convenient, and often necessary, to rely on simplified analytical expressions to reduce the computational burden and extend the range of feasible sample sizes. Baltagi et al. (2007) derived expressions for the inverse and determinant of the error covariance matrix:

$$\begin{aligned} \Sigma^{-1}_{\mathbf{SEMSRRE}} &= V_\psi^{-1} \otimes (B^\top B) + \frac{1}{d^2(1-\psi)^2}(V_\psi^{-1} J_T V_\psi^{-1}) \\ &\quad \otimes ([d^2(1-\psi)^2 \phi \mathrm{I}_N + (B^\top B)^{-1}]^{-1} - B^\top B) \\ |\Sigma_{\mathbf{SEMSRRE}}| &= |d^2(1-\psi)^2 \phi \mathrm{I}_N + (B^\top B)^{-1}| \cdot |(B^\top B)^{-1}|^{T-1}/(1-\psi^2)^N, \end{aligned}$$

where $\alpha = \sqrt{\frac{1+\psi}{1-\psi}}$ and $d^2 = \alpha^2 + (T-1)$. They can be plugged in the general likelihood (10.9) to estimate the **SAREMSRRE** model.

10.4.1.2 Serial and Spatial Correlation with KKP-Type Effects

As an alternative to the **SAREMSRRE** specification, Millo (2014) presents an extension of the **SEM2RE** errors a la Kapoor et al. (2007) to serial correlation in the remainder errors. As in the **SEM2RE** case, the random effects are spatially lagged together with the idiosyncratic ones, while the remainder errors ξ in turn are serially correlated:

$$\begin{aligned} y &= \lambda(\mathrm{I}_T \otimes W)y + Z\gamma + c \\ \epsilon &= (j_T \otimes \eta) + v \\ \epsilon &= \rho(\mathrm{I}_T \otimes W)\epsilon + \zeta \\ \zeta_t &= \psi\zeta_{t-1} + \xi_t \end{aligned}$$

This alternative specification assumes that individual effects follow the same spatial diffusion process as the idiosyncratic errors do. By analogy, it is termed **SAREM2SRRE**. Just as in the **SEM2RE** case, the error covariance is then again of the $B^\top \Omega B$ form (see Section 10.3.3.1), which simplifies computations considerably. In fact, the (scaled) error covariance for this model is:

$$\Sigma_{\mathbf{SEM2SRRE}} = (\phi J_T + V_\psi) \otimes (B^\top B)^{-1}$$

and, by the properties of Kronecker products, its inverse is

$$\Sigma^{-1}_{\mathbf{SEM2SRRE}} = (\phi J_T + V_\psi)^{-1} \otimes (B^\top B)$$

so that there is no need for the numerically demanding and unstable inversion of $B^\top B$.[14]

14 Models with serial and spatial correlation are often computationally cumbersome to estimate and can be challenging even for modern computers on moderate sample sizes. For an assessment of the practical computational limits of different specifications, see Millo (2014, Table 2).

Example 10.14 Serial and spatial correlation – `EvapoTransp` data set
Mountains are a crucial source of water for public, agricultural and hydropower use. Obojes et al. (2015) explore the effect of vegetation composition and structure on water balance on some high- elevation grasslands in the Alps, in order to infer the potential to influence the water balance of mountain areas through land management. In particular, they evaluate the consequences of the abandonment of mountain areas: leading to the proliferation of tall grasses and dwarf shrubs and therefore affecting the water balance. Of the different components of the water balance, evapotranspiration (ET) is the one most influenced by vegetation. They repeatedly measure the water balance of soil monoliths in deep seepage collectors in four experimental sites over three study areas, two in the French Alps, one in Switzerland, and one in Austria.

Persistence in both time and space is apparent in the data and attributed to small-scale features of the particular observation context. In order to account for these, they estimate a panel model with both spatial and serial correlation, using a distance-based, row-standardized weights matrix.

We replicate the results from the Austrian site, based on the original data, available as `EvapoTransp`, and weights matrix `etw`. There are 5 repeated measurements over 86 observation units.

```
data("EvapoTransp", package = "pder")
data("etw", package = "pder")
evapo <- et ~ prec + meansmd + potet + infil + biomass + plantcover +
  softforbs + tallgrass + diversity + matgram + dwarfshrubs + legumes
semsr.evapo <- spreml(evapo, data=EvapoTransp, w=etw,
                     lag=FALSE, errors="semsr")
summary(semsr.evapo)
ML panel with, AR(1) serial correlation, spatial error correlation

Call:
spreml(formula = evapo, data = EvapoTransp, w = etw, lag = FALSE,
    errors = "semsr")

Residuals:
   Min. 1st Qu.  Median    Mean 3rd Qu.    Max.
 -2.260  -0.500   0.021  -0.047   0.420   2.373

Error variance parameters:
    Estimate Std. Error t-value Pr(>|t|)
psi   0.1665     0.0482    3.45  0.00056 ***
rho   0.8665     0.0246   35.29  < 2e-16 ***

Coefficients:
             Estimate Std. Error t-value  Pr(>|t|)
(Intercept)  0.866041   0.562326    1.54    0.1235
prec        -0.129636   0.154338   -0.84    0.4009
meansmd      0.018968   0.004452    4.26   2.0e-05 ***
potet        0.551144   0.335828    1.64    0.1008
infil        0.023513   0.021876    1.07    0.2824
biomass      0.002335   0.000305    7.65   1.9e-14 ***
plantcover   0.019174   0.110332    0.17    0.8620
softforbs    0.132359   0.041463    3.19    0.0014 **
tallgrass    0.174540   0.054099    3.23    0.0013 **
```

```
diversity     0.040775    0.035790     1.14     0.2546
matgram      -0.029814    0.033040    -0.90     0.3669
dwarfshrubs   0.098405    0.054127     1.82     0.0691.
legumes      -0.016304    0.005591    -2.92     0.0035 **
---
Signif. codes:
0 '***' 0.001 '**' 0.01 '*' 0.05 '.' 0.1 ' ' 1
```

Although simple **OLS** would be consistent in this setting, the spatially and serially correlated **ML** model improves the precision of the estimates and leads to substantially different results. For example, in the spatial-serial error model, the coefficient on precipitation (`Prec`) is halved and not significant any more, with respect to what would result from **OLS** (reported below); analogously for potential evapotranspiration `potET`).

```
library("lmtest")
coeftest(plm(evapo, EvapoTransp, model="pooling"))

t test of coefficients:

              Estimate Std. Error t value Pr(>|t|)
(Intercept)  1.076993    0.168586     6.39  4.5e-10 ***
prec        -0.207954    0.037563    -5.54  5.5e-08 ***
meansmd      0.022749    0.006489     3.51   0.0005 ***
potet        0.767941    0.088380     8.69  < 2e-16 ***
infil        0.055648    0.031030     1.79   0.0736.
biomass      0.000104    0.000389     0.27   0.7900
plantcover   0.044657    0.157801     0.28   0.7773
softforbs    0.104305    0.058300     1.79   0.0743.
tallgrass    0.173013    0.078668     2.20   0.0284 *
diversity    0.016214    0.051333     0.32   0.7523
matgram     -0.069537    0.049001    -1.42   0.1566
dwarfshrubs  0.071451    0.077135     0.93   0.3548
legumes     -0.019447    0.008115    -2.40   0.0170 *
---
Signif. codes:
0 '***' 0.001 '**' 0.01 '*' 0.05 '.' 0.1 ' ' 1
```

The same happens for any specification omitting the spatial error term, like random effects or the serially correlated errors model, which can be obtained setting `errors` to `'sr'` (output not reported).

Controlling for spatial error correlation seems the key feature here, witness the large $\hat{\rho}$; nevertheless, despite the low magnitude of the serial correlation coefficient, omitting time persistence would still lead to substantially different results, namely to a false positive for the significance test on `DwarfShrubs`:

```
coeftest(spreml(evapo, EvapoTransp, w=etw, errors="sem"))

z test of coefficients:

              Estimate Std. Error z value Pr(>|z|)
(Intercept)  1.062540    0.566049     1.88  0.06050.
```

```
prec         -0.155642   0.153350   -1.01  0.31013
meansmd       0.017929   0.003962    4.52  6.0e-06 ***
potet         0.532574   0.327493    1.63  0.10390
infil         0.022110   0.019396    1.14  0.25430
biomass       0.002312   0.000286    8.08  6.3e-16 ***
plantcover    0.016307   0.097288    0.17  0.86689
softforbs     0.131949   0.036543    3.61  0.00031 ***
tallgrass     0.176606   0.047692    3.70  0.00021 ***
diversity     0.038389   0.031549    1.22  0.22369
matgram      -0.031006   0.029147   -1.06  0.28743
dwarfshrubs   0.104405   0.047821    2.18  0.02902 *
legumes      -0.016654   0.004929   -3.38  0.00073 ***
---
Signif. codes:
0 '***' 0.001 '**' 0.01 '*' 0.05 '.' 0.1 ' ' 1
```

10.4.2 Testing

Testing for either effect in the context of the spatially and serially correlated model with individual heterogeneity is performed within the same maximum likelihood framework used for estimation.

10.4.2.1 Tests for Random Effects, Spatial, and Serial Error Correlation

Baltagi et al. (2007) derive the joint, marginal, and conditional LM tests for the model with serial correlation.

They consider all possible combinations of joint, marginal and conditional tests:

- the joint test for $\rho = \psi = \sigma_\eta^2 = 0$, (*J*)
- the marginal tests for ρ, ψ, and σ_η^2 assuming in turn that the other two are zero (*M.1-3*)
- the joint tests for any combination of two of the parameters assuming the third one is zero (*M.4-6*)
- the marginal tests for ρ, ψ, and σ_η^2 assuming in turn that the other two may or may not be zero (*C.1-3*)
- the joint tests for any combination of two of the parameters assuming the third one may or may not be zero (*C.4-6*)

M.1-3 are well-established testing procedures in the literature (as observed in Baltagi et al., 2007). *M.1* (test for $\rho = 0$) is the LM test for spatial error correlation derived by Anselin (1988) in the context of a pooled model with no serial correlation or individual effects. On the other hand, *M.2* (test for $\psi = 0$) is analogous, for large T, to the well-known Breusch (1978), Godfrey (1978) serial correlation test. Finally, *M.3* is simply the Breusch and Pagan (1980) random effects test.

Baltagi et al. (2007, Appendix A.3) show that the test statistic for the joint hypothesis *M.4* ($\rho = \psi = 0$) assuming no random effects is simply the sum of the marginal tests *M.1* ($\rho = 0$) and *M.2* ($\psi = 0$). Additionally, *M.5* is the Baltagi et al. joint test outlined in section 10.3.4.1; and *M.6* is the joint test for random individual effects and serial correlation derived in Baltagi and Li (1995) (see section 4.3.2).

As a result of the previous discussion, we only consider the three-way joint test *J* and the one-way conditional tests *C.1-3*.

The corresponding null hypotheses are:

1) $H_0^a: \rho = \psi = \sigma_\eta^2 = 0$ under the alternative that at least one component is not zero (J)
2) $H_0^h: \rho = 0$, assuming $\psi \neq 0, \sigma_\eta^2 > 0$: test for spatial correlation, allowing for serial correlation and random individual effects (*C.1*)
3) $H_0^i: \psi = 0$, assuming $\rho \neq 0, \sigma_\eta^2 > 0$: test for serial correlation, allowing for spatial correlation and random individual effects (*C.2*)
4) $H_0^j: \sigma_\eta^2 = 0$, assuming $\rho \neq 0, \psi \neq 0$: test for random individual effects, allowing for spatial and serial correlation (*C.3*)

The joint LM test for H_0^a is given by:

$$\mathrm{LM}_j = \frac{NT^2}{2(T-1)(T-2)}[A^2 - 4AF + 2TF^2] + \frac{N^2T}{b}H^2 \tag{10.28}$$

where, $A = \hat{\epsilon}^\top (J_T \otimes \mathrm{I}_N)\hat{\epsilon}/\hat{\epsilon}^\top\hat{\epsilon} - 1$, $F = \hat{\epsilon}^\top (G_T \otimes \mathrm{I}_N)\hat{\epsilon}/2\hat{\epsilon}^\top\hat{\epsilon}$, $H = \hat{\epsilon}^\top (\mathrm{I}_T \otimes (W^\top + W))\hat{\epsilon}/2\hat{\epsilon}^\top\hat{\epsilon}$, $b = \mathrm{tr}(W + W^\top)^2/2$, G is a matrix with bidiagonal elements equal to one and $\hat{\epsilon}$ denotes OLS residuals. Under H_0^a, LM_J is distributed as χ_3^2.

The conditional *C.1* test for H_o^h gives rise to the following statistic, asymptotically distributed as χ_1^2 under H_0^h:

$$\mathrm{LM}_{\rho/\psi\eta} = \frac{\hat{D}(\rho)^2}{b(T - 2cg + c^2g^29)} \tag{10.29}$$

where

$$\hat{D}(\rho) = \frac{1}{2}\hat{\epsilon}^\top[V^{-1} - 2cV^{-1}J_TTV^{-1} + c^2[V^{-1}J_T]^2V^{-1}] \otimes (W^\top + W)\hat{\epsilon}$$

is the score vector (evaluated at the null), $\hat{\epsilon}$ a vector of ML residuals obtained from the estimation of the model with individual error components and serial correlation, $g = \frac{1}{\sigma_e^2}(1-\psi)2 + (T-2)(1-\psi)$, and b has been defined above.

The conditional *C.2* test for H_0^i is based on the following statistic, asymptotically distributed as χ_1^2 under the null:

$$\mathrm{LM}_{\psi/\rho\eta} = \hat{D}(\psi)^2 J_{33}^{-1} \tag{10.30}$$

where J_{33}^{-1} is the corresponding element of the information matrix,[15]

$$\begin{aligned}\hat{D}(\psi) = &-\frac{T-1}{T}(\hat{\sigma}_e^2\mathrm{tr}(Z(B^\top B)^{-1}) - N)\\ &+ \frac{\hat{\sigma}_e^2}{2}\hat{\epsilon}^\top[\frac{1}{\sigma_e^4}(E_TGE_T) \otimes (B^\top B) + \frac{1}{\sigma_\zeta^2}(\bar{J}_TGE_T) \otimes Z\\ &+ \frac{1}{\sigma_\zeta^2}(E_TG\bar{J}_T) \otimes Z + (\bar{J}_TG\bar{J}_T) \otimes Z(B^\top B)^{-1}Z]\hat{\epsilon}\end{aligned} \tag{10.31}$$

with $Z = [T\sigma_\eta^2\mathrm{I}_N + \sigma_e^2(B^\top B)^{-1}]$ the score evaluated at the null and $\hat{\epsilon}$ the vector of ML residuals from the estimation of a panel model with individual error components and serial correlation. Both g and b assume the same expression as before, while as usual $E_T = \mathrm{I}_T - \bar{J}_T$.

The conditional *C.3* test for H_o^j is based on the following statistic:

$$\mathrm{LM}_{\eta/\rho\psi} = \hat{D}(\sigma_\eta^2)J_{22}^{-1} \tag{10.32}$$

15 For the expression of the information matrix see Baltagi et al. (2007, Eq. 3.10)

where

$$\hat{D}(\sigma_\eta^2) = -\frac{\text{tr}(V^{-1}J_T)}{2}\text{tr}(B^\top B) + \frac{1}{2\sigma_e^4}\hat{\epsilon}^\top[V_\psi^{-1}J_T V_\psi^{-1} \otimes (B^\top B)^2]\hat{\epsilon}$$

is the score evaluated at the null, J_{22}^{-1} is the corresponding element of the information matrix[16] and $\hat{\epsilon}$ is the vector of estimated residuals from the **ML** estimation of a panel model with spatially and serially correlated errors but no individual error components. The $\text{LM}_{\eta/\rho\psi}$ test statistic is asymptotically distributed as χ_1^2 under H_0^j:

Example 10.15 Conditional BSJK tests – `RiceFarms` data set

We now address the issue of serial correlation in the remainder errors of the `RiceFarms` model. In other words, we check whether persistence characteristics in the output of an individual farm have effectively been accounted for by the individual effects, which in previous examples have proved significant, statistical evidence favoring the random hypothesis. On the spatial side, there has been ample evidence of spatial effects of the SEM type. For all this, on one hand a joint test is guaranteed to reject; on the other, tests for each single "effect" (spatial or serial correlation, or individual effects) will have to account for the possible presence of one or both of the others.

The tests are performed specifying a `formula` and a `data.frame`, the spatial weights `listw` to be employed and the `test`. Although we are here particularly interested in the `'C.2'` test, for the sake of comparison we perform both the joint test `'J'` and all three conditional tests. As observed, the joint test will use the **OLS** residuals, while the others those from the appropriate restricted specification, e.g, the *C*.2 those from a **SEMRE** model.

```
bsjk.LM <- matrix(ncol = 4, nrow = 2)
tests <- c("J", paste("C", 1:3, sep = "."))
dimnames(bsjk.LM) <- list(c("LM test", "p-value"),
                          tests)
for(i in tests) {
    mytest <- bsjktest(riceprod, data = RiceFarms, index = "id",
                       listw = ricelw, test = i)
    bsjk.LM[1, i] <- mytest$statistic
    bsjk.LM[2, i] <- mytest$p. value
    }
round(bsjk.LM, 6)
            J   C.1      C.2  C.3
LM test 319.5 371.5 11.894431 75.8
p-value   0.0   0.0  0.000563  0.0
```

All tests reject the respective null hypotheses: the joint and the *C*.1 (spatial effects) most forcefully and then the *C*.3 (random effects). The *C*.2 test rejects less forcefully; still it provides evidence for some serial correlation in the remainder errors of the rice production equation *after* controlling for spatial and individual random effects.

Example 10.16 Spatial and serial correlation – `RiceFarms` data set

The result from the *C*.2 test, although not very sharp, warrants an investigation into the serial correlation issue. To this end, we estimate the full SEMSRRE model, visualizing only the significance table for the error components. t-statistics are expected to mimic the results of the

16 See Baltagi et al. (2007, Section 3.4)

asymptotically equivalent LM tests closely: still there is more information to be extracted from the encompassing model, i.e., the magnitudes, and hence the substantial importance, of the estimated parameters.

```
semsrre.rice <- spreml(riceprod, data = Rice,
                       w=riceww, lag = FALSE, errors = "semsrre")
round(summary(semsrre.rice)$ErrCompTable, 6)
    Estimate Std. Error t-value Pr(>|t|)
phi   0.2500    0.05904   4.234 0.000023
psi   0.1250    0.04092   3.054 0.002259
rho   0.6136    0.04617  13.291 0.000000
```

Spatial error correlation is confirmed as the statistically strongest effect, with the now familiar large coefficient. Both individual effects and serial correlation play minor roles: the variance ratio of the random effects over the idiosyncratic errors ϕ is about one fourth; the estimated serial correlation coefficient is but 0.13.

10.4.2.2 Spatial Lag vs Error in the Serially Correlated Model

Testing for spatial lag vs spatial error in a model allowing for random effects and/or serially correlated errors can be done via the Wald approach, from the encompassing specification.[17]

Example 10.17 Spatial and serial correlation – `EvapoTransp` data set

In the evapotranspiration example, spatial correlation in the errors comes as a consequence of the particular experimental environment: in other words, it is imposed over the alternative of spatial lag dependence as a theoretical *a priori* of the researchers. In fact, it seems difficult to come up with reasons why the *outcome*, actual evapotranspiration, at one site should influence that of neighboring sites, while it is quite natural to expect that measurement errors at nearby sites be correlated. As a statistical check, we estimate the encompassing specification with SAR, SEM, and serial error correlation, reporting only the relevant coefficient tables for the error variance parameters and the spatial lag coefficient:

```
saremsrre.evapo <- spreml(evapo, data = EvapoTransp,
                          w = etw, lag = TRUE, errors = "semsr")
summary(saremsrre.evapo)$ARCoefTable
       Estimate Std. Error t-value Pr(>|t|)
lambda   -0.322     0.2804  -1.148   0.2508
round(summary(saremsrre.evapo)$ErrCompTable, 6)
    Estimate Std. Error t-value Pr(>|t|)
psi   0.1679    0.04820   3.483 0.000496
rho   0.9000    0.02902  31.006 0.000000
```

The statistical evidence from estimation backs the *a priori* considerations: unlike the spatial error coefficient $\hat{\rho}$, the estimate $\hat{\lambda}$ of the spatial lag is not significant.

17 The encompassing specification has many parameters; therefore, it is quite complicated to estimate and prone to numerical problems. See Millo (2014, 5.1.5). Likelihood optimization may yield corner solutions; or the numerical evaluation of the Hessian matrix, which produces the standard errors, can fail for some parameters. This is common when the model is overspecified, most often when the "problematic" parameters are close to 0, i.e., not significant.

Bibliography

A. Acconcia, G. Corsetti, and S. Simonelli. Mafia and public spending: evidence on the fiscal multiplier from a quasi-experiment. *American Economic Review*, 104:2185–2209, 2014.

A.D. Acemoglu, S. Johnson, J.A. Robinson, and P. Yared. Income and democracy. *American Economic Review*, 98(3):808–842, 2008.

S. Alan, B.E. Honoré, L. Hu, and S. Leth-Petersen. Estimation of panel data regression model with two-sided censoring or truncation. *Journal of Econometric Methods*, 3 (1):1–20, 2013.

C. Alonso-Borrego and M. Arellano. Symmetrically normalized instrumental-variable estimation using panel data. *Journal of Business and Economic Statistics*, 17 (1):36–49, 1999.

R.M. Alvarez, G. Garrett, and P. Lange. Government partisanship, labor organization, and macroeconomic performance. *The American Political Science Review*, 85 (2):539–556, 1991.

T. Amemiya. The estimation of the variances in a variance–components model. *International Economic Review*, 12:1–13, 1971.

T. Amemiya and T.E. MaCurdy. Instrumental-variable estimation of an error-components model. *Econometrica*, 54 (4):869–80, July 1986.

A.L. Andersen, D.D. Lassen, and L.H.W. Nielsen. Late budgets. *American Economic Journal, Economic Policy*, 4 (4):1–40, 2012.

T.W. Anderson and C. Hsiao. Formulation and estimation of dynamic models using panel data. *Journal of Econometrics*, 18:47–82, 1982.

J.D. Angrist and W.K. Newey. Over-identification tests in earnings functions with fixed effects. *Journal of Business & Economic Statistics*, 9 (3):317–323, 1991.

L. Anselin. *Spatial Econometrics: Methods and Models*, volume 4. Springer, 1988.

L. Anselin, A.K. Bera, R. Florax, and M.J. Yoon. Simple diagnostic tests for spatial dependence. *Regional Science and Urban Economics*, 26 (1):77–104, 1996.

L. Anselin, J. Le Gallo, and H. Jayet. Spatial panel econometrics. In L. Matyas and P. Sevestre, editors, *The Econometrics of Panel Data, Fundamentals and Recent Developments in Theory and Practice (3rd Edition)*, pages 624 – 660. Springer-Verlag, Berlin Heidelberg, 2008.

M. Arellano. Computing robust standard errors for within-groups estimators. *Oxford bulletin of Economics and Statistics*, 49 (4):431–434, 1987.

M. Arellano. *Panel Data Econometrics*. Oxford University press, 2003.

M. Arellano and S. Bond. Some tests of specification for panel data: Monte carlo evidence and an application to employment equations. *Review of Economic Studies*, 58:277–297, 1991.

M. Arellano and O. Bover. Another look at the at the instrumental variables estimation of error components. *Journal of Econometrics*, 68:29–51, 1995.

R.B. Avery. Error components and seemingly unrelated regressions. *Econometrica*, 45:199–209, 1977.

D. Bailey and J.N. Katz. Implementing panel corrected standard errors in r: The pcse package. *Journal of Statistical Software*, 42 (CS1):1–11, 2011.

Panel Data Econometrics with R, First Edition. Yves Croissant and Giovanni Millo.

Companion website: www.wiley.com/go/croissant/data-econometrics-with-R

P. Balestra and M. Nerlove. Pooling cross-section and time-series data in the estimation of dynamic models: The demand for natural gas. *Econometrica*, 34:585–612, 1966.

P. Balestra and J. Varadharajan-Krishnakumar. Full information estimations of a system of simultaneous equations with error components. *Econometric Theory*, 3:223–246, 1987.

B.H. Baltagi. On seemingly unrelated regressions with error components. *Econometrica*, 48:1547–1551, 1980.

B.H. Baltagi. Simultaneous equations with error components. *Journal of Econometrics*, 17:21–49, 1981.

B.H. Baltagi. Estimating an economic model of crime using panel data from north carolina. *Journal of Applied Econometrics*, 21 (4), May - June 2006.

B.H. Baltagi. Narrow replication of Serlenga and Shin (2007) Gravity models of intra-eu trade: Application of the ccep-ht estimation in heterogeneous panels within unobserved common time-specific factors. *Journal of Applied Econometrics*, 25:505–506, 2012.

B.H. Baltagi. *Econometric Analysis of Panel Data*. John Wiley and Sons ltd, 5th edition, 2013.

B.H. Baltagi and J.M. Griffin. Pooled estimators vs. their heterogeneous counterparts in the context of dynamic demand for gasoline. *Journal of Econometrics*, 77 (2):303–327, 1997.

B.H. Baltagi and J.M. Griffin. The econometrics of rational addiction: The case of cigarettes. *Journal of Business & Economic Statistics*, 19 (4):449–454, 2001.

B.H. Baltagi and S. Khanti-Akom. On efficient estimation with panel data: An empirical comparison of instrumental variables estimators. *Journal of Applied Econometrics*, 5 (4), Oct. - Dec. 1990.

B.H. Baltagi and Q. Li. A lagrange multiplier test for the error components model with incomplete panels. *Econometric Reviews*, 9:103–107, 1990.

B.H. Baltagi and Q. Li. A joint test for serial correlation and random individual effects. *Statistics and Probability Letters*, 11:277–280, 1991.

B.H. Baltagi and Q. Li. A note on the estimation of simultaneous equations with error components. *Econometric Theory*, 8 (01):113–119, March 1992.

B.H. Baltagi and Q. Li. Testing AR(1) against MA(1) disturbances in an error component model. *Journal of Econometrics*, 68:133–151, 1995.

B.H. Baltagi and N. Pinnoi. Public capital stock and state productivity growth: further evidence from an error components model. *Empirical Economics*, 20: 351–359, 1995.

B.H. Baltagi, Y.J. Chang, and Q. Li. Monte Carlo results on several new and existing tests for the error components model. *Journal of Econometrics*, 54: 95–120, 1992.

B.H. Baltagi, J.M. Griffin, and W. Xiong. To pool or not to pool: Homogeneous versus heterogeneous estimators applied to cigarette demand. *The Review of Economics and Statistics*, 82 (1):117–126, 2000.

B.H. Baltagi, S.H. Song, and B.C. Jung. The unbalanced nested error component regression model. *Journal of Econometrics*, 101:357–381, 2001.

B.H. Baltagi, G. Bresson, J.M. Griffin, and A. Pirotte. Homogeneous, heterogeneous or shrinkage estimators? Some empirical evidence from French regional gasoline consumption. *Empirical Economics*, 28 (4):795–811, 2003a.

B.H. Baltagi, S. Heun Song, and W. Koh. Testing panel data regression models with spatial error correlation. *Journal of Econometrics*, 117 (1):123–150, 2003b.

B.H. Baltagi, S.H. Song, B.C. Jung, and W. Koh. Testing for serial correlation, spatial autocorrelation and random effects using panel data. *Journal of Econometrics*, 140 (1):5–51, 2007.

B.H. Baltagi, G. Bresson, and A. Pirotte. To pool or not to pool? In *The Econometrics of Panel Data*, pages 517–546. Springer, 2008.

B.H. Baltagi, P. Egger, and M. Pfaffermayr. A generalized spatial panel data model with random effects. *Econometric Reviews*, 32:650–685, 2013.

P. Bardhan and D. Mookherjee. Determinants of redistributive politics: An empirical analysis of land reform in West Bengal, India. *American Economic Review*, 100 (4):1572–1600, 2010.

D. Bates and M. Maechler. *Matrix: Sparse and Dense Matrix Classes and Methods*, 2016. URL https://CRAN.R-project.org/package=Matrix. R package version 1.2-4.

S. Bazzi. Wealth heterogeneity and the income elasticity of migration. *American Economic Journal, Applied Economics*, 9 (2):219–255, 2017.

N. Beck and J.N. Katz. What to do (and not to do) with time-series cross-section data. *American Political Science Review*, 89 (03):634–647, 1995.

N. Beck, J.N. Katz, R.M. Alvarez, G. Garrett, and P. Lange. Government partisanship, labor organization, and macroeconomic performance: A corrigendum. *American Political Science Review*, 87 (04):945–948, 1993.

K.P. Bell and N.E. Bockstael. Applying the generalized-moments estimation approach to spatial problems involving micro-level data. *The Review of Economics and Statistics*, 82 (1):72–82, 2000.

A.K. Bera and M.J. Yoon. Specification testing with locally misspecified alternatives. *Econometric Theory*, 9 (04):649–658, 1993.

A.K. Bera, W. Sosa-Escudero, and M. Yoon. Tests for the error component model in the presence of local misspecification. *Journal of Econometrics*, 101:1–23, 2001.

A.K. Bera, G. Montes-Rojas, and W. Sosa-Escudero. Testing under local misspecification and artificial regressions. *Economics Letters*, 104 (2):66–68, 2009.

A. Bhargava, L. Franzini, and W. Narendranathan. Serial correlation and the fixed effects model. *Review of Economic Studies*, 49:533–554, 1982.

R. Bivand. *spdep: Spatial Dependence: Weighting Schemes, Statistics and Models*, 2008. R package version 0.4-17.

R. Blundell and S. Bond. Initital conditions and moment restrictions in dynamic panel data models. *Journal of Econometrics*, 87:115–143, 1998.

R. Blundell and S. Bond. GMM estimation with persistent panel data: An application to production functions. *Econometric Reviews*, 19 (3):321–340, 2000.

S.R. Bond. Dynamic panel data models: A guide to micro data methods and practice. *Portugese Economic Journal*, 1:141–162, 2002.

S.R. Bond, A. Hoeffler, and J. Temple. GMM estimation of empirical growth models. *CEPR Discussion Paper*, 3048, 2001.

J. Brandts and D.J. Cooper. A change would do you good... An experimental study on how to overcome coordination failure in organizations *American Economic Review*, 96 (3):669–693, 2006.

A. Brender and A. Drazen. How do budget deficits and economic growth affect reelection prospects? Evidence from a large panel of countries. *American Economic Review*, 98(5):2203–2220, 2008.

T.S. Breusch. Testing for autocorrelation in dynamic linear models. *Australian Economic Papers*, 17 (31):334–355, 1978.

T.S. Breusch and A.R. Pagan. The Lagrange multiplier test and its applications to model specification in econometrics. *Review of Economic Studies*, 47: 239–253, 1980.

T.S. Breusch, G.E. Mizon, and P. Schmidt. Efficient estimation using panel data. *Econometrica*, 57 (3):695–700, May 1989.

P. Burridge. On the Cliff-Ord test for spatial correlation. *Journal of the Royal Statistical Society. Series B (Methodological)*, 42 (1):107–108, 1980.

A.C. Cameron and P.K. Trivedi. *Regression Analysis of Count Data*. Cambridge, 1998.

A.C. Cameron, J.B. Gelbach, and D.L. Miller. Robust inference with multiway clustering. *Journal of Business & Economic Statistics*, 29 (2),2011.

A.C. Case. Spatial patterns in household demand. *Econometrica*, 59 (4): 953–965, 1991.

F. Caselli, G. Esquivel, and F. Lefort. Reopening the convergence debate: A new look at cross-country growth empirics. *Journal of Economic Growth*, 1: 363–389, 1996.

S.B. Caudill, J.M. Ford, and D.L. Kaserman. Certificate-of-need regulation and the diffusion of innovations: A random coefficient model. *Journal of Applied Econometrics*, 10 (1), 1995.

S.G. Cecchetti. The frequency of price adjustment: A study of the newsstand prices of magazines. *Journal of Econometrics*, 31:255–274, 1986.

G. Chamberlain. Analysis of covariance with qualitative data. *Review of Economic Studies*, 47:225–238, 1980.

G. Chamberlain. Multivariate regression models for panel data. *Journal of Econometrics*, 18:5–46, 1982.

J.M. Chambers. *Programming with Data: A guide to the S Language*. Springer, 1998.

G. Charness and M. Villeval. Cooperation and competition in intergenerational experiments in the field and the laboratory. *American Economic Review*, 99 (3):956–978, 2009.

X. Chen, S. Lin, and W.R. Reed. A Monte Carlo evaluation of the efficiency of the PCSE estimator. *Applied Economics Letters*, 17 (1):7–10, 2009.

M. Cincer. Patents, R&D, and technological spillovers at the firm level: Some evidence from econometric count models for panel data. *Journal of Applied Econometrics*, 12 (3), 1997.

J. Coakley, A.M. Fuertes, and R. Smith. Unobserved heterogeneity in panel time series models. *Computational Statistics & Data Analysis*, 50 (9):2361–2380, 2006.

A. Cohen and L. Einav. The effects of mandatory seat belt laws on driving behavior and traffic fatalities. *The Review of Economics and Statistics*, 85 (4):828–843, November 2003.

C. Cornwell and P. Rupert. Efficient estimation with panel data: An empirical comparison of instrumental variables estimators. *Journal of Applied Econometrics*, 3:149–155, 1988.

C. Cornwell and W.N. Trumbull. Estimating the economic model of crime with panel data. *Review of Economics and Statistics*, 76:360–366, 1994.

C. Cornwell, P. Schmidt, and D. Wyhowski. Simultaneous equations and panel data. *Journal of Econometrics*, 51 (1–2):151–181, 1992.

A. Cottrell and R. Lucchetti. *Gretl User's Guide*, May 2007. URL http://gretl.sourceforge.net/.

Y. Croissant. *pglm: panel generalized linear model*, 2017. URL http://www.r-project.org. R package version 0.2-0.

Y. Croissant and G. Millo. Panel data econometrics in R: The plm package. *Journal of Statistical Software*, 27 (2):1–43, 2008.

Y. Croissant and G. Millo. *pder: Panel Data Econometrics with R*, 2017. URL http://www.r-project .org. R package version 1.0-0.

Y. Croissant and A. Zeileis. *truncreg: Truncated Gaussian Regression Models*, 2016. URL https:// CRAN.R-project.org/package=truncreg. R package version 0.2-4.

R.E. De Hoyos and V. Sarafidis. Testing for cross–sectional dependence in panel–data models. *The Stata Journal*, 6 (4):482–496, 2006.

N. Debarsy and C. Ertur. Testing for spatial autocorrelation in a fixed effects panel data model. *Regional Science and Urban Economics*, 40 (6):453–470, 2010.

J.C. Driscoll and A.C. Kraay. Consistent covariance matrix estimation with spatially dependent panel data. *Review of Economics and Statistics*, 80 (4): 549–560, 1998.

D.M. Drukker. Testing for serial correlation in linear panel–data models. *The Stata Journal*, 3 (2):168–177, 2003.

V. Druska and W.C. Horrace. Generalized moments estimation for spatial panel data: Indonesian rice farming. *American Journal of Agricultural Economics*, 86 (1):185–198, 2004.

M. Eberhardt, C. Helmers, and H. Strauss. Do spillovers matter when estimating private returns to R&D? *Review of Economics and Statistics*, 95 (2):436–448, 2013.

P. Egger and M. Pfaffermayr. Distance, trade, and FDI: A Hausman-Taylor SUR approach. *Journal of Applied Econometrics*, 19(2):227–46, 2004.

M. El-Gamal and H. Inanoglu. Inefficiency and heterogeneity in Turkish banking: 1990-2000. *Journal of Applied Econometrics*, 20 (5):641–664, 2005.

J.P. Elhorst. Specification and estimation of spatial panel data models. *International Regional Science Review*, 26 (3):244–268, 2003.

J.P. Elhorst. Serial and spatial error correlation. *Economics Letters*, 100 (3): 422–424, 2008.

J.P. Elhorst. Applied spatial econometrics: Raising the bar. *Spatial Economic Analysis*, 5 (1):9–28, 2010.

J.P. Elhorst. Spatial panel data models. In *Spatial Econometrics*, pages 37–93. Springer, 2014.

J.P. Elhorst and S. Fréret. Evidence of political yardstick competition in france using a two-regime spatial Durbin model with fixed effects. *Journal of Regional Science*, 49 (5):931–951, 2009.

J.P. Elhorst and K. Zigova. Competition in research activity among economic departments: Evidence by negative spatial autocorrelation. *Geographical Analysis*, 46 (2):104–125, 2014.

H.S. Farber, D. Silverman, and T. von Wachter. Determinants of callbacks to job applications: An audit study. *American Economic Review*, 106 (5):314–318, 2016.

B. Fingleton. A generalized method of moments estimator for a spatial panel model with an endogenous spatial lag and spatial moving average errors. *Spatial Economic Analysis*, 3 (1):27–44, 2008.

R.J.G.M. Florax, H. Folmer, and S.J. Rey. Specification searches in spatial econometrics: The relevance of Hendry's methodology. *Regional Science and Urban Economics*, 33 (5):557–579, 2003.

K.J. Forbes. A reassessment of the relation between inequality and growth. *American Economic Review*, 90 (4):869–887, september 2000.

J. Fox and S. Weisberg. *An R Companion to Applied Regression*. Sage, Thousand Oaks CA, second edition, 2011. URL http://socserv.socsci.mcmaster.ca/jfox/Books/Companion.

R.J. Franzese and J.C. Hays. Strategic interaction among EU governments in active labor market policy-making: subsidiarity and policy coordination under the European employment strategy. *European Union Politics*, 7 (2):167–189, 2006.

R.J. Franzese and J.C. Hays. Spatial econometric models of cross-sectional interdependence in political science panel and time-series-cross-section data. *Political Analysis*, 15 (2):140–164, 2007.

R.J. Franzese and J.C. Hays. Interdependence in comparative politics: Substance, theory, empirics, substance. *Comparative Political Studies*, 41 (4-5):742–780, 2008.

K.A. Froot. Consistent covariance matrix estimation with cross-sectional dependence and heteroskedasticity in financial data. *Journal of Financial and Quantitative Analysis*, 24 (03):333–355, 1989.

J.L. Furman and S. Stern. Climbing atop the shoulders of giants: The impact of institutions on cumulative research. *American Economic Review*, 101 (5): 1933–1963, august 2011.

R. Furrer and S.R. Sain. spam: A sparse matrix R package with emphasis on MCMC methods for Gaussian Markov random fields. *Journal of Statistical Software*, 36 (10):1–25, 2010. URL http://www.jstatsoft.org/v36/i10/.

L.G. Godfrey. Testing against general autoregressive and moving average error models when the regressors include lagged dependent variables. *Econometrica*, 46 (6):1293–1301, 1978.

C. Gourieroux, A. Holly, and A. Monfort. Likelihood ratio test, Wald test, and Kuhn–Tucker test in linear models with inequality constraints on the regression parameters. *Econometrica*, 50:63–80, 1982.

C.W.J. Granger and P. Newbold. Spurious regressions in econometrics. *Journal of Econometrics*, 2 (2):111–120, 1974.

W.H. Greene. *Econometric Analysis*. Prentice Hall, 5th edition, 2003.

D.A. Griffith and G. Arbia. Detecting negative spatial autocorrelation in georeferenced random variables. *International Journal of Geographical Information Science*, 24 (3):417–437, 2010.

A. Hall. Testing for a unit root in time series with pretest data-based model selection. *Journal of Business & Economic Statistics*, 12 (4):461–470, 1994.

L.P. Hansen. Large sample properties of generalized method of moments estimators. *Econometrica*, 50:1029–1054, 1982.

J.A. Hansman and D.A. Wise. Social experimentation, truncated distributions and efficient estimation. *Econometrica*, 45 (4):919–938, may 1976.

M.N. Harris, L. Matyas, and P. Sevestre. Dynamic models for short panels. In Laszlo Matyas and Patrick Sevestre, editors, *The Econometrics of Panel Data*, pages 249–278. Springer, 2008.

D. Harrison and D.L. Rubinfeld. Hedonic housing prices and the demand for clean air. *Journal of Environmental Economics and Management*, 5:81–102, 1978.

J. Hausman, B.H. Hall, and Z. Griliches. Patents and R&D: Is there a lag? *International Economic Review*, pages 265–283, 1986.

J.A. Hausman. Specification tests in econometrics. *Econometrica*, 46: 1251–1271, 1978.

J.A. Hausman and W.E. Taylor. Panel data and unobservable individual effects. *Econometrica*, 49:1377–1398, 1981.

J.A. Hausman, B.H. Hall, and Z. Griliches. Econometric models for count data with and application to the patents–R&D relationship. *Econometrica*, 52: 909–938, 1984.

A. Henningsen. *censReg: Censored Regression (Tobit) Models*, 2017. URL https://CRAN.R-project.org/package=censReg. R package version 0.5-26.

A. Henningsen and O. Toomet. maxlik: A package for maximum likelihood estimation in R. *Computational Statistics*, 26 (3):443–458, 2011. doi: 10.1007/s00180-010-0217-1. URL http://dx.doi.org/10.1007/s00180-010-0217-1.

M. Hlavac *stargazer: LaTeX code for well-formatted regression and summary statistics tables*. Harvard University, Cambridge, USA, 2013. URL http://CRAN.R-project.org/package=stargazer. R package version 3.0.1.

S. Holly, M.H. Pesaran, and T. Yamagata. A spatio-temporal model of house prices in the USA. *Journal of Econometrics*, 158 (1):160–173, 2010.

D. Holtz-Eakin, W. Newey, and H.S. Rosen. Estimating vector autoregressions with panel data. *Econometrica*, 56:1371–1395, 1988.

Y. Honda. Testing the error components model with non–normal disturbances. *Review of Economic Studies*, 52:681–690, 1985.

B.E. Honoré. Trimmed LAD and least squares estimation of truncated and censored regression models with fixed effects. *Econometrica*, 60 (3), may 1992.

B.E. Honoré. Nonlinear models with panel data. *Portuguese Economic Journal*, 1 (2):163–179, 2002.

W.C. Horrace and P. Schmidt. Confidence statements for efficiency estimates from stochastic frontier models. *Journal of Productivity Analysis*, 7:257–282, 1996.

W.C. Horrace and P. Schmidt. Multiple comparisons with the best, with economic applications. *Journal of Applied Econometrics*, 15 (1):1–26, 2000.

T. Hothorn, K. Hornik, M.A. van De Wiel, and A. Zeileis. A Lego system for conditional inference. *The American Statistician*, 60 (3), 2006.

C. Hsiao. *Analysis of Panel Data*. Cambridge University Press, Cambridge, 2003.

C. Hsiao and M.H. Pesaran. Random coefficient models. In *The Econometrics of Panel Data*, pages 185–213. Springer, 2008.

M.M. Hutchison and I. Noy. How bad are twins? Output costs of currency and banking crises. *Journal of Money, Credit and Banking*, 4:725–752, august 2005.

K.S. Im, M.H. Pesaran, and Y. Shin. Testing for unit roots in heterogeneous panels. *Journal of Econometrics*, 115(1):53–74, 2003.

C.H. Jackson. Multi-state models for panel data: The msm package for R. *Journal of Statistical Software*, 38 (8):1–29, 2011. URL http://www.jstatsoft.org/v38/i08/.

G. Kapetanios, M.H. Pesaran, and T. Yamagata. Panels with non-stationary multifactor error structures. *Journal of Econometrics*, 160 (2):326–348, 2011.

M. Kapoor, H.H. Kelejian, and I.R. Prucha. Panel data models with spatially correlated error components. *Journal of Econometrics*, 140 (1):97–130, 2007.

H.H. Kelejian and I.R. Prucha. A generalized moments estimator for the autoregressive parameter in a spatial model. *International Economic Review*, 40 (2):509–533, 1999.

A.S. Kessler, N.A. Hansen, and C. Lessman. Interregional redistribution and mobility in federations: A positive approach. *The Review of Economic Studies*, 78:1345–78, 2011.

M.S. Khan and M.D. Knight. Import compression and export performance in developing countries. *Review of Economics and Statistics*, 70 (2):315–321, 1988.

N.M. Kiefer. Estimation of fixed effect models for time series of cross-sections with arbitrary intertemporal covariance. *Journal of Econometrics*, 14 (2): 195–202, 1980.

T. Kinal and K. Lahiri. A computational algorithm for multiple equation models with panel data. *Economic Letters*, 34:143–146, 1990.

T. Kinal and K. Lahiri. On the estimation of simultaneous-equations error-components models with an application to a model of developing country foreign trade. *Journal of Applied Econometrics*, 8:81–92, 1993.

M.L. King and P.X. Wu. Locally optimal one–sided tests for multiparameter hypotheses. *Econometric Reviews*, 33:523–529, 1997.

J.F. Kiviet. On bias, inconsistency, and efficiency of various estimators in dynamic panel data models. *Journal of Econometrics*, 68:53–78, 1995.

C. Kleiber and A. Zeileis. *Applied Econometrics with R*. Springer-Verlag, New York, 2008. URL http://CRAN.R-project.org/package=AER. ISBN 978-0-387-77316-2.

R. Koenker and A. Zeileis. On reproducible econometric research. *Journal of Applied Econometrics*, 24 (5):833–847, 2009.

S.C. Kumbhakar. Estimation of cost efficiency with heteroscedasticity: An application to electric utilities. *Journal of the Royal Statistical Society, series D*, 45:319–335, 1996.

C.E. Landry, A. Lange, J.A. List, M.K. Price, and N.G. Rupp. Is a donor in hand better than two in the bush? Evidence from a natural field experiment. *American Economic Review*, 100 (3):958–983, 2012.

L. Lee and J. Yu. Spatial panels: Random components versus fixed effects. *International Economic Review*, 53 (4):1369–1412, November 2012.

L.F. Lee and J. Yu. Estimation of spatial autoregressive panel data models with fixed effects. *Journal of Econometrics*, 154 (2):165–185, 2010a.

L.F. Lee and J. Yu. Some recent developments in spatial panel data models. *Regional Science and Urban Economics*, 40 (5):255–271, 2010b.

P. Leifeld. texreg: Conversion of statistical model output in R to LaTeX and html tables. *Journal of Statistical Software*, 55 (8):1–24, 2013. URL http://www.jstatsoft.org/v55/i08/.

F. Leisch. Sweave: Dynamic generation of statistical reports using literate data analysis. In *Compstat*, pages 575–580. Springer, 2002.

A. Levin, C.F. Lin, and C.S.J. Chu. Unit root tests in panel data: Asymptotic and finite sample properties. *Journal of Econometrics*, 108:1–24, 2002.

R. Levine, N. Loayza, and T. Beck. Financial intermediation and growth: Causality and causes. *Journal of Monetary Economics*, 46:31–77, 2000.

K-Y Liang and S.L. Zeger. Longitudinal data analysis using generalized linear models. *Biometrika*, 73 (1):13–22, 1986.

T. Lumley and A. Zeileis. sandwich: Model–robust standard error estimation for cross–sectional, time series and longitudinal data. R package version 2.0-2, 2007. URL http://CRAN.R-project .org.

G.S. Maddala and S. Wu. A comparative study of unit root tests with panel data and a new simple test. *Oxford Bulletin of Economics and Statistics*, 61: 631–52, 1999.

J. Magnus. Maximum likelihood estimation of the GLS model with unknown parameters in the disturbance covariance matrix. *Journal of Econometrics*, 7: 281–312, 1978.

J. Mairesse and B.H Hall. Estimating the productivity of research and development in French and US manufacturing firms: An exploration of simultaneity issues with GMM methods. In K. Wagner and B. Van-Ark, editors, *International Productivity Differences and their Explanations*, pages 285–315. Elsevier Science, 1996.

E. Meredith and J.S. Racine. Towards reproducible econometric research: The Sweave framework. *Journal of Applied Econometrics*, 24 (2):366–374, 2009.

S. Michalopoulos and E. Papaioannou. The long-run effects of the scramble for Africa. *American Economic Review*, 106 (7):1802–1848, 2016.

G. Millo. Maximum likelihood estimation of spatially and serially correlated panels with random effects. *Computational Statistics & Data Analysis*, 71: 914–933, 2014.

G. Millo. Narrow replication of 'A spatio-temporal model of house prices in the USA' using R. *Journal of Applied Econometrics*, 30 (4):703–704, 2015.

G. Millo. A simple randomization test for spatial correlation in the presence of common factors and serial correlation. *Regional Science and Urban Economics*, 66:28–38, 2017a.

G. Millo. Robust standard error estimators for panel models: A unifying approach. *Journal of Statistical Software*, 82 (3):1–27, 2017b.

G. Millo and G. Piras. splm: Spatial panel data models in R. *Journal of Statistical Software*, 47 (1):1–38, 2012.

B.R. Moulton. Random group effects and the precision of regression estimates. *Journal of Econometrics*, 32 (3):385–397, 1986.

B.R. Moulton. An illustration of a pitfall in estimating the effects of aggregate variables on micro units. *The Review of Economics and Statistics*, 72 (2): 334–338, 1990.

Y. Mundlak. Empirical production function free of management bias. *Journal of Farm Economics*, 43 (1):44–56, 1961.

Y. Mundlak. On the pooling of time series and cross section data. *Econometrica*, 46 (1):69–85, 1978.

A. Munnell. Why has productivity growth declined? Productivity and public investment. *New England Economic Review*, pages 3–22, 1990.

W. Murphy. *fiftystater: Map Data to Visualize the Fifty U.S. States with Alaska and Hawaii Insets*, 2016. URL https://CRAN.R-project.org/package=fiftystater. R package version 1.0.1.

J. Mutl and M. Pfaffermayr. The Hausman test in a Cliff and Ord panel model. *Econometrics Journal*, 14 (1):48–76, 2011.

M. Nerlove. Further evidence on the estimation of dynamic economic relations from a time–series of cross–sections. *Econometrica*, 39:359–382, 1971.

W.K. Newey and K.D. West. A simple, positive semi-definite, heteroskedasticity and autocorrelation consistent covariance matrix. *Econometrica*, 55 (3): 703–08, 1987.

S.J. Nickell. Biases in dynamic models with fixed effects. *Econometrica*, 49: 1417–1426, 1981.

W. Oberhofer and J. Kmenta. A general procedure for obtaining maximum likelihood estimates in generalized regression models. *Econometrica*, 42 (3): 579–590, 1974.

N. Obojes, M. Bahn, E. Tasser, J. Walde, N. Inauen, E. Hiltbrunner, P. Saccone, J. Lochet, J.C. Clément, S. Lavorel, et al. Vegetation effects on the water balance of mountain grasslands depend on climatic conditions. *Ecohydrology*, 8 (4):552–569, 2015.

R.W. Parks. Efficient estimation of a system of regression equations when disturbances are both serially and contemporaneously correlated. *Journal of the American Statistical Association*, 62 (318):500–509, 1967.

S. Peltzman. The effects of automobile safety regulation. *Journal of Political Economy*, 83 (4):677–725, August 1975.

R.D. Peng. Reproducible research in computational science. *Science (New York, Ny)*, 334 (6060):1226–1227, 2011.

M.H. Pesaran. General diagnostic tests for cross section dependence in panels. CESifo Working Paper Series, 1229, 2004.

M.H. Pesaran. Estimation and inference in large heterogeneous panels with a multifactor error structure. *Econometrica*, 74 (4):967–1012, 2006.

M.H. Pesaran. A simple panel unit root test in the presence of cross-section dependence. *Journal of Applied Econometrics*, 22 (2):265–312, 2007.

M.H. Pesaran and R. Smith. Estimating long-run relationships from dynamic heterogeneous panels. *Journal of Econometrics*, 68 (1):79–113, 1995.

M.H. Pesaran and E. Tosetti. Large panels with common factors and spatial correlation. *Journal of Econometrics*, 161 (2):182–202, 2011.

M.A. Petersen. Estimating standard errors in finance panel data sets: Comparing approaches. *Review of Financial Studies*, 22 (1):435–480, 2009.

P.C.B. Phillips and H.R. Moon. Linear regression limit theory for nonstationary panel data. *Econometrica*, 67 (5):1057–1111, 1999.

P.C.B. Phillips and D. Sul. Dynamic panel estimation and homogeneity testing under cross section dependence. *The Econometrics Journal*, 6 (1):217–259, 2003.

J. Pinheiro, D. Bates, S. DebRoy, D. Sarkar, and R Core Team. *nlme: Linear and Nonlinear Mixed Effects Models*, 2017. URL https://CRAN.R-project.org/package=nlme. R package version 3.1-131.

J. Powell. Symmetrically trimmed least squares estimators for tobit models. *Econometrica*, 54:1435–1460, 1986.

R Core Team. *foreign: Read Data Stored by 'Minitab', 'S', 'SAS', 'SPSS', 'Stata', 'Systat', 'Weka', 'dBase', ...*, 2017. URL https://CRAN.R-project.org/package=foreign. R package version 0.8-69.

J. Racine and R. Hyndman. Using R to teach econometrics. *Journal of Applied Econometrics*, 17 (2):175–189, 2002.

J.S. Racine. Rstudio: A platform-independent IDE for R and Sweave. *Journal of Applied Econometrics*, 27 (1):167–172, 2012.

C. Raux, S. Souche, and Y. Croissant. How fair is pricing perceived to be? An empirical study. *Public Choice*, 139(1):227–240, 2009.

D. Roodman. How to do xtabond2: An introduction to difference and system GMM in Stata. *The Stata Journal*, 9:86–136, 2009a.

D. Roodman. A note on the theme of too many instruments. *Oxford Bulletin of Economics and Statistics*, 71:135–158, 2009b.

A.J. Rossini, R.M. Heiberger, R.A. Sparapani, M. Maechler, and K. Hornik. Emacs Speaks Statistics: A multiplatform, multipackage development environment for statistical analysis. *Journal of Computational and Graphical Statistics*, 2004.

V. Sarafidis and T. Wansbeek. Cross-sectional dependence in panel data analysis. *Econometric Reviews*, 31 (5):483–531, 2012.

J.D. Sargan. The estimation of economic relationships using instrumental variables. *Econometrica*, 26:393–415, 1958.

H. Schaller. A re-examination of the q theory of investment using US firm data. *Journal of Applied Econometrics*, 5(4):309–325, 1990.

L. Serlenga and Y. Shin. Gravity models of intra-EU trade: application of the CCEP-HT estimation in heterogeneous panels with unobserved common time-specific factors. *Journal of Applied Econometrics*, 22:361–381, 2007.

J.H. Stock. Asymptotic properties of least squares estimators of cointegrating vectors. *Econometrica*, 55 (5):1035–1056, 1987.

J.H. Stock and M.W. Watson. *Introduction to Econometrics*. Pearson/Addison Wesley Boston, 2007.

P.A.V.B. Swamy. Efficient inference in a random coefficient regression model. *Econometrica*, 38:311–323, 1970.

P.A.V.B. Swamy and S.S Arora. The exact finite sample properties of the estimators of coefficients in the error components regression models. *Econometrica*, 40:261–275, 1972.

T. Tantau. *The TikZ and PGF Packages*, 2013. URL http://sourceforge.net/projects/pgf/.

T. Therneau. *bdsmatrix: Routines for Block Diagonal Symmetric Matrices*, 2014. URL https://CRAN.R-project.org/package=bdsmatrix. R package version 1.3-2.

T.M. Therneau and P.M. Grambsch. *Modeling Survival Data: Extending the Cox Model*. Springer, New York, 2000. ISBN 0-387-98784-3.

S. Theußl and A. Zeileis. Collaborative software development using R-Forge. *The R Journal*, 1 (1):9–14, May 2009.

S.B. Thompson. Simple formulas for standard errors that cluster by both firm and time. *Journal of Financial Economics*, 99 (1):1–10, 2011.

J. Tobin. Estimation of relationships for limited dependent variables. *Econometrica*, 26 (1):24–36, 1958.

J. Tobin. A general equilibrium approach to monetary theory. *Journal of Money, Credit and Banking*, 1:15–29, 1969.

F. Vella and M. Verbeek. Whose wages do unions raise? A dynamic model of unionism and wage rate determination for young men. *Journal of Applied Econometrics*, 13:163–183, 1998.

W.N. Venables and B.D. Ripley. *Modern Applied Statistics with S*. Springer, New York, fourth edition, 2002. URL http://www.stats.ox.ac.uk/pub/MASS4. ISBN 0-387-95457-0.

T.D. Wallace and A. Hussain. The use of error components models in combining cross section with time series data. *Econometrica*, 37 (1):55–72, 1969.

H. White. A heteroskedasticity-consistent covariance matrix estimator and a direct test for heteroskedasticity. *Econometrica*, 48 (4):817–838, 1980.

H. White. *Advances in statistical analysis and statistical computing, vol. 1*, chapter Instrumental variables analogs of generalized least squares estimators. Mariano, R.S., 1986.

H. Wickham. *ggplot2: Elegant Graphics for Data Analysis*. Springer-Verlag New York, 2009. ISBN 978-0-387-98140-6. URL http://ggplot2.org.

H. Wickham and R. Francois. *dplyr: A Grammar of Data Manipulation*, 2016. URL https://CRAN.R-project.org/package=dplyr. R package version 0.5.0.

J.L. Willis. Magazine prices revisited. *Journal of Applied Econometrics*, 21 (3): 337–344, 2006.

F. Windmeijer. A finite sample correction for the variance of linear efficient two–step GMM estimators. *Journal of Econometrics*, 126:25–51, 2005.

J.M. Wooldridge. *Econometric Analysis of Cross–Section and Panel Data*. MIT press, 2010.

Y. Xie. *Dynamic Documents with R and knitr*. Chapman and Hall/CRC, Boca Raton, Florida, 2nd edition, 2015. URL http://yihui.name/knitr/. ISBN 978-1498716963.

A.T. Yalta and R. Lucchetti. The GNU/Linux platform and freedom respecting software for economists. *Journal of Applied Econometrics*, 23 (2):279–286, 2008.

A.T. Yalta and A.Y. Yalta. Gretl 1.6. 0 and its numerical accuracy. *Journal of Applied Econometrics*, 22 (4):849–854, 2007.

A.T. Yalta and A.Y. Yalta. Should economists use open source software for doing research? *Computational Economics*, 35 (4):371–394, 2010.

A. Zeileis. Econometric computing with HC and HAC covariance matrix estimators. *Journal of Statistical Software*, 11 (10):1–17, 2004. URL http://www.jstatsoft.org/v11/i10/.

A. Zeileis. Object-oriented computation of sandwich estimators. *Journal of Statistical Software*, 16 (9):1–16, 2006a. URL http://www.jstatsoft.org/v16/i09/.

A. Zeileis. Implementing a class of structural change tests: An econometric computing approach. *Computational Statistics & Data Analysis*, 50 (11): 2987–3008, 2006b.

A. Zeileis and Y. Croissant. Extended model formulas in R: Multiple parts and multiple responses. *Journal of Statistical Software*, 34 (XYZ):1–12, 2010. URL http://www.jstatsoft.org/v34/iXYZ/.

A. Zeileis and T. Hothorn. Diagnostic checking in regression relationships. R *News*, 2 (3):7–10, 2002. URL http://CRAN.R-project.org/doc/Rnews/.

A. Zellner. An efficient method of estimating seemingly unrelated regressions and tests of aggregation bias. *Journal of the American Statistical Association*, 57:500–509, 1962.

Index

General index

Akaike information criteria 205
Amemiya and MaCurdy estimator 147, 148, 153, 154
Amemiya estimator 36, 58, 68, 77
Anderson and Hsiao estimator 167, 168, 172
Angrist and Newey test 93, 95
asymptotic least squares estimator 93
augmented Dickey-Fuller regression 204–207
auto-regressive process 21, 101, 261

Baltagi and Li test 97–98, 101
Baltagi, Song and Koh test 269–272
Baltagi, Song, Jung and Koh test 281–284
Bayes' theorem 238
Bera, Sosa-Escudero and Yoon test 97, 98
between estimator 28, 29
between transformation 24, 26, 35
binomial model 211, 213
block-diagonal matrix 17, 54, 170, 171
Breusch, Mizon and Schmidt estimator 147, 148
Breusch-Godfrey test 97, 101
Breusch-Pagan test 84–88, 105, 269

censored model 211
Chamberlain test 90–95
Cholesky decomposition 67, 156
Cobb-Douglas functional form 77
cointegration 207–209
common correlated effects 196–198, 200, 207, 209, 249, 254, 255
conditional logit model 219–222
constained least squares 65
constrained estimator 93
contiguity matrix 250
count data 236–243
cross-sectional and timewise correlation consistent covariance matrix 115, 117, 119
cross-sectional augmented regression 207–209
cross-sectional dependence 104–108, 207, 208, 247–251
cross-sectional heteroscedasticity and serial correlation consistent covariance matrix 111, 115, 117, 119
cross-sectionally augmented Im, Pesaran and Shin test 207–209

Dickey-Fuller test 204
differenced generalised method of moments estimator 168–172
distance matrix 250
Durbin-Watson test 101
dynamic model 161–184

endogeneity 139–159
error component 144–146, 148, 149
error components instrumental variables estimator 143
error components three stage least squares 156–158
error components two stage least squares 146

F test 84, 86–88
feasible generalized least squares 123, 128, 129
first difference transformation 2, 103, 120, 121, 132, 136, 137
first generation unit root tests 204, 206
fixed effects 101, 102, 120, 121, 130–132, 135–137, 261, 269

Panel Data Econometrics with R, First Edition. Yves Croissant and Giovanni Millo.

Companion website: www.wiley.com/go/croissant/data-econometrics-with-R

fixed effects censored model 229–233
fixed effects model 30, 55–56, 93
fixed effects Negbin model 239–243
fixed effects Poisson model 237–239
fixed effects truncated model 227–229
Frisch-Waugh theorem 30, 53, 55

general feasible generalized least squares estimator 17, 127–137
generalized least squares 17, 20, 31, 33–35, 38–45, 47, 48, 51, 54, 58, 71, 74, 89, 90, 93, 99, 100, 122, 127–133, 135, 140, 141, 144, 146, 149, 150, 155, 156, 159, 163, 190, 263, 264, 267
generalized linear model 211
generalized method of moments 168, 171–174, 176, 177, 180, 182, 183, 185
generalized moments estimation 254, 267–269
Gourieroux, Holly and Monfort test 86, 88

Hausman and Taylor estimator 36, 146, 151, 153, 154
Hausman test 90, 125, 149, 150, 152
heteroscedasticity and autocorrelation consistent covariance matrix 110
heteroscedasticity and cross-sectional correlation consistent covariance matrix 111, 115–117, 119
heteroscedasticity consistent covariance matrix 120
Honda test 86, 87

idempotent matrix 25, 30, 54, 141
Im, Pesaran and Shin test 205, 207
incidental parameter problem 212
instrument proliferation 172–174
instrumental variable estimator 140, 166
instrumental variables estimator 140–159

Kapoor, Kelejian and Prucha estimator 261, 267, 268, 278
King and Wu test 86, 88
Kronecker product 24

Lagrange multiplier test 18, 84–88, 97, 98, 101, 105, 247, 260, 269–274, 281–283
Lagrangian function 65
least absolute deviations 229, 231
least squares dummy variables 2, 10
Levin, Lin and Chu test 205
likelihood ratio test 18, 99, 100, 276, 277
logit 211

Maddala and Wu test 206
maximum likelihood estimator 71–74, 95, 99, 166, 212, 226–227, 254, 258, 262, 267, 269, 271, 272, 275–277, 280, 282, 283
mean groups 190–192, 197, 198, 200, 209, 254
measurement error 139
multinomial logit model 239
Mundlak model 89, 90

Negbin model 211, 236–237
neighborhood matrix 250
Nerlove estimator 37
nested error components model 74–80
Newey-West robust covariance matrix 117
nonstationarity 200–209

omitted variable 139
ordinal model 211, 214
orthogonal decomposition 25
orthogonal deviations 166

panel corrected standard errors 122, 123, 128
partitioned matrix 29
Pesaran cross-sectional dependence test 105
Poisson model 211, 236
poolability tests 192–194
pooled common correlated effects 198–200, 209
pooling estimator 27–28
power of a matrix 54
probit 211
purchasing power parity 124, 125

quadratic form 58–60, 75, 76, 93, 169

random coefficients model 187–192
random effects 97–99, 101, 104, 106, 120, 121, 129, 261, 264–269, 272, 275–278, 283
random effects binomial model 214–217
random effects censored model 234
random effects model 56–57
random effects Negbin model 240

random effects ordinal model 217–219
random effects Poisson model 239–240
random effects truncated model 233–234, 236
randomized W test 247–250
robust covariance matrix 178–179
robust Lagrange multiplier test 273, 275

Sargan test 180
scaled Lagrange multiplier 105, 247
Schwarz information criteria 204
second generation unit root tests 207, 209
seemingly unrelated regressions 64–71, 91, 128, 154, 155
serially autoregressive random effect 277, 278
simultaneity 139
sparse matrix 17
spatial correlation consistent 116, 117, 120, 127, 131, 132, 135–137
spatial error model 21, 255, 256, 261–263, 265–269, 272–278, 283, 284
spatial lag and spatial error model 261, 268, 272, 275, 277, 278
spatial lags 250–257
spatially autoregressive model 21, 254–256, 261–264, 265, 267, 268, 272–277, 284
sufficient statistic 212
Swamy and Arora estimator 37, 58, 68, 77
Swamy estimator 187–189
symmetrical trimmed estimator 225–226
system generalized method of moments estimator 174–177

three stage least squares 155–158
trace of a matrix 29, 30, 59
translog functional form 68
trimmed estimator 212
truncated and censored model 223–236
truncated model 211
two stage least squares 141, 142, 144–149, 154, 156–159, 268
two-ways error component model 47–49, 54–64, 85

unbalanced panel 53–64, 86, 105
unit root tests 201–209

vector autoregressive model 183

Wald test 14, 18, 102, 179, 275
Wallace and Hussain estimator 36, 58, 77
weak instruments 174
White robust covariance matrix 111, 115–117, 119
within estimator 29, 164–168
within instrumental variables estimator 141–143
within model 145
within transformation 25, 26, 35
Wooldridge unobserved effects test 95
Wooldridge *within*-based test 102
Wooldridge first-difference-based test 103

Functions

AER
  tobit 211
MASS
  glm.nb 211
  polr 211, 218
base
  attach 10
  cbind 9
  crossprod 9
  detach 10
  diff 5
  mean 10
  read.table 8
  sapply 39, 43, 49
  solve 9, 17
  summary 14
car
  lht 15, 16
  linearHypothesis 15, 123, 124
censReg
  censReg 211
default
  summary 193
dplyr
  mutate 69
graphics
  plot 14
lmtest
  bgtest 102
  coeftest 4, 112, 117, 123, 124, 251
  dwtest 102
  waldtest 123, 124

msm
deltamethod 191
nlme
gls 99, 100
lme 99
pglm
ordinal 218
pglm 73, 216, 218
plm
aneweytest 95
Between 41, 94
between 94
cipstest 208, 209
cortab 107
ercomp 38, 43
index 32
mtest 182
pbgtest 101, 102
pbltest 97, 98
pbsytest 98
pcce 198, 200, 255
pcdtest 105–107, 207, 249
pdata.frame 31, 77
pdim 31, 60
pdwtest 101, 102
pFtest 86
pggls 129–131
pgmm 171
phtest 93, 126, 132
piest 94
pldv 212, 232
plm 4, 5, 16, 31, 70, 77, 112, 114, 120, 130, 131, 142, 148, 157, 164, 165, 171, 216, 251, 259, 275
plmtest 87, 97
pmg 190, 197, 198, 255
pooltest 192
purtest 206
pvcm 187, 188, 192, 193
pwartest 103
pwfdtest 103
sargan 180
vcovDC 117
vcovG 117
vcovNW 117, 200
vcovSCC 117
Within 41, 274
sandwich
vcovHAC 120
vcovHC 15, 16, 102, 113, 120, 179, 200
spdep
lagsarlm 254
nb2mat 247
nblag 247
splm
bsktest 271
pcdtest 247
pmg 207, 249
rwtest 249
slag 253, 255
slmtest 273
spgm 268
spml 259, 268, 275
spreml 255, 256, 259, 275
stats
coef 6, 15, 16
glm 211, 215, 216, 218
lag 164
lm 3, 4, 9, 33, 100, 112, 113, 164, 215
vcov 15, 16, 179
survival
clogit 222
texreg
screenreg 80, 216, 219
truncreg
truncreg 211

Data

AER
Fatalities 3, 9, 14
pcse
agl 123
pder
Callbacks 244
CoordFailure 244
DemocracyIncome 161, 164, 165, 167, 171, 177, 179, 180, 182
DemocracyIncome25 46, 173
Dialysis 188
Donor 234
etw 279
EvapoTransp 279, 284
FinanceGrowth 183

ForeignTrade 42, 148, 157
GiantsShoulders 240, 241
HousePricesUS 107, 190, 193, 197, 198, 206–208, 245, 248, 254
IncomeMigrationH 244
IncomeMigrationV 244
IneqGrowth 183
LandReform 243
LateBudgets 231
Mafia 159
MagazinePrices 220, 222
RDPerfComp 184
RDSpillovers 105, 126, 135, 191, 200
Reelection 215, 216
RegIneq 184
ScrambleAfrica 243
SeatBelt 142
Seniors 243
Solow 183
TexasElectr 44, 68
Tileries 6, 15, 16, 60
TobinQ 31, 37, 41, 48
TradeEU 151
TradeFDI 159
TurkishBanks 44
TwinCrises 159

pglm
Fairness 218
PatentsRD 244
PatentsRDUS 244
UnionWage 49, 244

plm
Cigar 252
Cigarette 251
Crime 159
EmplUK 102, 103, 129, 131, 132, 183
Grunfeld 99, 126
Hedonic 113
Produc 77, 79, 111, 117, 118, 122, 124
Snmesp 183
Wages 159

splm
RiceFarms 73, 79, 86, 93, 96, 98, 101, 104, 133, 256, 258, 259, 264, 266, 268, 271, 274–276, 283
riceww 256

Packages

AER 211
car 123, 124
censReg 211
dplyr 69, 241
fiftystater 245
foreign 8
Formula 142, 167
ggplot2 241, 245
lmtest 101, 112, 123
MASS 211, 218
Matrix 17
MaxLik 17
msm 191
nlme 17, 99
pcse 123
pder 31, 142, 148, 151, 159, 182–184, 215, 222, 241, 243, 244
pglm 13, 73, 211, 218, 244
plm 5, 13, 14, 16, 17, 31, 79, 95, 108, 109, 120, 159, 164, 171, 182, 183, 187, 212, 255
sandwich 120
spam 17
spdep 247, 254
splm 13, 268
stargazer 50
survival 222
texreg 80, 153, 216
truncreg 211

Programming Language and Software

C 17
Emacs 13
ESS 13
FORTRAN 17
Gretl 8
knitr 12
R 1, 3, 5–15, 17, 32, 118, 120, 164, 211, 245, 254
RStudio 13
S 14
Sweave 12
tikz xiv

Printed and bound by CPI Group (UK) Ltd, Croydon, CR0 4YY
12/05/2024
14500523-0005